Color in Computer Vision

Wiley–IS&T Series in Imaging Science and Technology

Series Editor: **Michael A. Kriss**

Consultant Editor: **Lindsay W. MacDonald**

Reproduction of Colour (6th Edition) / R. W. G. Hunt

Colour Appearance Models (2nd Edition) / Mark D. Fairchild

Colorimetry: Fundamentals and Applications / Noburu Ohta and Alan R. Robertson

Color Constancy / Marc Ebner

Color Gamut Mapping / Ján Morovič

Panoramic Imaging: Sensor-Line Cameras and Laser Range-Finders / Fay Huang, Reinhard Klette and Karsten Scheibe

Digital Color Management (2nd Edition) / Edward J. Giorgianni and Thomas E. Madden

The JPEG 2000 Suite / Peter Schelkens, Athanassios Skodras and Touradj Ebrahimi (Eds.)

Color Management: Understanding and Using ICC Profiles / Phil Green (Editor)

Fourier Methods in Imaging / Roger L. Easton, Jr.

Measuring Colour (4th Edition) / R.W.G. Hunt and M.R. Pointer

The Art and Science of HDR Imaging / John McCann and Alessandro Rizzi

Color in Computer Vision: Fundamentals and Applications / Theo Gevers, Arjan Gijsenij, Joost van de Weijer, and Jan-Mark Geusebroek

Published in Association with the Society for Imaging Science and Technology

Color in Computer Vision

Fundamentals and Applications

Theo Gevers

Intelligent Systems Lab. Amsterdam,
University of Amsterdam (The Netherlands)
and
Computer Vision Center,
Universitat Autònoma de Barcelona (Spain)

Arjan Gijsenij

Intelligent Systems Lab. Amsterdam,
University of Amsterdam (The Netherlands)

Joost van de Weijer

Computer Vision Center,
Universitat Autonòma de Barcelona (Spain)

Jan-Mark Geusebroek

Intelligent Systems Lab. Amsterdam,
University of Amsterdam (The Netherlands)

A John Wiley & Sons, Inc., Publication

Cover image: Jasper van Turnhout
Cover design: Michael Rutkowski

Published by John Wiley & Sons, Inc., Hoboken, New Jersey
Published simultaneously in Canada

Library of Congress Cataloging-in-Publication Data:

Color in computer vision : fundamentals and applications / Theo Gevers ... [et al.].
p. cm.
Includes bibliographical references and index.
ISBN 978-0-470-89084-4 (pbk.)
1. Computer vision. 2. Color vision. 3. Color photography. I. Gevers, Theo.
TA1634.C637 2012
006.3′7—dc23

2012000650

ISBN: 9780470890844

10 9 8 7 6 5 4 3 2 1

To my parents, Dick and Wil

—Theo Gevers

To my wife Petra

—Arjan Gijsenij

To Line

—Joost van de Weijer

To my wife Astrid and our daughters Nora and Ellen

—Jan-Mark Geusebroek

Contents

Preface .... **xv**

1 Introduction **1**
1.1 From Fundamental to Applied 2
1.2 Part I: Color Fundamentals 3
1.3 Part II: Photometric Invariance 3
1.3.1 Invariance Based on Physical Properties 4
1.3.2 Invariance By Machine Learning 4
1.4 Part III: Color Constancy 4
1.5 Part IV: Color Feature Extraction 5
1.5.1 From Luminance to Color 5
1.5.2 Features, Descriptors, and Saliency 6
1.5.3 Segmentation 6
1.6 Part V: Applications 7
1.6.1 Retrieval and Visual Exploration 7
1.6.2 Color Naming 7
1.6.3 Multispectral Applications 8
1.7 Summary 9

PART I Color Fundamentals **11**

2 Color Vision **13**
2.1 Introduction 13
2.2 Stages of Color Information Processing 14
2.2.1 Eye and Optics 14
2.2.2 Retina: Rods and Cones 14
2.2.3 Ganglion Cells and Receptive Fields 16
2.2.4 LGN and Visual Cortex 16

2.3 Chromatic Properties of the Visual System 18
2.3.1 Chromatic Adaptation 18
2.3.2 Human Color Constancy 18
2.3.3 Spatial Interactions 20
2.3.4 Chromatic Discrimination and Color Deficiency 23
2.4 Summary 24

3 Color Image Formation 26
3.1 Lambertian Reflection Model 28
3.2 Dichromatic Reflection Model 29
3.3 Kubelka–Munk Model 32
3.4 The Diagonal Model 34
3.5 Color Spaces 36
3.5.1 *XYZ* System 36
3.5.2 RGB System 38
3.5.3 Opponent Color Spaces 40
3.5.4 Perceptually Uniform Color Spaces 41
3.5.5 Intuitive Color Spaces 42
3.6 Summary 44

PART II Photometric Invariance 47

4 Pixel-Based Photometric Invariance 49
4.1 Normalized Color Spaces 50
4.2 Opponent Color Spaces 52
4.3 The HSV Color Space 52
4.4 Composed Color Spaces 53
4.4.1 Body Reflectance Invariance 53
4.4.2 Body and Surface Reflectance Invariance 55
4.5 Noise Stability and Histogram Construction 58
4.5.1 Noise Propagation 58
4.5.2 Examples of Noise Propagation through Transformed Colors ... 60
4.5.3 Histogram Construction by Variable Kernel Density Estimation 61
4.6 Application: Color-Based Object Recognition 64
4.6.1 Dataset and Performance Measure 64
4.6.2 Robustness Against Noise: Simulated Data 65
4.7 Summary 68

5 Photometric Invariance from Color Ratios 69
5.1 Illuminant Invariant Color Ratios 71
5.2 Illuminant Invariant Edge Detection 73
5.3 Blur-Robust and Color Constant Image Description 74

5.4 Application: Image Retrieval Based on Color Ratios . . . 77
5.4.1 Robustness to Illuminant Color . . . 77
5.4.2 Robustness to Gaussian Blur . . . 78
5.4.3 Robustness to Real-World Blurring Effects . . . 78
5.5 Summary . . . 80

6 Derivative-Based Photometric Invariance . . . 81
6.1 Full Photometric Invariants . . . 84
6.1.1 The Gaussian Color Model . . . 84
6.1.2 The Gaussian Color Model by an RGB Camera . . . 88
6.1.3 Derivatives in the Gaussian Color Model . . . 89
6.1.4 Differential Invariants for the Lambertian Reflection Model . . . 90
6.1.5 Differential Invariants for the Dichromatic Reflection Model . . . 95
6.1.6 Summary of Full Color Invariants . . . 98
6.1.7 Geometrical Color Invariants in Two Dimensions . . . 100
6.2 Quasi-Invariants . . . 101
6.2.1 Edges in the Dichromatic Reflection Model . . . 101
6.2.2 Photometric Variants and Quasi-Invariants . . . 103
6.2.3 Relations of Quasi-Invariants with Full Invariants . . . 104
6.2.4 Localization and Discriminative Power of Full and Quasi-Invariants . . . 108
6.3 Summary . . . 111

7 Photometric Invariance by Machine Learning . . . 113
7.1 Learning from Diversified Ensembles . . . 114
7.2 Temporal Ensemble Learning . . . 119
7.3 Learning Color Invariants for Region Detection . . . 120
7.4 Experiments . . . 124
7.4.1 Error Measures . . . 124
7.4.2 Skin Detection: Still Images . . . 125
7.4.3 Road Detection in Video Sequences . . . 129
7.5 Summary . . . 134

PART III Color Constancy . . . 135

8 Illuminant Estimation and Chromatic Adaptation . . . 137
8.1 Illuminant Estimation . . . 139
8.2 Chromatic Adaptation . . . 141

9 Color Constancy Using Low-level Features . . . 143
9.1 General Gray-World . . . 143
9.2 Gray-Edge . . . 146

9.3 Physics-Based Methods 150
9.4 Summary 151

10 Color Constancy Using Gamut-Based Methods 152
10.1 Gamut Mapping Using Derivative Structures 155
10.1.1 Diagonal-Offset Model 155
10.1.2 Gamut Mapping of Linear Combinations of Pixel Values 155
10.1.3 N-Jet Gamuts 157
10.2 Combination of Gamut Mapping Algorithms 157
10.2.1 Combining Feasible Sets 159
10.2.2 Combining Algorithm Outputs 159
10.3 Summary 160

11 Color Constancy Using Machine Learning 161
11.1 Probabilistic Approaches 161
11.2 Combination Using Output Statistics 162
11.3 Combination Using Natural Image Statistics 163
11.3.1 Spatial Image Structures 164
11.3.2 Algorithm Selection 165
11.4 Methods Using Semantic Information 167
11.4.1 Using Scene Categories 167
11.4.2 Using High-Level Visual Information 169
11.5 Summary 171

12 Evaluation of Color Constancy Methods 172
12.1 Data Sets 172
12.1.1 Hyperspectral Data 173
12.1.2 RGB Data 173
12.1.3 Summary 174
12.2 Performance Measures 175
12.2.1 Mathematical Distances 176
12.2.2 Perceptual Distances 176
12.2.3 Color Constancy Distances 177
12.2.4 Perceptual Analysis 178
12.3 Experiments 180
12.3.1 Comparing Algorithm Performance 181
12.3.2 Evaluation 182
12.4 Summary 185

PART IV Color Feature Extraction 187

13 Color Feature Detection 189
13.1 The Color Tensor 191
13.1.1 Photometric Invariant Derivatives 193

13.1.2 Invariance to Color Coordinate Transformations 195
13.1.3 Robust Full Photometric Invariance 196
13.1.4 Color-Tensor-Based Features 197
13.1.5 Experiment: Robust Feature Point Detection and Extraction 204
13.2 Color Saliency 205
13.2.1 Color Distinctiveness 207
13.2.2 Physics-Based Decorrelation 208
13.2.3 Statistics of Color Images 211
13.2.4 Boosting Color Saliency 212
13.2.5 Evaluation of Color Distinctiveness 214
13.2.6 Repeatability 215
13.2.7 Illustrations of Generality 218
13.3 Conclusions 218

14 Color Feature Description 221
14.1 Gaussian Derivative-Based Descriptors 225
14.2 Discriminative Power 229
14.3 Level of Invariance 235
14.4 Information Content 236
14.4.1 Experimental Results 242
14.5 Summary 243

15 Color Image Segmentation 244
15.1 Color Gabor Filtering 245
15.2 Invariant Gabor Filters Under Lambertian Reflection 247
15.3 Color-Based Texture Segmentation 247
15.4 Material Recognition Using Invariant Anisotropic Filtering 249
15.4.1 MR8-NC Filterbank 253
15.4.2 MR8-INC Filterbank 254
15.4.3 MR8-LINC Filterbank 255
15.4.4 MR8-SLINC Filterbank 255
15.4.5 Summary of Filterbank Properties 256
15.5 Color Invariant Codebooks and Material-Specific Adaptation 256
15.6 Experiments 258
15.6.1 Material Classification by Color Invariant Codebooks 258
15.6.2 Color–Texture Segmentation of Material Images 260
15.6.3 Material Classification by Adaptive Color Invariant Codebooks 262
15.7 Image Segmentation by Delaunay Triangulation 263
15.7.1 Homogeneity Based on Photometric Color Invariance 264
15.7.2 Homogeneity Based on a Similarity Predicate 265
15.7.3 Difference Measure 265
15.7.4 Segmentation Results 267
15.8 Summary 268

PART V Applications **269**

16 Object and Scene Recognition **271**
16.1 Diagonal Model 272
16.2 Color SIFT Descriptors 273
16.3 Object and Scene Recognition 276
16.3.1 Feature Extraction Pipelines 276
16.3.2 Classification 277
16.3.3 Image Benchmark: PASCAL Visual Object Classes Challenge 278
16.3.4 Video Benchmark: Mediamill Challenge 279
16.3.5 Evaluation Criteria 279
16.4 Results 280
16.4.1 Image Benchmark: PASCAL VOC Challenge 280
16.4.2 Video Benchmark: Mediamill Challenge 282
16.4.3 Comparison 283
16.5 Summary 285

17 Color Naming **287**
17.1 Basic Color Terms 288
17.2 Color Names from Calibrated Data 291
17.2.1 Fuzzy Color Naming 293
17.2.2 Chromatic Categories 294
17.2.3 Achromatic Categories 298
17.2.4 Fuzzy Sets Estimation 300
17.3 Color Names from Uncalibrated Data 304
17.3.1 Color Name Data Sets 306
17.3.2 Learning Color Names 307
17.3.3 Assigning Color Names in Test Images 311
17.3.4 Flexibility Color Name Data Set 312
17.4 Experimental Results 313
17.5 Conclusions 316

18 Segmentation of Multispectral Images **318**
18.1 Reflection and Camera Models 319
18.1.1 Multispectral Imaging 319
18.1.2 Camera and Image Formation Models 319
18.1.3 White Balancing 320
18.2 Photometric Invariant Distance Measures 321
18.2.1 Distance between Chromaticity Polar Angles 321
18.2.2 Distance between Hue Polar Angles 322
18.2.3 Discussion 325
18.3 Error Propagation 325
18.3.1 Propagation of Uncertainties due to Photon Noise 325
18.3.2 Propagation of Uncertainty 326

18.4 Photometric Invariant Region Detection by Clustering 328
18.4.1 Robust K-Means Clustering 328
18.4.2 Photometric Invariant Segmentation 329
18.5 Experiments 330
18.5.1 Propagation of Uncertainties in Transformed Spectra 331
18.5.2 Photometric Invariant Clustering 334
18.6 Summary 338

Citation Guidelines 339

References 341

Index 363

Preface

Visual information is our most natural source of information and communication. Apart from human vision, visual information plays a vital and indispensable role in society and is the nucleus of current communication frameworks such as the World Wide Web and mobile phones. With the ever-growing production, use, and exploitation of digital visual information (e.g., documents, websites, images, videos, and movies), a visual overflow will occur, and hence demands are urgent for the (automatic) understanding of visual information. Moreover, as digital visual information is nowadays available in color format, there is the irreversible necessity for the understanding of visual *color* information. Computer vision deals with the understanding of visual information. Although color became a central topic in various disciplines (ranging from mathematics and physics to the humanities and art) quite early on, in the field of computer vision it has emerged only recently. We take on the challenge of providing a substantial set of tools for image understanding from a color perspective. The central topic of this book is to present color theories, representation models, and computational methods that are essential for image understanding in the field of computer vision.

The idea to make this book was born when the authors were sitting on a terrace overlooking the Amstel River. The rich artistic history of Amsterdam, the river, and that sunny day gave us the inspiration for discussing the role of color in art, in life, and eventually in computer vision. There, we decided to do something about the lack of textbooks on color in computer vision. We agreed that the most productive and pleasant way to reflect our findings on this topic was to write this book together. A book in which color is taken as a valuable collaborative source of synergy between two research fields: *color science* and *computer vision*. The book is the result of more than 10 years of research experience of all four authors who worked closely together (as PhDs, postdocs, professors, colleagues, and eventually friends) on the same topic of color computer vision at the University of Amsterdam. Because of this long-term collaboration among the authors, our research on color computer vision is a tight connection of color theories, color image processing methods, machine learning, and applications in the field of computer vision, such as image segmentation, understanding, and search. Even though many of the chapters in the book have their origin as a journal article, we ascertained that our work is rewritten and trimmed down. This process, the long-term collaboration, and many discussions resulted in a book in which a uniform style has emerged and in which the material represents the best of us.

The book is a valuable textbook for graduate students, researchers, and professionals in the field of computer vision, computer science, color, and engineering. The book covers upper-level undergraduate and graduate courses and can also be used in more advanced courses such as postgraduate tutorials. It is a good reference for anyone, including those in industry, interested in the topic of color and computer vision. A prerequisite is a basic knowledge of image processing and computer vision. Further, a general background in mathematics is required, such as linear algebra, calculus, and probability theory. Some of the material in this book has been presented as part of graduate and postgraduate courses at the University of Amsterdam. Also, part of the material has been presented at conference tutorials and short courses at image processing conferences (International Conference on Image Processing (ICIP) and International Conference on Pattern Recognition (ICPR)), computer vision conferences (Computer Vision and Pattern Recognition (CVPR) and the International Conference on Computer Vision (ICCV)), and color conferences (Colour in Graphics, Imaging, and Vision (CGIV) and conferences organized by the International Society for Optics and Photonics (SPIE)). Computer vision contains more topics than what we have presented in this book. The emphasis is on image understanding. However, the topic of image understanding has been taken as the path along which we were able to present our work. Although the material represents our view on color in computer vision, our sincere intention was to include all relevant research. Therefore, we believe this book is one of the first extensive works on color in computer vision to be published with over 360 citations.

This book consists of five parts. The topics range from (low-level) color image formation to (intermediate-level) color invariant feature extraction and color image processing to (high-level) semantic descriptors for object and scene recognition. The topics are treated from low-level to high-level processing and from fundamental to more applied research. Part I contains the (color) fundamentals of the book. This part presents the concept of trichromatic color processing and the similarity between human and computer vision systems. Furthermore, the basics are provided on the color image formation. Reflection models that describe the imaging process, the interplay between light and matter, and how photometric conditions influence the RGB values in an image are presented. In Part II, we consider the research area of extracting color invariant information. We build detailed models of the color image formation process and design mathematical methods to infer the quantities of interest. Pixel-based and derivative-based photometric invariance are discussed. An overview is given on the computation of both photometric invariance and differential information. Part III contains an overview on color constancy. Computational methods are presented to estimate the illumination. An evaluation of color constancy methods is given on large-scale datasets. The problem of how to select and combine different methods is addressed. A statistical approach is taken to quantify the priors of unknowns in noisy data to infer the best possible estimate of the illumination from the visual scene. Feature detection and color descriptors are discussed in Part IV. Color image processing tools are provided. An algebraic (vector-based) approach is taken to extend scalar-signal to vector-signal processing. Computational methods are introduced to extract a variety of local image features, such as circle detectors, curvature estimation, and optical flow. Finally, in Part V, different applications are presented, such as image segmentation, object recognition, color naming, and image retrieval.

This book comes with a large amount of supplementary material, which can be found at

- http://www.colorincomputervision.com

Here you can find

- Software implementations of many of the methods presented in the book.
- Datasets and pointers to public image datasets.
- Slides corresponding to the material covered in the book.
- Slides of new material presented at tutorials at conferences.
- Pointers to workshops and conferences.
- Discussions on current developments, including latest publications.

Our policy is to make our software and datasets available as a contribution to the research community. Also, in case you want to share your software or dataset, please drop us a line so we can add a pointer to it on our website. If you have any suggestions for improving the book, please send us an e-mail. We want to keep the book accurate as much as possible.

Finally, we thank all the people who have worked with us over the years and shared their passion for research and color with us.

Arnold Smeulders at the University of Amsterdam is one of the best researchers we had the opportunity to work with. He was heading the group during the time we paved the way for this book. His insatiable passion for research and lively debates have been a source of inspiration to all of us. We enjoyed working with him.

We are very grateful to Marcel Lucassen who contributed Chapter 2 to this book. Furthermore, his thorough proofreading and enthusiasm were indispensable for the quality of the book. It is a fortune to have him as a human (color) vision scientist amidst us. It was certainly a pleasure to work with him. We are indebted to Jan van Gemert for his proofreading and Frank Aldershoff for LaTeX and Mathematica issues.

We are also grateful to NWO (Dutch Organisation for Scientific Research), who granted Theo Gevers with a VICI (#639.023.705) with the same title of this book ‘‘Color in Computer Vision’’ and Jan-Mark Geusebroek with a VENI. These grants were valuable for this book.

While working at the University of Amsterdam, we had the opportunity to collaborate with many wonderful colleagues. We want to thank Arnold Smeulders for his work on Chapters 6 and 13, Rein van de Boomgaard for Chapter 6, Gertjan Burghouts for Chapters 14 and 15, Koen van de Sande and Cees Snoek for their help on Chapter 16, and Harro Stokman for Chapter 18. Furthermore, we thank the following persons: Virginie Mes, Roberto Valenti, Marcel Worring, Dennis Koelma, and all other members of the ISIS group.

At the Computer Vision Center (Universitat Autònoma de Barcelona), we thank José Álvarez and Antonio López for their contribution to Chapter 7. Further, we are indebted to Robert Benavente, Maria Vanrell, and Ramon Baldrich for their contribution to Chapter 17. At the LEAR team in INRIA rhône Alpes, France, we thank Cordelia Schmid, Jakob Verbeek, and Diane Larlus for their help with Chapters 5 and 17. We also appreciate the contribution of Andrew Bagdanov at the Media Integration and Communication Center in Florence, Italy. Furthermore, Joost van de Weijer acknowledges the support of the Spanish Ministry of Science and Innovation in Madrid, Spain, in particular for funding the Consolider MIPRCV project and for providing him with the Ramon y Cajal Fellowship.

Lastly, we will always remember that this book would not have been possible without our families and loved ones whose energy and love inspired us to make our work colorful and worthwhile.

October 2011
Amsterdam, The Netherlands

Theo Gevers
Arjan Gijsenij
Joost van de Weijer
Jan-Mark Geusebroek

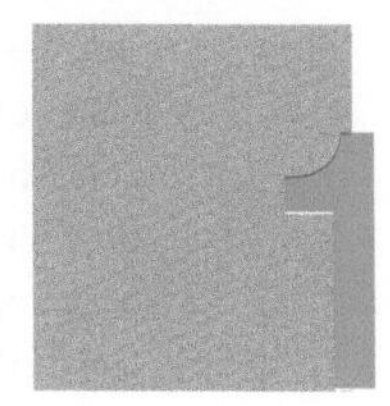

1 Introduction

Color is one of the most important and fascinating aspects of the world surrounding us. To comprehend the broad characteristics of color, a range of research fields has been actively involved, including physics (light and reflectance modeling), biology (visual system), physiology (perception), linguistics (cultural meaning of color), and art.

From a historical perspective, covering more than 400 years, prominent researchers contributed to our present understanding of light and color. Snell and Descartes (1620–1630) formulated the law of light refraction. Newton (1666) discovered various theories on light spectrum, colors, and optics. The perception of color and the influence on humans has been studied by Goethe in his famous book "Farbenlehre" (1840). Young and Helmholtz (1850) proposed the trichromatic theory of color vision. Work on light and color resulted in quantum mechanics elaborated by Max Planck, Albert Einstein, and Niels Bohr. In art (industrial design), Albert Munsell (1905) invented the theory on color ordering in his "A Color Notation." Further, the value of the biological and therapeutic effects of light and color have been analyzed, and views on color from folklore, philosophy, and language have been articulated by Schopenhauer, Hegel, and Wittgenstein.

Over the last decades, with the technological advances of printers, displays, and digital cameras, an explosive growth in the diversity of needs in the field of color computer vision has been witnessed. More and more, the traditional gray value imaginary is replaced by color systems. Moreover, today, with the growth and popularity of the World Wide Web, a tremendous amount of visual information, such as images and videos, has become available. Hence, nowadays, all visual data is available in color. Furthermore, (automatic) image understanding is becoming indispensable to handle large amount of visual data. Computer vision deals with image understanding and search technology for the management of

Color in Computer Vision: Fundamentals and Applications, First Edition.
Theo Gevers, Arjan Gijsenij, Joost van de Weijer, and Jan-Mark Geusebroek.

large-scale pictorial datasets. However, in computer vision, the use of color has been only partly explored so far.

This book determines the use of color in computer vision. We take on the challenge of providing a substantial set of color theories, computational methods, and representations, as well as data structures for image understanding in the field of computer vision. Invariant and color constant feature sets are presented. Computational methods are given for image analysis, segmentation, and object recognition. The feature sets are analyzed with respect to their robustness to noise (e.g., camera noise, occlusion, fragmentation, and color trustworthiness), expressiveness, discriminative power, and compactness (efficiency) to allow for fast visual understanding. The focus is on deriving semantically rich color indices for image understanding. Theoretical models are presented to express semantics from both a physical and a perceptual point of view.

1.1 From Fundamental to Applied

The aim of this book is to present color theories and techniques for image understanding from (low level) basic color image formation to (intermediate level) color invariant feature extraction and color image processing to (high level) learning of object and scene recognition by semantic detectors. The topics, and corresponding chapters, are organized from low level to high level processing and from fundamental to more applied research. Moreover, each topic is driven by a different research area using color as an important stand-alone research topic and as a valuable collaborative source of information bridging the gap between different research fields (Fig. 1.1).

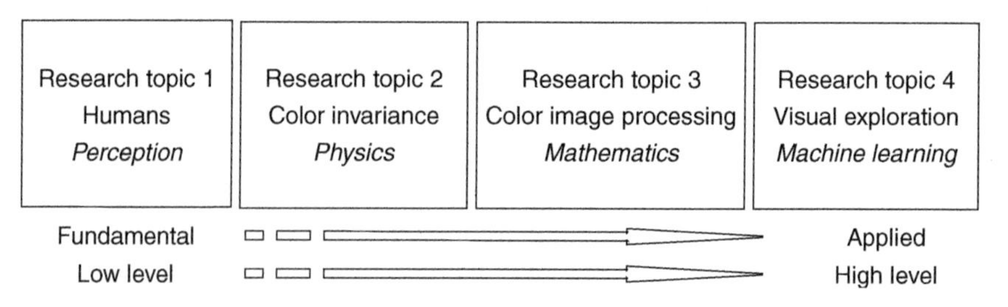

Figure 1.1 The different topics are organized from low level to high level processing and from fundamental to more applied research. Each topic is driven by a different research area from human perception, physics, and mathematics to machine learning.

The book starts with the explanation of the mechanisms of *human color perception*. Understanding the human visual pathway is crucial for computer vision systems, which aim to describe color information in such a way that it is relevant to humans.

Then, *physical* aspects of color are studied, resulting in reflection models from which photometric invariance is derived. Photometric invariance is important for computer vision, as it results in color measurements that are independent of accidental imaging conditions such as a change in camera viewpoint or a variation in the illumination.

A *mathematical* perspective is taken to cope with the difference between gray value (scalar) and color (vector) information processing, that is, the extension of single-channel signal to multichannel signal processing. This mathematical approach will result in a sound way to perform color processing to obtain (low level) computational methods for (local) feature computation (e.g., color derivatives), descriptors (e.g., SIFT), and image segmentation. Furthermore, based on both mathematical and physical fundamentals, color image feature extraction is presented by integrating differential operators and color invariance.

Finally, color is studied in the context of *machine learning*. Important topics are color constancy, photometric invariance by learning, and color naming in the context of object recognition and video retrieval. On the basis of the multichannel approach and color invariants, computational methods are presented to extract salient image patches. From these salient image patches, color descriptors are computed. These descriptors are used as input for various machine learning methods for object recognition and image classification.

The book consists of five parts, which are discussed next.

1.2 Part I: Color Fundamentals

The observed color of an object depends on a complex set of imaging conditions. Because of the similarity in trichromatic color processing between humans and computer vision systems, in Chapter 2, an outline on human color vision is provided. The different stages of color information processing along the human visual pathway are presented. Further, important chromatic properties of the visual system are discussed such as chromatic adaptation and color constancy. Then, to provide insights in the imaging process, in Chapter 3, the basics on color image formation are presented. Reflection models are introduced describing the imaging process and how photometric changes, such as shadows and specularities, influence the RGB values in an image. Additionally, a set of relevant color spaces are enumerated.

1.3 Part II: Photometric Invariance

In computer vision, invariant descriptions for image understanding are relatively new but quickly gaining ground. The aim of photometric invariant features is to compute image properties of objects irrespective of their recording conditions.

This comes, in general, at the loss of some discriminative power. To arrive at invariant features, the imaging process should be taken into account.

In Chapters 4–6, the aim is to extract color invariant information derived from the physical nature of objects in color images using reflection models. Reflection models are presented to model dull and gloss materials, as well as shadows, shading, and specularities. In this way, object characteristics can be derived (based on color/texture statistics) for the purpose of image understanding. Physical aspects are investigated to model and analyze object characteristics (color and texture) under different viewing and illumination conditions. The degree of invariance should be tailored to the recording circumstances. In general, a color model with a very wide class of invariance loses the power to discriminate among object differences. Therefore, in Chapter 6, the aim is to select the tightest set of invariants suited for the expected set of nonconstant conditions.

1.3.1 Invariance Based on Physical Properties

As discussed in Chapter 4, most of the methods to derive photometric invariance are using 0th order photometric information, that is, pixel values. The effect of the reflection models on higher-order- or differential-based algorithms remained unexplored for a long time. The drawbacks of the photometric invariant theory (i.e., the loss of discriminative power and deterioration of noise characteristics) are inherited by the differential operations. To improve the performance of differential-based algorithms, the stability of photometric invariants can be increased through the noise propagation analysis of the invariants. In Chapters 5 and 6, an overview is given on how to advance the computation of both photometric invariance and differential information in a principled way.

1.3.2 Invariance By Machine Learning

While physical-based reflection models are valid for many different materials, it is often difficult to model the reflection of complex materials (e.g., with nonperfect Lambertian or dielectrical surfaces) such as human skin, cars, and road decks. Therefore, in Chapter 7, we also present techniques to estimate photometric invariance by machine learning models. On the basis of these models, computational methods are studied to derive the (in)sensitivity of transformed color channels to photometric effects obtained from a set of training samples.

1.4 Part III: Color Constancy

Differences in illumination cause measurements of object colors to be biased toward the color of the light source. Humans have the ability of color constancy; they tend to perceive stable object colors despite large differences in illumination. A similar color constancy capability is necessary for various computer vision applications such as image segmentation, object recognition, and scene classification.

In Chapters 8–10, an overview is given on computational color constancy. Many state-of-the-art methods are tested on different (freely) available datasets. As color constancy is an underconstrained problem, color constancy algorithms are based on specific imaging assumptions. These assumptions include the set of possible light sources, the spatial and spectral characteristics of scenes, or other assumptions (e.g., the presence of a white patch in the image or that the averaged color is gray). As a consequence, no algorithm can be considered as universal. With the large variety of available methods, the inevitable question, that is, how to select the method that induces the equivalence class for a certain imaging setting, arises. Furthermore, the subsequent question is how to combine the different algorithms in a proper way. In Chapter 11, the problem of how to select and combine different methods is addressed. An evaluation of color constancy methods is given in Chapter 12.

1.5 Part IV: Color Feature Extraction

We present how to extend luminance-based algorithms to the color domain. One requirement is that image processing methods do not introduce new chromaticities. A second implication is that for differential-based algorithms, the derivatives of the separate channels should be combined without loss of derivative information. Therefore, the implications on the multichannel theory are investigated, and algorithmic extensions for luminance-based feature detectors such as edge, curvature, and circular detectors are given. Finally, the photometric invariance theory described in earlier parts of the book is applied to feature extraction.

1.5.1 From Luminance to Color

The aim is to take an algebraic (vector based) approach to extend scalar-signal to vector-signal processing. However, a vector-based approach is accompanied by several mathematical obstacles. Simply applying existing luminance-based operators on the separate color channels, and subsequently combining them, will fail because of undesired artifacts.

As a solution to the opposing vector problem, for the computation of the color gradient, the color tensor (structure tensor) is presented. In Chapter 13, we give a review on color-tensor-based techniques on how to combine derivatives to compute local structures in color images in a principled way. Adaptations of the tensor lead to a variety of local image features, such as circle detectors, curvature estimation, and optical flow.

1.5.2 Features, Descriptors, and Saliency

Although color is important to express saliency, the explicit incorporation of color distinctiveness into the design of image feature detectors has been largely ignored. To this end, we give an overview on how color distinctiveness can be explicitly incorporated in the design of color (invariant) representations and feature detectors. The approach is based on the analysis of the statistics of color derivatives. Furthermore, we present color descriptors for the purpose of object recognition. Object recognition aims to detect high level semantic information present in images and videos. The approach is based on salient visual features and using machine learning to build concept detectors from annotated examples. The choice of features and machine learning algorithms is of great influence on the accuracy of the concept detector. Features based on interest regions, also known as *local features*, consist of an interest region detector and a region descriptor. In contrast to the use of intensity information only, we will present both interest point detection (Chapter 13) and region description (Chapter 14), see Figure 1.2.

Figure 1.2 Visual exploration is based on the paradigm to divide the images into meaningful parts from which features are computed. Salient point detection is applied first from which color descriptors are computed. Then, machine learning is applied to provide classifiers for object recognition.

1.5.3 Segmentation

In computer vision, texture is considered as all what is left after color and local shape have been considered or it is given in terms of structure and randomness. Many common textures are composed of small textons usually too large in number to be perceived as isolated objects. In Chapter 15, we give an overview on powerful features based on natural image statistics or general principles from surface physics in order to classify a large number of materials by their texture. On the basis of their textural nature, different materials and concepts containing certain types of material can be identified (Fig. 1.3). For features at the level of (entire) objects, the aim is to aggregate pieces of local visual information to characteristic geographical arrangements of (possibly missing)

Figure 1.3 On the basis of their textural nature, different materials and concepts containing certain types of material can be identified.

parts. The objective is to find computational models to combine individual observations of an object's appearance under the large number of variations in that appearance.

1.6 Part V: Applications

In the final part of the book, we emphasize on the importance of color in several computer vision applications.

1.6.1 Retrieval and Visual Exploration

In Chapter 16, we follow the state-of-the-art object recognition paradigm consisting of a learning phase and a (runtime) classification phase (Fig. 1.4). The learning module consists of color feature extraction and supervised learning strategies. Color descriptors are computed at salient points in the image by different point detectors (Fig. 1.2). The learning part is executed offline. The runtime classification part takes an image or video as an input from which features are extracted. Then, the classification scheme will provide a probability to what class of concepts the query image/video belongs to (people, mountain, or cars). A concept is defined as a *material* (e.g., grass, brick, or sand, as illustrated in Fig. 1.3a) or as an object (e.g., car, bike, or person, as illustrated in Fig. 1.3b), an event (explosion, crash, etc.), or a scene (e.g., mountain, beach, or city), see Figure 1.5.

1.6.2 Color Naming

Color names are linguistic labels that humans attach to colors. We use them routinely and seemingly without effort to describe the world around us. They have been primarily studied in the fields of visual psychology, anthropology, and linguistics. One of the most influential works in color naming is the linguistic study of Berlin and Kay on basic color terms. In Chapter 17, color names are

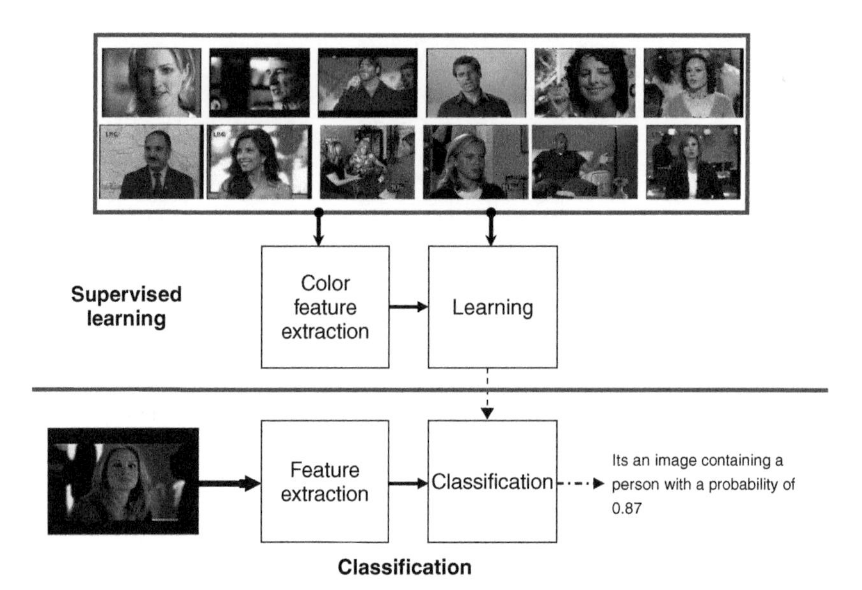

Figure 1.4 First, during training, features are extracted and objects/scenes are learned offline by giving examples of different concepts (e.g., people, buildings, mountains) as the input to a learning system (in this case pictures containing people). Then, during online recognition, features are extracted from the incoming image/video and provided to the classification system to result in a probability of being one of the concepts.

Figure 1.5 TRECVID concepts and corresponding key frames.

presented in the context of image retrieval. This allows for searching objects in images by a certain color name.

1.6.3 Multispectral Applications

Finally, in Chapter 18, we give an overview on multispectral imaginary and applications to segmentation and detection. In fact, techniques are presented to detect regions in multispectral images. To obtain robustness against noise, noise propagation is adopted.

1.7 Summary

Visual information (images and video) is one of the most valuable sources of information. In fact, it is the core of current technologies such as the Internet and mobile phones. The immense stimulus of the use and exploitation of digital visual information demands for advanced knowledge representations, learning systems, and image understanding techniques. As all digital information is nowadays available in color (documents, images, videos, and movies), there is an increasing demand for the use and understanding of color information.

Although color has been proved to be a central topic in various disciplines, it has only been partly explored so far in computer vision, which this book resolves. The central topic of this book is to present color theories, color representation models, and computational methods, which are essential for visual understanding in the field of computer vision. Color is taken as the merging topic between different research areas such as mathematics, physics, machine learning, and human perception. Theoretical models are studied to express color semantics from both a physical and a perceptual point of view. These models are the foundations for visual exploration, which are tested in practice.

PART I

COLOR FUNDAMENTALS

2 Color Vision

By Marcel P. Lucassen

2.1 Introduction

For any vision system, color vision is possible only when two or more light sensors sample the spectral energy distribution of the incoming light in different ways. In animal life, several instantiations of this principle are found, some of them even using parts of the electromagnetic spectrum not visible to the human eye. Human color vision is basically trichromatic, involving three types of cone photoreceptors in the retinae of our eyes. According to a number of reports, however, some women may possess tetrachromatic vision involving four photoreceptor types. Less than three functional sensors—*color deficiency*—is a well-known phenomenon in humans, often erroneously termed as *color blindness*. But apart from these two anomalies, ''normal'' color vision starts with the absorption of light in three cone types. Responses arising from these cones are combined in retinal ganglion cells to form three opponent channels: one achromatic (black–white) and two chromatic channels (red–green and yellow–blue). Retinal ganglion cells send off pulselike signals through the optic nerve to the visual cortex, where the perception of color eventually takes place. With the advances in neural imaging techniques, vision researchers have learned much about the specific locations of information processing in the visual cortex. How this eventually results in the perception of color and associated color phenomena in the context of other perceptual attributes such as shape and motion is largely unknown. This chapter describes the basic

Color in Computer Vision: Fundamentals and Applications, First Edition.
Theo Gevers, Arjan Gijsenij, Joost van de Weijer, and Jan-Mark Geusebroek.

building blocks of the visual pathway and provides some grip on the factors that affect the fascinating process of color vision.

2.2 Stages of Color Information Processing

2.2.1 Eye and Optics

Color vision starts with light that enters our eyes. At the cornea, a very sensitive part of our eyes, the incoming light is refracted. The diameter of the pupil, the hole in the iris through which light enters the eye, is dependent on the light intensity. Iris muscles cause the dilation and contraction of the pupil, which thereby regulates the amount of light entering the eye ball by a factor of about 10–30, depending on the exact minimum and maximum pupil diameters. Adjustment of the lens curvature by the lens muscles is the process known as *accommodation* and ensures the projection of a sharply focused image on the retina at the back of the eye ball. Unfortunately, because of the chromatic aberration of the lens it is not possible to have a focused image for all wavelengths simultaneously. This explains why red text on a blue background or vice versa can appear blurry and difficult to read. Blue and red are associated with the lower and upper ends of the visible wavelength spectrum, implying that when we focus on one, the other will be out of focus.

2.2.2 Retina: Rods and Cones

The retina contains two kinds of light-sensitive cells, rods and cones, named after their basic shapes. Each retina holds about 100 million photoreceptors, roughly 95 million rods and 5 million cones. At low light levels (<0.01 cd/m^2), our vision is *scotopic* and served by rod activity only. In pure scotopic vision we sense differences in the light–dark dimension, but color vision is not possible. Also, visual acuity is poor. At intermediate light levels (0.01–1 cd/m^2) our vision is *mesopic*, in which both rods and cones are active. In mesopic light conditions color discrimination is poor. At light levels above 1 cd/m^2 our vision becomes *photopic*, where cone activity is best and allows for good color discrimination.

The spatial distribution of rods and cones along the retina is not uniform. Where cone density is high, rod density is low, and vice versa. Usually the visual field is divided into a central area (having high cone density) and a peripheral area (high rod density). Cone density is at maximum (around 150,000–200,000 cones/mm^2) in a tiny spot central to the retina, the *fovea*, which allows us to perform high acuity tasks such as reading, and provides the best color discrimination. A yellow macular pigment covers the fovea and may serve to maintain high visual acuity because it filters out the blurry short wavelength light that is scattered in the ocular media. At the very heart of the fovea, an area known as the *foveola*, no S-cones are present at all, which causes small blue objects to be invisible to the S-cone system (Fig. 2.1c). This phenomenon is known as *small-field tritanopia*, a color

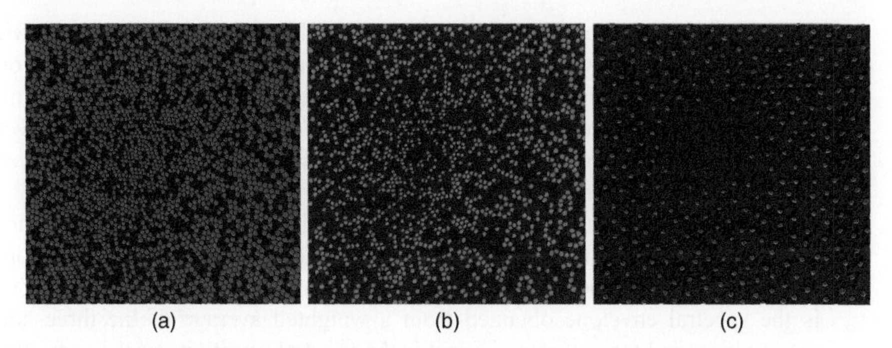

Figure 2.1 Cone mosaic at the central fovea, showing (a) L-cones, (b) M-cones, and (c) S-cones. The area shown is approximately 0.3×0.3 mm and is rod-free. The labeling in red, green, and blue refers to the spectral region where the cones have their maximum sensitivity. Note the different number of cones and the absence of S-cones in the center. *Source:* Figures adapted from Reference 1.

vision deficiency for objects subtending visual angles smaller than $0.35°$. The three cone types (L, M, S) occur in different numbers, in L:M:S ratios of about 60:30:5 although these numbers may vary considerably from person to person.

The three cone types have peak sensitivities at different wavelengths and are sensitive to the long-wave (L), middle-wave (M), and short-wave (S) portions of the wavelength spectrum. In Figure 2.2, the spectral sensitivities of the cone types are shown. Note that the sensitivities of the L- and M-cones are largely overlapping whereas the S-cones are spectrally more isolated. Owing to the spectral overlap, at each wavelength there exists a unique combination of L, M, S sensitivities. However, wavelength information is lost in the process that determines the cone responses. For each cone type, the response is obtained by

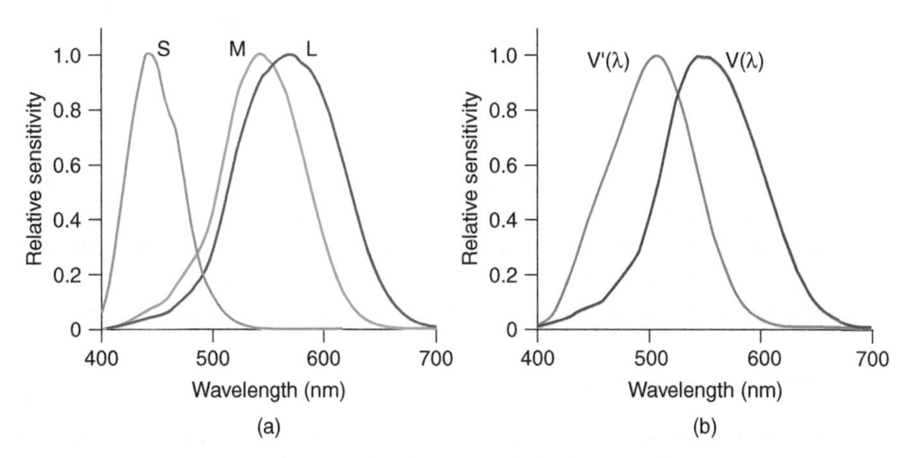

Figure 2.2 (a) Relative spectral sensitivity of the three cone types. (b) Spectral luminous efficiency functions $V(\lambda)$ for photopic vision and $V'(\lambda)$ for scotopic vision, with sensitivities normalized to their maximum. *Source:* Data for 2° observer, after Reference 2.

summing up the wavelength-by-wavelength product of the light spectrum with the spectral sensitivity over the spectral window, resulting in three numbers (one for each cone type). The perceived color of an object is determined by the relative magnitude of these three numbers that the object "produces," but not exclusively so. The visual system also makes spatial comparisons, which make the perceived color of an object dependent on neighboring colors as well.

A quantity often used in vision is the spectral luminous efficiency function, which is denoted by the symbol $V(\lambda)$ for photopic vision and $V'(\lambda)$ for scotopic vision. It represents the spectral sensitivity of the eye. For photopic vision, $V(\lambda)$ is the spectral envelope obtained from a weighted average of the three cone sensitivities, and for scotopic vision it is the spectral sensitivity of the rods. Note that the latter is shifted toward the blue end of the spectrum.

2.2.3 Ganglion Cells and Receptive Fields

If each photoreceptor were to be connected to individual brain cells, one can imagine that a neural cable of considerable thickness would be required. It makes sense therefore that, before signals are sent to the brain, the output signals of the cones are spatially pooled and combined. Also, from an information theory point of view it makes sense to compress the amount of visual information, given the limited bandwidth of the visual pathway [3]. The rods and cones are connected to subsequent layers of horizontal cells, bipolar cells, amacrine cells, and ganglion cells. Interestingly, the incoming light has to first pass these layers in reverse order to reach the layer containing the photoreceptors. The incoming light and the nerve signals thus travel in opposite directions. All neurons have inputs and outputs forming a complex structure in the retinal layer. The output of a neuron is influenced by inputs that can be excitatory (stimulating the output) or inhibitory (suppressing the output). The horizontal and amacrine cells make it possible to combine information from photoreceptors at different spatial locations. A single ganglion cell may thus receive inputs from many photoreceptors. The area on the retina that contributes to the stimulation of a ganglion cell is known as the *receptive field*. Likewise, neural cells along the visual pathway also have their receptive fields, but these are not necessarily equal to the receptive fields of ganglion cells. The axons of the ganglion cells together form the *optic nerve*, the connection between the eyes and the brain. When excited, the ganglion cells will fire sharply peaked output signals (pulses or spikes) to the optic nerve. To summarize, the light that is initially absorbed in the cone photoreceptors is transformed to electrical pulse signals that encode the visual information.

2.2.4 LGN and Visual Cortex

The next processing stage upstream the visual pathway to consider is the *lateral geniculate nucleus*, or LGN in short. It is the place where two streams of visual information meet: one stream coming from the left part of the visual field (projected on the right part of each retina) and another coming from the right part of the visual field (projected on the left part of each retina). The LGN can be thought of as a relay station, where signals from the retina pass and are sent to the primary visual cortex (V1) in the back of the head. The left and right ''halves'' of V1 thus receive information from the right and left halves of the visual field, respectively. Properties of cells within the LGN are very much like those of the retinal ganglion cells, including their receptive field organization. Important for the understanding of the (color) vision process is the notion of opponent cells, usually in a center-surround configuration. The so-called *on-cells* are excited by light stimulation in the central part of the receptive field, whereas they are inhibited by stimulation in the outer part of it (surrounding the center). *Off-cells* have the opposite spatial characteristics, that is, inhibition by light stimulation in the center of the receptive field and excitation in the surround. Cells with a center-surround configuration play an important role in vision, since they are capable of detecting spatial transitions in light intensity (such as edges) and color. Two types of chromatic cone opponent cells have been reported, sometimes called red–green and blue–yellow cells [4, 5]. Such cells compare signals from different cone types. In the case of the red–green on-cell, abbreviated to *red-on*, the cell is excited by stimulation of the L-cones and inhibited by the stimulation of the M-cones.

From LGN, nerve signals are sent to the visual cortex, which can be thought of as divided in a number of functionally distinct areas (V1–V5). The idea is that cells within such an area are predominantly responsible for analyzing different properties of the retinal image, such as shape, motion, orientation, and color [6]. Area V4 is considered an area that is specialized in color processing, although its role as ''color center'' is under debate. A recent review of the research of the past 25 years on cortical processing of color signals has put more emphasis on the role of area V1 [7]. Since the different areas in visual cortex are interconnected and feature both forward and backward loops, it is indeed hard to imagine that a single brain area would take care of all the color processing. We have also learned that color cannot be considered as a completely isolated visual property, since it is always in interaction with shape, texture, contrast, and so on, which thus would require information exchange between specialized brain areas. It is clear, however, that the visual information in one area depends on the presence of information in a preceding area. Opponent cells were found in LGN and also in V1. Another type of opponent cells, double opponent cells, was found in the primary visual cortex. These cells are capable of both spatial and chromatic opponency and are optimally excited when the color in the center of the receptive field is the opposite color from the one in the surround. And to make it even more complex, these cells also show temporal opponent characteristics [8]. Using noninvasive imaging techniques such as PET (positron emission tomography) and fMRI (functional magnetic resonance imaging), many studies have reported on the mapping of brain activity, and many will follow. This will hopefully lead to a more complete understanding of the processes underlying color vision and perception, and how it integrates into higher order processes involving, for instance, emotion and behavior.

2.3 Chromatic Properties of the Visual System

2.3.1 Chromatic Adaptation

The dynamic range of the human visual system is very impressive, covering a light intensity range of about 10^{12}. This is achieved by adaptation to the ambient light level, a process in which the sensitivity to light is adjusted. Two variants of adaptation we are commonly aware of are *light adaptation* and *dark adaptation*, occurring whenever we change from a low light intensity to a high light intensity situation or vice versa. Light adaptation is a relatively fast process, in the order of seconds, whereas dark adaptation takes minutes to complete. Perhaps somewhat less noticeable is the process of chromatic adaptation, in which the sensitivities of the primary color channels (L, M, S) are individually adjusted. This has the effect of white-balancing because any color dominance is counterbalanced by the sensitivity readjustments. Chromatic adaptation is a continuous and spatially localized process, which may bring specific appearance effects when making eye movements after a period of fixation. Studies into the temporal characteristics of chromatic adaptation have shown that the underlying visual processes are characterized by both a fast and a slow component and are located at the receptor level as well as the cortical level [9, 10]. Figure 2.3 demonstrates the effect of chromatic adaption.

(a)

(b)

Figure 2.3 Demonstration of chromatic adaptation (inspired by the work of John Sadowski). Stare at the black dot in the image (a) for about 20 s, without blinking or moving your eyes. Then quickly look at the black spot in the center of the image (b). The image will appear as having natural colors for a brief period because of the aftereffect of chromatic adaptation.

2.3.2 Human Color Constancy

The spectral distribution of daylight changes during the day. Despite these changes, the color appearance of objects is remarkably stable, a phenomenon known as *color constancy*. Grass remains green throughout the day, whereas from a physical point of view the more reddish light toward the end of the day would predict the grass to appear brownish. Color constancy is considered a basic property of the visual system and has been intensively studied in the past few decades. There exist different approaches to solving the problem of color constancy, which focus on the question of how to disentangle the product of illumination and surface reflection that enters our eye. Reviews of human color constancy studies are presented by Smithson [11] and Foster [12]. An overview of the computational approach to color constancy by illuminant estimation is presented in Chapter 8. Contrary to what the term *constancy* may suggest, there is abundant psychophysical evidence, coming from different experimental paradigms, showing that human color constancy is *not* perfect. The degree of color constancy can be quantified using a constancy index ranging between 0 (no constancy at all) and 1 (perfect constancy). Foster [12] tabulated values for the constancy index for some 30 different experimental studies, showing widely varying values. Imperfect constancy implies that a change in the color of the illuminant is not fully discounted for by the visual system, which results in noticeable shifts in object colors. Figure 2.4 presents a demonstration of color constancy. Figure 2.4b shows the original scene, and Figure 2.4a shows a simulated change in the color of the global illuminant acting on the whole image. Although we easily perceive the global shift toward a purplish color, the fruit colors stay reasonably constant. If, on the other hand, the simulated change in the illuminant is locally restricted to the apple in the center of the fruit basket, color constancy is lost and the apple appears purple. This demonstrates the different effects of local versus global changes in the illumination.

2.3.2.1 Human Color Constancy by Ratios How can we explain the different appearances of the apple in the images (a) and (c) in Figure 2.4 while the physical light distributions reflected from the apples are identical? The key to the explanation is the fact that for the global change in illumination, ratios across object boundaries within the individual L-, M-, S-cone signals stay the same, whereas for the local illuminant change these ratios change. The latter results in

Figure 2.4 (a) Global change in illumination, (b) original image (standard image from ISO 12640:1997), and (c) local change in illumination. Note the very different appearance of the color of the apple for the global and the local illuminant change, although physically they are identical.

the perception of a completely different color, as if the apple had been replaced by a different object. Ratios across borders or edges also play an important role in the *retinex theory* [13, 14]. According to the theory, the visual system independently processes three images, each image belonging to one cone type (L, M, or S). Within each cone image, lightness values (so-called *designators*) are calculated from spatial comparisons of the reflectance at a specific point to the maximum reflectance in the image. The combination of the three lightness values occupies a point in a three-dimensional space and determines the color. Retinex theory was shown to correlate well with visual perception and received a lot of attention from vision researchers (both in a positive and a negative way). Hurlbert [15] showed that several other lightness algorithms, all having the retinex algorithm as their precursor, are formally connected by one and the same mathematical formula. We refer to Chapter 5 where the role of color ratios for computational color constancy is discussed.

2.3.2.2 Human Color Constancy by Chromatic Adaptation An alternative explanation of color constancy has a physiological basis. A well-known and often used chromatic adaptation model is the *coefficient rule* of von Kries [16]. It states that the sensitivities of the three cone types are regulated by cone-specific gain factors that are inversely proportional to the level of cone stimulation. To illustrate, let us assume that we are in a room in which we adapt to neutral (white) illumination that stimulates the L-, M-, and S-cones in equal amounts. Within the room are several colored objects and also a white object. Now we change the room illumination from neutral toward blue such that the S-cone system is stimulated twice as much, whereas the L- and M-cone stimulation remains unaffected. According to the von Kries coefficient law, the sensitivity of the S-cone system will be reduced by a factor of 2 to effectively rebalance the L-, M-, S-cone stimulation. For the white object, which takes on the illuminant color, this will result in unchanged cone stimulations, implying that von Kries adaptation permits perfect color constancy for the white object. For the colored objects in the room, however, perfect color constancy is not guaranteed because the interaction between the illuminant spectrum and the surface reflectance may result in S-cone ratios being different from 2.

Helson [17] proposed an adaptation model in which the visual system is adapted to a medium gray level. Objects with reflectances above that of the adaptation level take on the color of the illuminant, whereas objects with reflectances below that of the adaptation level take on the complementary color. This effect is known as the *Helson–Judd Effect*.

2.3.3 Spatial Interactions

The perceived color of an object is determined not only by the light coming from that object but also by the light coming from neighboring objects in the scene. Colors seen in complete isolation, such as a patch of color on a black background presented on a color display, can appear as if they are self-luminous and emit light.

When put in context of other colors, however, the appearance is different and dependent on the exact definition of the surrounding colors. Two important spatial interactions are mentioned here, which influence color perception, *contrast*, and *assimilation*. In contrast effects, the difference between a color and its surround is enhanced so that the two will look more different. The effect can be interpreted as an induction effect, whereby the color complementary to that of the surround is induced into the center. Different surrounds may give dramatically different effects, as demonstrated in Figure 2.5.

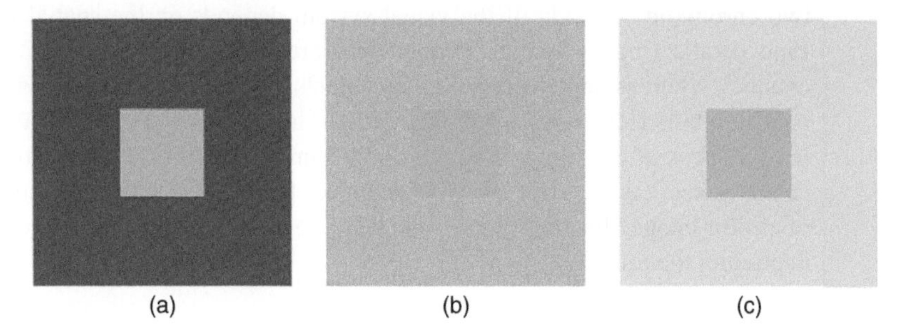

Figure 2.5 Simultaneous color contrast: the center squares are physically identical but appear different because of a difference in surround color.

The effect of assimilation, on the other hand, is the opposite of the contrast effect because with assimilation the difference between a color region and the adjacent color appears smaller. This leads to the perception that the color seems to be shifted toward that of the surrounding color. Figure 2.6 demonstrates how

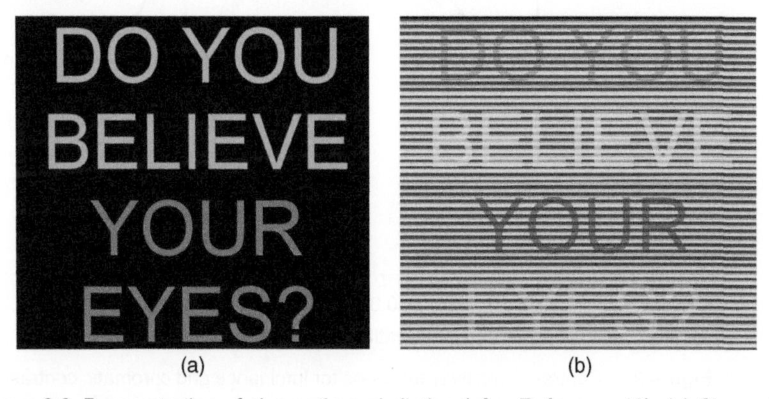

Figure 2.6 Demonstration of chromatic assimilation (after Reference 18). (a) Shows four lines of text, the first two and the last two having the same color. When placed on differently colored backgrounds and "behind" thin colored stripes, the color of the stripes seems to spread into the color of the words. Physically, the colors of the text in (a) and the uncovered parts of the text in (b) are identical.

the perceived color of text may change completely. It appears that the color of the stripes covering the text spreads out into the text. In other words, the surrounding color induces its color into the target color.

The demonstrations in Figures 2.5 and 2.6 are dependent on viewing distance, or more precisely, on the visual angles that the details subtend on the retina. We already mentioned that the number of S-cones is much less than that of the L- and M-cones; therefore they sample the retinal image at a lower spatial resolution. This has consequences also for the spatial resolution of the blue–yellow channel. Figure 2.7 shows how the contrast sensitivity of the achromatic channel and the two chromatic channels of the visual system depends on the spatial frequency. Fine details (higher spatial frequencies) are best detected by the luminance channel, whereas the two chromatic channels are better equipped to detect more coarse details (lower spatial frequencies). This property of the visual system is used successfully in image compression techniques. Since the chromatic channels cannot detect (at a certain viewing distance) the high spatial frequency contents of a color image, this information can be removed or compressed without visually degrading the image.

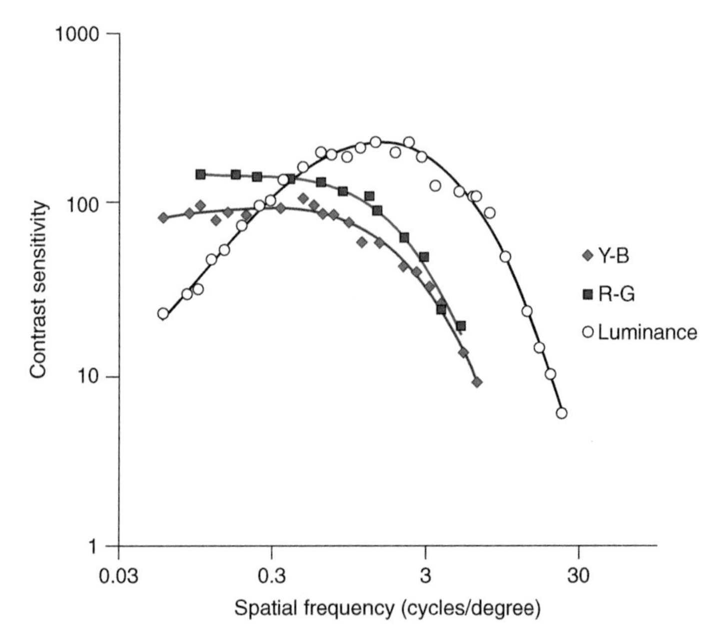

Figure 2.7 Contrast sensitivity functions for luminance and chromatic contrast, as a function of spatial frequency. *Source:* Replotted from Figures 7 and 9 in Reference 19. Solid lines represent fits to the data. Note the difference between the low pass characteristic of the chromatic channels and the bandpass characteristic of the achromatic channel.

Spatial effects can occur only when some form of spatial comparison is performed by the visual system. We already noted the importance of center-surround cells for vision because they allow the detection of intensity and color edges. Mathematically, these edge detectors are obtained by taking spatial derivatives, as presented in Chapter 6.

2.3.4 Chromatic Discrimination and Color Deficiency

A number of studies have focused on the question of how many colors can be perceived by humans. There is no single answer to this question, since it depends on the criteria used for counting discriminable colors. As a result, estimates vary from order 10^3 to 10^6. If we go out to buy a can of red paint to match the color of a tomato we saw earlier that day, chances are very high that the two colors will not match. Humans are far better in seeing differences between colors (relative color) than in memorizing absolute colors. Early measurements of chromatic discrimination thresholds [20] have laid the basis for the developments of a perceptually uniform color space (CIELAB), and the derivation of mathematical formulae to quantify color differences [21]. The latter are abundantly used in industry.

There exist various tests to measure someone's chromatic discrimination ability. Even for normal trichromats, people with ''normal'' color vision, this ability may change from person to person. There are different ways in which color vision may be impaired; usually the distinction is made between *acquired* and *congenital* color vision deficiencies. Aging causes the ocular media to become more yellow, which reduces color discrimination along the yellow–blue axis of color space [22]. Some diseases, alcohol consumption [23], medication, and drugs [24] can negatively affect color vision abilities. These are examples of acquired color vision deficiencies. With congenital deficiencies, abnormalities in the photopigments are inherited and are already present at birth. This affects about 8% of men and 0.45% of women. The spectral sensitivities of the photopigments can differ from normal trichromats in many different ways. The terms *protan*, *deutan*, and *tritan* are used to indicate that the L-, M-, and S-cone, respectively, are abnormal. We can indicate the severeness of this abnormality by a number ranging between 0 (cone type missing) and 1 (normal). If the abnormality is somewhere in between 0 and 1, we speak of *anomalous trichromats*. If one cone pigment is missing, only two functional cone types are left, resulting in dichromatic color vision. Depending on the cone type that is lacking (L, M, or S), *dichromats* are characterized as *protanopes*, *deuteranopes*, or *tritanopes*. Color discrimination for dichromats is strongly reduced as illustrated in Figure 2.8.

It is mistaken belief that color-deficient people are not able to see color, as the term *color blind* would suggest. What is meant is that they are less well able to discriminate colors; some colors are confused, which can be graphically shown in color space (Fig. 2.9). Colors located on the so-called *confusion lines* cannot be distinguished, and hence appear equal. For the different types of deficiency, the confusion lines originate in different copunctal points.

Figure 2.8 (a) Original image. (b) Simulated appearance for a deuteranope (missing the M-cone photopigment). Simulated image obtained with the TNO color deficiency simulator.

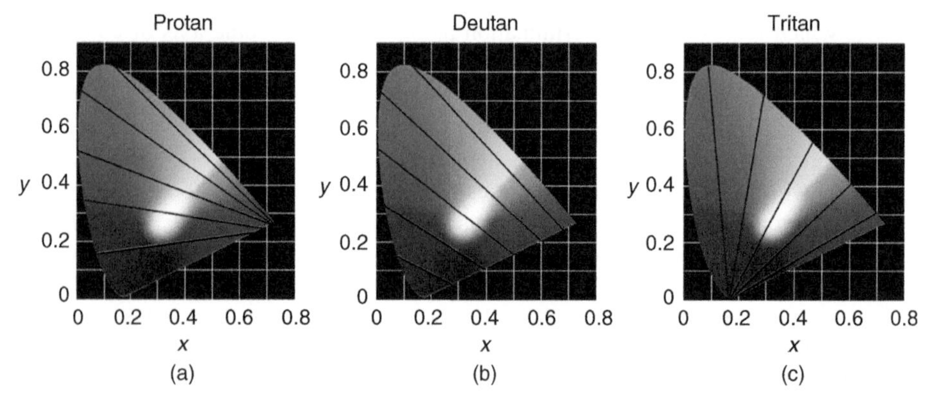

Figure 2.9 CIE 1931 x, y chromaticity space showing confusion lines for a protan, deutan, and tritan. Colors located on such confusion lines are not distinguished by color deficients.

2.4 Summary

The different stages of color information processing along the human visual pathway have been highlighted. Color vision begins with the absorption of light in the three cone types at the retinal level. Cone responses are spatially compared and transformed to three opponent color signals (one achromatic and two chromatic), traveling along the optic nerve from LGN to the visual cortex, where the perception of color eventually takes place. We discuss important chromatic properties of the visual system, such as chromatic adaptation and color constancy, which provide demonstrations of spatial interactions and finally take a look at color deficiency.

3 Color Image Formation

The image formation process described in this chapter involves three processes (illumination, material reflection, and detection/observation) interacting to generate the final color image. The process starts with light, which illuminates the visual scene. Light is described as electromagnetic radiation of a certain intensity, consisting of particles (photons) containing energy of certain wavelengths, each photon traveling in a certain direction. When many of the photons travel in the same direction, the light is directed and forms a beam of light. When all photons travel in a random direction, the light is diffuse. Light is typically emitted by light sources. A light source can be characterized by the way the light bundle is directed and by the emitted spectra of photons over the wavelengths. When more photons of short wavelength are emitted relative to the long wavelengths, the color of the light source is bluish. When more photons of long wavelengths are emitted, the color is reddish. For candle light and halogen illumination, the emitted spectra follow that of a so-called black body radiator [25], for which the smooth emitted spectra can be uniquely characterized by a single number, being the temperature of the radiator. As many natural light sources emit spectra that are similar in color to such a black body radiator, the color of a light source is defined by the ''correlated color temperature,'' that is, the temperature of a black body radiator at which a similar color is perceived. However, keep in mind that there are many nonnatural light sources (such as fluorescent light) that might have a color quite similar to black body radiators, but with a spectrum very different from that of the smooth blackbody radiator.

The second process in image formation involves materials. Materials in the scene interact with the incoming light, causing its reflection (Fig. 3.1). Materials absorb photons, reflecting only part of the light hitting the material. In case of ''white'' materials, most of the photons are reflected. For ''black'' materials, most of the photons are absorbed. Hence, in a certain way, materials modulate the light

Color in Computer Vision: Fundamentals and Applications, First Edition.
Theo Gevers, Arjan Gijsenij, Joost van de Weijer, and Jan-Mark Geusebroek.

intensity. Furthermore, the amount of light particles absorbed by a material may depend on the wavelength of the photons. Depending on the material properties, photons of certain wavelengths may be absorbed, whereas others are reflected. In that case, the light is spectrally modulated by such a material, causing it to appear, for example, reddish when all middle and short wavelengths are absorbed, and others reflected. The Kubelka–Munk theory presented in Section 3.3 models such effects in detail.

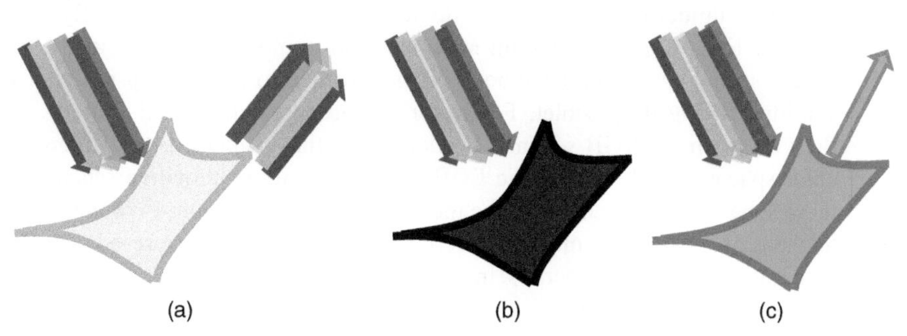

Figure 3.1 Example of white light interacting with material. White light contains energy over all wavelengths of the visual spectrum. (a) When interacting with a perfect white surface all light is reflected. (b) In the case of a perfect black object the light is absorbed. (c) In case of a blue material only the spectrum representing blue light is reflected and the other wavelengths are absorbed by the material.

Besides the absorption inside the material, there is an effect when light hits the material. When light hits the material, light changes medium by going from air through the material boundary. As such, part of the light is reflected at the ‘‘interface’’ between air and the material, which causes Fresnel reflection or, the term we use in this book, interface reflection. Interface reflection is responsible for specularities (also called *highlights*) on objects. This aspect is taken into account in the dichromatic reflection model in Section 3.2.

In the third process in image formation, light is recorded or observed by a camera or the eye. Here, the photons are registered, by integrating the energy over a certain bandwidth, a certain spatial area, and for a certain period of time. For the eye (Section 2.2.1), the integration is performed over three spectral broadbands, covering the short-, middle-, and long-range wavelengths of the visual spectrum. Integration time is around 50 ms, and the visual acuity depends on the position of the retina at which the light falls. Acuity is highest for the central (fovea) area of the retina, and falls off toward the periphery. Color cameras more or less mimic the temporal and spectral characteristics of the eye, being composed to record three color bands in about 50 ms, and often have a uniform spatial resolution in the order of megapixels.

The modeling of the interactions between light, materials, and the observation process together is the main goal of reflection models. Such models aim at a simplification of the physics involved in order to understand certain aspects of the process. Different models make different simplifications, thereby being more suitable under different conditions, or more suitable for different mathematical frameworks, than alternative models. In this chapter, we discuss the most relevant models used in computer vision.

Besides reflection modeling, one also needs to quantize the resulting color information. As computers deal with numbers, the registration of the spectral information at each pixel of the camera should be condensed into numbers. Any arbitrary numbering scheme could do, as long as each number uniquely defines a color. For example, a coding scheme could number the colors in the same order as they appear in a rainbow, starting with zero being deep red and, say, 1 million being deep violet. For historical reasons, commercial cameras yield their results in an RGB scheme. Section 3.5 in this chapter deals with common color spaces, which reorder the RGB information into alternative schemes. All these schemes essentially describe the same color information. However, similar to the color formation models, a certain scheme might be more advantageous to highlight certain properties in color information than another scheme. For example, the well-known hue-saturation-value model decomposes the RGB values into an orthogonal color scheme, where the (achromatic) intensity information is independent of the chromatic information. As seen in this chapter, there are many color coding schemes, each with its own advantages and disadvantages when used for computer vision.

3.1 Lambertian Reflection Model

Many computer vision applications are based on the assumption of Lambertian reflectance, which means that the intensity of light reflected by the surface is independent of the viewing angle. The surface luminance is said to be isotropic. Materials that have this property are called *matte materials*. Examples of Lambertian reflectances are chalk, paper, and unfinished wood.

Consider that the illumination of the scene is given by, $e(\lambda, \mathbf{x})$, where λ is the wavelength and $\mathbf{x}$ is the spatial position in the image. Often we assume the spectral distribution of the light source to be spatially uniform across the scene. In that case, we write $e(\lambda)$. The reflected energy (i.e., radiance) from the surface E is given by

$$E(\lambda, \mathbf{x}) = m^b(\mathbf{x})\, s(\lambda, \mathbf{x})\, e(\lambda, \mathbf{x}), \tag{3.1}$$

where s is the surface albedo that describes the spectral reflectance properties of the material. The geometric dependence of the reflectance is described by the term m^b and depends on the light source direction and surface orientation according to $m^b = cos(\alpha)$, where α is the angle between the surface normal and the illumination direction. $\mathbf{x}$ denotes the spatial coordinates of the image, and we apply bold face to indicate vectors.

The measured observation values, $\mathbf{f}^{\mathbf{RGB}} = (R, G, B)$, of the camera with spectral sensitivities $\rho^c(\lambda)$, $c \in \{R, G, B\}$, are modeled by integrating over the visible spectrum ω,

$$f^c(\mathbf{x}) = \int_\omega E(\lambda, \mathbf{x})\, \rho^c(\lambda)\, \mathrm{d}\lambda \tag{3.2}$$

$$= m^b(\mathbf{x}) \int_\omega s(\lambda, \mathbf{x})\, e(\lambda)\, \rho^c(\lambda)\, \mathrm{d}\lambda. \tag{3.3}$$

This can also be written in vectors as

$$\mathbf{f}(\mathbf{x}) = m^b(\mathbf{x})\, \mathbf{c}^{\mathbf{b}}(\mathbf{x}), \tag{3.4}$$

where the body reflectance

$$\mathbf{c}^{\mathbf{b}}(\mathbf{x}) = \int_\omega s(\lambda, \mathbf{x})\, e(\lambda) \rho^c(\lambda)\, \mathrm{d}\lambda. \tag{3.5}$$

The Lambertian model predicts that the pixels on a single colored object lie on a line passing through the origin of the RGB cube. Note that for many materials the Lambertian assumption does not hold in the strict sense. For example, materials might be glossy, causing specularities at some spots on the material. Also for these materials, the Lambertian assumption can be a good approximation since often the specularities only occupy a small part of the objects. However, there are better approximations of material properties in these cases, as discussed in the following text.

3.2 Dichromatic Reflection Model

The Lambertian model does not include reflections such as specularities (highlights). The *dichromatic reflection model* (DRM) includes the interface reflection or Fresnel reflection, which allows the anisotropic reflections of specularities.

The DRM is proposed by Shafer [26] and is, besides Lambert's law, one of the most popular reflection models in computer vision. The model focuses on the color aspects of light reflection and has only limited usage for geometry recovery of scenes. The model assumes a single light source in the scene. It separates reflectance into surface body reflectance and interface reflectance. The model is valid for the class of inhomogeneous materials, which covers a wide range of materials such as wood, paints, papers, and plastics (but excludes homogeneous materials such as metals). The DRM is the summation of the body reflectance (superscript b) and the interface reflectance (superscript i):

$$f^c(\mathbf{x}) = m^b(\mathbf{x}) \int_\omega s(\lambda, \mathbf{x}) e(\lambda) \rho^c(\lambda) \, \mathrm{d}\lambda + m^i(\mathbf{x}) \int_\omega i(\lambda) e(\lambda) \rho^c(\lambda) \, \mathrm{d}\lambda. \quad (3.6)$$

Note that for $m^i(\mathbf{x}) = \mathbf{0}$, this equation is equal to Equation 3.3. We will assume neutral interface reflection (NIR), meaning that the Fresnel reflectance i is independent of λ. Accordingly, we will omit i in further equations. The geometric dependence of the reflectance is described by the terms m^b and m^i, which depend on the viewing angle, light source direction, and surface orientation.

In many cases, we can assume the illumination in a scene to be white, and hence $e(\lambda) = i$ is constant. This can be obtained by, for example, white balancing or estimation of the light source as discussed in Part III of this book. Removing the dependence on $e(\lambda)$ yields

$$f^c(\mathbf{x}) = m^b(\mathbf{x}) \int_\omega s(\lambda, \mathbf{x}) \rho^c(\lambda) \, \mathrm{d}\lambda + m^i(\mathbf{x}) \int_\omega \rho^c(\lambda) \, \mathrm{d}\lambda, \quad (3.7)$$

where the constant factor i is incorporated in the geometrical terms m^b and m^i. When we further assume that the area under the sensitivity functions ρ is approximately the same, called the *integral white condition*, that is, $\int_\lambda \rho_R(\lambda)\mathrm{d}\lambda = \int_\lambda \rho_G(\lambda)\mathrm{d}\lambda = \int_\lambda \rho_B(\lambda)\mathrm{d}\lambda = 1$, the equation simplifies to

$$f^c(\mathbf{x}) = m^b(\mathbf{x}) \int_\omega s(\lambda, \mathbf{x}) \rho^c(\lambda) \, \mathrm{d}\lambda + m^i(\mathbf{x}). \quad (3.8)$$

An insightful way to interpret the DRM is by representing it in vector notation. Then we can write Equation 3.6 as

$$\mathbf{f}(\mathbf{x}) = m^b(\mathbf{x}) \, \mathbf{c^b}(\mathbf{x}) + m^i(\mathbf{x}) \, \mathbf{c^i}(\mathbf{x}), \quad (3.9)$$

and Equation 3.8 as

$$\mathbf{f}(\mathbf{x}) = m^b(\mathbf{x}) \, \mathbf{c^b}(\mathbf{x}) + m^i(\mathbf{x}). \quad (3.10)$$

The reflection of the light consists of two parts: (i) the body reflection part $m^b(\mathbf{x}) \, \mathbf{c}^b$, which describes the light that is reflected after interaction with the surface albedo, and (ii) the interface reflection $m^i(\mathbf{x})\mathbf{c}^i$, which describes the part of the light that is immediately reflected at the surface, causing specularities. Both parts consist of a geometrical part dependent on the location in the scene, and a spectral part dependent on the spectral wavelength. The dichromatic reflection

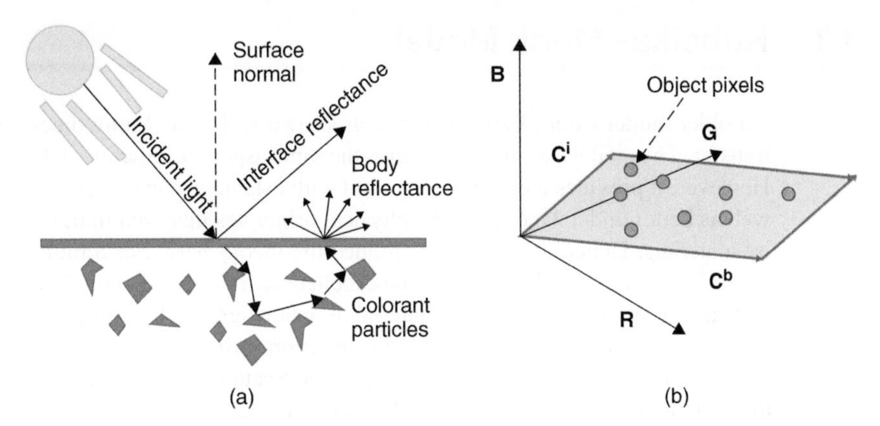

Figure 3.2 Interaction of light with material according to the dichromatic reflection model. (a) The reflected light consists of two parts, the body reflection and the interface reflection. (b) The dichromatic reflection model predicts that the pixel values of a single colored object lie on a parallelogram formed by the vectors of the body and the interface reflectance.

model projects pixel values of a single colored object $\mathbf{f}(\mathbf{x})$ onto a parallelogram (Fig. 3.2). The position on the parallelogram is determined by the amount of body reflectance and the amount of interface reflectance.

When the assumptions made by the original DRM are not met, more complex reflectance models are required. One such case is the presence of ambient light, that is, light coming from all directions. Ambient light is present in outdoor scenes where next to the dominant illuminant, that is, the sun, there is diffuse light coming from the sky. Similarly, it is present in indoor situations where diffuse light is caused by reflection from walls and ceilings. Shafer [26] models the diffuse light, $\mathbf{a}$, by a third term

$$\mathbf{f}(\mathbf{x}) = m^b(\mathbf{x})\,\mathbf{c}^\mathbf{b}(\mathbf{x}) + m^i(\mathbf{x})\,\mathbf{c}^\mathbf{i}(\mathbf{x}) + \mathbf{a}. \tag{3.11}$$

Later work improved the modeling [27] and showed that the ambient term results in an object-color-dependent offset, which could be crucial in handling the case of colored shadows.

The original application to which the DRM was applied was the separation of shading from specularities [26]. The specularities, being dependent on scene incidental events such as viewpoint and surface normal, could be removed to simplify color image understanding. The removal of specularities allowed for improved segmentation algorithms [29, 28]. Throughout this book, we see several applications of the DRM, such as color constancy, photometric invariant feature computation, and color image segmentation.

3.3 Kubelka–Munk Model

An older model from physics is the well-known Kubelka–Munk theory of light transfer. The model essentially captures the same aspects of Shafer's DRM model. However, it provides a better background with established experimental work, as well as better understanding of the physical principles involved in light reflection by materials. Hence, we shortly introduce the theory here. For a more involved explanation, we refer the reader to the excellent work by Judd and Wyszecki [30].

Transfer of light through a material is characterized by three fundamental processes: absorption, scattering, and emission. Absorption is the process by which radiant energy is transformed into heat. Scattering is the process by which the radiant energy is diffused toward different directions. Emission is the process by which new radiant energy is created (not considered in this book). The Kubelka–Munk theory models the effect of these processes under the assumption of an one-dimensional light flux, thereby implying isotropic scattering within the material [25, 30–32]. Under this assumption, the material layer (i.e., the object surface) is characterized by a wavelength-dependent scatter coefficient and absorption coefficient. The class of materials for which the theory is useful ranges from dyed paper and textiles, opaque plastics, paint films, liquids, up to enamel and dental silicate cements. The model may be applied to both reflecting and transparent materials.

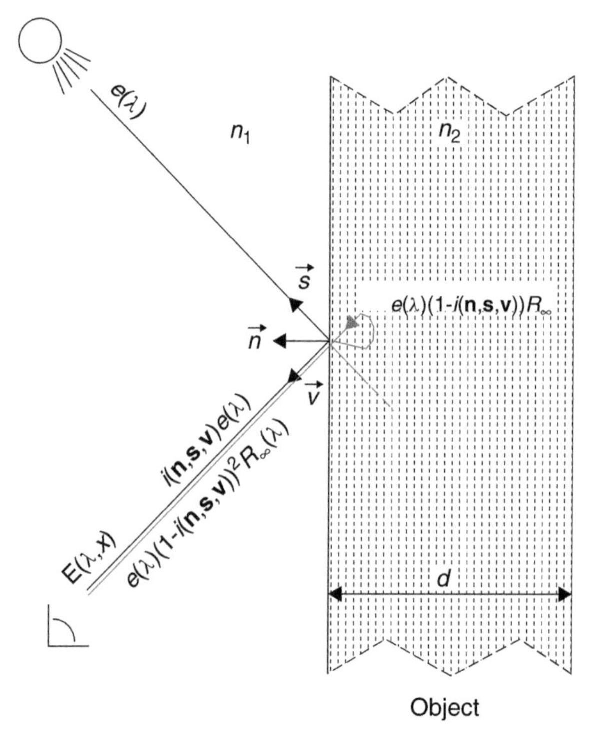

Figure 3.3 The various aspects involved in the Kubelka–Munk model.

Consider a homogeneously colored material patch of uniform thickness d and infinitesimal area (Fig. 3.3) characterized by its absorption coefficient $K_a(\lambda)$ and scatter coefficient $K_s(\lambda)$. When illuminated by light with spectral distribution $e(\lambda)$, light scattering within the material causes diffuse body reflection, while Fresnel interface reflection occurs at the surface boundaries. Fresnel reflection at the back can be neglected when the thickness of the layer is such that further increase in thickness does not affect the reflected color. In that case, we consider the material to have infinite optical thickness.

The incident light is partly reflected at the front surface and partly enters the material, is isotropically scattered, and a part again passes the front surface boundary. The reflected spectrum in the viewing direction $\mathbf{v}$, ignoring secondary scattering after internal boundary reflection, is given by

$$E(\lambda) = e(\lambda)(1 - i(\lambda, \mathbf{n}, \mathbf{s}, \mathbf{v}))^2 R_\infty(\lambda) + e(\lambda) i(\lambda, \mathbf{n}, \mathbf{s}, \mathbf{v}), \tag{3.12}$$

where $\mathbf{n}$ is the surface patch normal and $\mathbf{s}$ is the direction of the illumination source. Further, i is the Fresnel interface reflection coefficient in the viewing direction $\mathbf{v}$. The body reflectance $R_\infty(\lambda) = a(\lambda) - b(\lambda)$ depends on the absorption and scattering coefficients $K_a(\lambda)$ and $K_s(\lambda)$ by

$$a(\lambda) = 1 + \frac{K_a(\lambda)}{K_s(\lambda)}, \tag{3.13}$$

$$b(\lambda) = \sqrt{a(\lambda)^2 - 1}. \tag{3.14}$$

Note that the body reflectance $R_\infty(\lambda)$ is equivalent to the surface reflectance $s(\lambda)$ discussed earlier.

Assuming Neutral Interface Reflectance (NIR) (Section 3.2), the reflection model of Equation 3.12 simplifies as follows:

$$E(\lambda) = e(\lambda)(1 - i(\mathbf{n}, \mathbf{s}, \mathbf{v}))^2 R_\infty(\lambda) + e(\lambda) i(\mathbf{n}, \mathbf{s}, \mathbf{v}). \tag{3.15}$$

Using this model, perfectly diffuse surfaces are modeled by substituting $i(\cdot) = 0$, while perfect mirroring of the light source is modeled by substituting $i(\cdot) = 1$. In reality, $i(\cdot)$ will assume a value somewhere in between 0 and 1, which results in the spectral color $E(\lambda)$, which is an additive mixture of the color of the light source and the perfectly diffuse body reflectance color, similar as in the DRM model.

Because the 3D coordinates are projected onto a 2D image plane, the vectors $\mathbf{n}$, $\mathbf{s}$, and $\mathbf{v}$ depend on the position in the image. The energy of the incoming spectrum at spatial location $\mathbf{x}$ on the image plane is then related to

$$E(\lambda, \mathbf{x}) = e(\lambda, \mathbf{x})(1 - i(\mathbf{x}))^2 R_\infty(\lambda, \mathbf{x}) + e(\lambda, \mathbf{x}) i(\mathbf{x}), \tag{3.16}$$

where the spectral distribution at each point $\mathbf{x}$ is generated off a specific material patch. When the following substitutions are made

$$c^b(\lambda, \mathbf{x}) = e(\lambda, \mathbf{x}) R_\infty(\lambda, \mathbf{x}), \tag{3.17}$$

$$c^i(\lambda, \mathbf{x}) = e(\lambda, \mathbf{x}), \tag{3.18}$$

$$m^b(\mathbf{x}) = (1 - i(\mathbf{x}))^2, \tag{3.19}$$

$$m^i(\mathbf{x}) = i(\mathbf{x}), \tag{3.20}$$

Equation 3.16 reduces to the dichromatic reflection model as proposed by Shafer [26]:

$$E(\lambda, \mathbf{x}) = m^b(\mathbf{x}) c^b(\lambda, \mathbf{x}) + m^i(\mathbf{x}) c^i(\lambda, \mathbf{x}), \tag{3.21}$$

and hence

$$f^c(\mathbf{x}) = \int_\omega E(\lambda, \mathbf{x}) \, \rho^c(\lambda) \, d\lambda. \tag{3.22}$$

However, do note that the coefficients m^b and m^i are dependent on each other, as can be derived from the above equation. Further simplification is obtained by assuming only matte, or dull surfaces, for which specular reflection is negligible, that is, $i(\mathbf{x}) \approx 0$, and Equation 3.16 reduces to the Lambertian model for diffuse body reflection:

$$E(\lambda, \mathbf{x}) = e(\lambda, \mathbf{x}) R_\infty(\lambda, \mathbf{x}). \tag{3.23}$$

The Kubelka–Munk theory also generalizes the case of transmission of light. Assuming the material layer is of limited optical thickness, light will leave the material at the side opposite of the one it entered. In that case, the absorption and scattering within the material cause an exponential decay of the light intensity, resulting in the well-known Beer–Lambert equation

$$E(\lambda, \mathbf{x}) = e(\lambda, \mathbf{x}) \exp\left\{-d(\mathbf{x}) c(\mathbf{x}) \alpha(\lambda, \mathbf{x}))\right\}, \tag{3.24}$$

where d is the local thickness of the layer, c is the concentration of the colorant particles, and α indicates the absorption and scattering coefficient for the colorant particles. Again, e is the emitted spectrum of the illuminant. The law plays an important role in, for example, transmissive light microscopy.

3.4 The Diagonal Model

The colors in a scene change significantly with variations in the light source color. Part III of this book is dedicated to estimating the light source of a scene. Here, we shortly discuss the diagonal model that predicts the change in the camera responses under variations of the illuminant.

The *diagonal transform* or *von Kries Model* [33] is given by

$$\begin{pmatrix} R_c \\ G_c \\ B_c \end{pmatrix} = \begin{pmatrix} a & 0 & 0 \\ 0 & b & 0 \\ 0 & 0 & c \end{pmatrix} \begin{pmatrix} R_u \\ G_u \\ B_u \end{pmatrix}, \tag{3.25}$$

or in short

$$\mathbf{f}_c = \mathbf{D}_{u,c}\mathbf{f}_u, \tag{3.26}$$

where $\mathbf{f}_u$ is the image taken under an unknown light source, $\mathbf{f}_c$ is the same image transformed, so that it appears as if it was taken under the canonical illuminant, and $\mathbf{D}_{u,c}$ is a diagonal matrix that maps colors that are taken under an unknown light source u to their corresponding colors under the canonical illuminant c. Often a white illuminant such as D65 is used as the canonical reference illuminant.

The diagonal model can be derived by assuming dirac delta functions for the camera sensitivities $\rho^c(\lambda) = \delta(\lambda_c)$. If we substitute this into Equation 3.3, then

$$f^c(\mathbf{x}) = m^b(\mathbf{x})\, s(\lambda_c, \mathbf{x})\, e(\lambda_c), \tag{3.27}$$

at wavelength λ_c. If we consider two different illuminants e^1 and e^2, then the relation between the two is given by

$$\frac{f_1^c(\mathbf{x})}{f_2^c(\mathbf{x})} = \frac{m^b(\mathbf{x})\, s(\lambda_c, \mathbf{x})\, e^1(\lambda_c)}{m^b(\mathbf{x})\, s(\lambda_c, \mathbf{x})\, e^2(\lambda_c)} = \frac{e^1(\lambda_c)}{e^2(\lambda_c)} \tag{3.28}$$

and hence the relation between the camera responses under the two illuminants can be modeled by $\mathbf{f}_1 = \mathbf{D}_{1,2}\mathbf{f}_2$, where $\mathbf{D}_{1,2}$ is a diagonal matrix.

To include a diffuse light term, Finlayson et al. [34] extended the diagonal model with an offset $(o1, o2, o3)$, resulting in the diagonal-offset model:

$$\begin{pmatrix} R_c \\ G_c \\ B_c \end{pmatrix} = \begin{pmatrix} a & 0 & 0 \\ 0 & b & 0 \\ 0 & 0 & c \end{pmatrix} \begin{pmatrix} R_u \\ G_u \\ B_u \end{pmatrix} + \begin{pmatrix} o1 \\ o2 \\ o3 \end{pmatrix}. \tag{3.29}$$

Although this model is merely an approximation of illuminant change and might not accurately be able to model photometric changes, it is widely accepted as a color correction model and is at the base of many color constancy algorithms. Several improvements of the diagonal transformation exist, such as changing the color basis [35] and applying spectral sharpening [36].

3.5 Color Spaces

After light is reflected off an object surface, it can be detected and "measured" by a human observer or by a color camera. The light that reaches the sensor (either eye or camera) is the result of the interaction between the spectral power distribution of the light source $e(\lambda)$ and the spectral reflectance distribution of the object $s(\lambda)$ and is modeled using any of the equations for E in the previous section. These distributions can be transformed into actual color signals as follows:

$$f^c(\mathbf{x}) = \int_\omega E(\lambda, \mathbf{x})\rho^c(\lambda)\mathrm{d}\lambda, \quad (3.30)$$

where the simplest model, being the Lambertian model, obtains the tristimulus values f^c by integrating the product of the three components at each wavelength of the visible spectrum ω. The three components are the spectral power distribution of the light source $e(\lambda)$, the reflectance distribution of the object $s(\lambda)$, and the sensitivity function $\rho^c(\lambda)$ of the sensor. Trichromacy theory indicates that three channels are required to generate the full range of human visible colors, so three camera sensitivities need to be defined to specify the sensitivity of the sensor to the incoming spectral power distribution.

3.5.1 *XYZ* System

From Equation 3.30 it becomes obvious that the camera sensitivities significantly influence the final color values. As one would like cameras to "perceive" spectra similarly as the human eye does, a specification of the human sensitivities is required. However, the human visual system in operation can only be probed as a black box, to which one can ask questions such as "are these two colors similar?" This is exactly how knowledge about the sensitivities of the eye is obtained [37]. Two panels are shown, one illuminated by a test light source of arbitrary color and the other illuminated by a mixture of three primary light sources. Now, an observer is asked to change the intensities of the primary light sources until the two test panels appear equivalent. That is, one cannot distinguish a color edge in between the opposing panels. The intensity values of the three primary light sources (the *tristimulus values*) now indicate a "color matching pair." When choosing three monochromatic (small-band) light sources for the primaries, and taking monochromatic test light sources of various wavelengths, the resulting intensity values per test wavelength yield the so-called *color matching functions*. These color matching functions essentially describe the human chromatic response. Unfortunately, for some wavelengths in the above-described experiment, no satisfactory color match could be obtained. In that case, the experimenter had to move one of the primaries from one panel, to add to the test light source at the other panel. This effectively resulted in negative values of the color matching functions. To circumvent this, the Commission Internationale de l'Éclairage (CIE) introduced three imaginary primaries, X, Y, and Z, that result in only positive tristimulus values of the associated color matching functions.

To match the human visual system as closely as possible, the CIE introduced two standards: the CIE 1931 2° standard observer (shortened to the 2° standard observer) and the CIE 1964 10° standard observer (shortened to 10° standard observer). The first should be used to model an observer with a narrow field of view (~0.4 in at reading distance of 10 in), while the latter standard corresponds to visual matching of larger sampling (~1.9 in at reading distance of 10 in).

With three color matching functions of the standard observer, three numbers (called the *tristimulus values*) can be computed equivalent to the response of a standard observer:

$$X = \int_\lambda E(\lambda, \mathbf{x})\bar{x}(\lambda)\mathrm{d}\lambda, \tag{3.31}$$

$$Y = \int_\lambda E(\lambda, \mathbf{x})\bar{y}(\lambda)\mathrm{d}\lambda, \tag{3.32}$$

$$Z = \int_\lambda E(\lambda, \mathbf{x})\bar{z}(\lambda)\mathrm{d}\lambda, \tag{3.33}$$

where $\bar{x}(\lambda)$, $\bar{y}(\lambda)$, and $\bar{z}(\lambda)$ are the CIE color matching functions of either the 2° CIE standard observer or the 10° CIE standard observer (Fig. 3.4). These *XYZ* values can be converted to chromaticity coordinates to describe the chromaticity of the color

$$x = \frac{X}{X+Y+Z}, \tag{3.34}$$

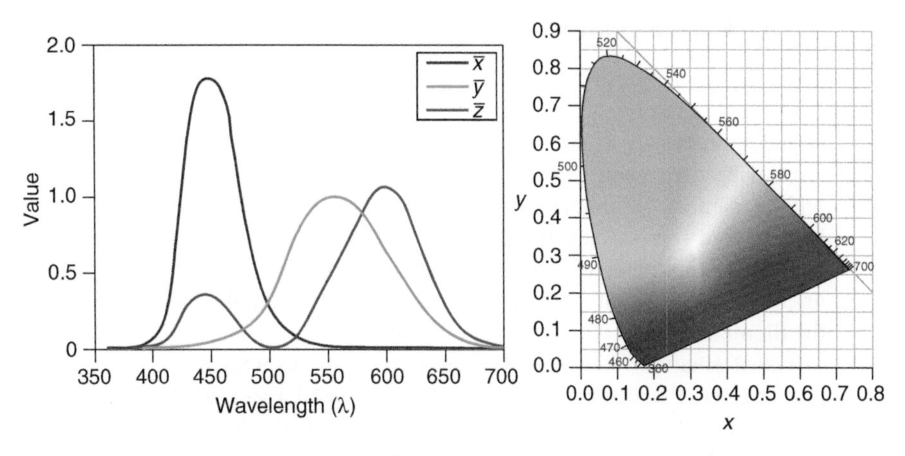

Figure 3.4 (a) Color matching functions of the CIE 1931 2° standard observer. (b) CIE 1931 *xy* chromaticity diagram. The horseshoe shape indicates the gamut of visible colors in the *xy*-plane. On the outside curve (the spectral locus) the wavelengths are indicated.

$$y = \frac{Y}{X + Y + Z}, \tag{3.35}$$

$$z = \frac{Z}{X + Y + Z}. \tag{3.36}$$

Since the intensity of the color is factored out and the sum of these chromaticity values equals unity, only two chromaticity values are sufficient to describe the color. However, in order to retain full information on the color as well as the intensity, it is often specified in terms of two chromaticity channels x and y, as well as the intensity channel Y, resulting in xyY color space. The visible spectrum forms a horseshoe shape in the xy-plane as can be seen in Figure 3.4. Values outside the horseshoe shape are not visible for humans. Conversion from xyY to the original XYZ coordinates is specified as follows:

$$X = \frac{xY}{y}, \tag{3.37}$$

$$Y = Y, \tag{3.38}$$

$$Z = \frac{(1 - x - y)Y}{y}. \tag{3.39}$$

Using Equations 3.31–3.36, precise numerical values can be assigned to the color sensation of a standard observer in an objective manner. In fact, the XYZ system introduced by CIE is the scientific basis of objective color measurement. In the next section, color systems (RGB) are derived to express colors for use in displays and digital cameras.

3.5.2 RGB System

The television industry has opted for RGB sensitivities according to a ''standard'' that is reasonably well able to match the human eye. However, as several manufacturers had slightly different photo sensors, the RGB sensitivity curves are device dependent. This is immediately obvious when recording an image with a camera (illuminant not known, or in best case estimated, primaries known) and reproducing the image on a monitor (known but different illuminant and primaries), or printing it on paper (known but different primaries, unknown ''illuminant''—being the paper's reflectance function combined with the illuminant under which the paper is observed).

Transformations between XYZ and RGB are expressed by a set of color primaries (xyY coordinates), a reference white and a γ-correction function. The γ-correction function is used for visualization of the linear color values on a nonlinear display device and is usually specified as a power law expression:

$$\mathbf{f}_{\text{out}} = \mathbf{f}_{\text{in}}^{\gamma}, \tag{3.40}$$

where γ is chosen to match the display device. All *RGB* color spaces are defined for a specific device (or set of devices) and are defined with a specific value for γ. The only notable difference is the *sRGB* space, which is discussed later.

The reference white is the color of a nominally white object color stimulus. When this value is unknown, which it usually is, an assumption must be made. Since the color primaries are partially dependent on the value of the reference white, the transformation between *XYZ* and *RGB* is expressed as combination of color primaries and reference white, where the latter is usually defined by referring to one of the standard CIE illuminants (e.g., D65, D50, C, etc.). Many different sets of color matching functions have been proposed to be able to visualize colors on displays, resulting in different RGB standards, for example, NTSC-RGB, PAL-RGB, and sRGB. Given these data, the conversion between *XYZ* and *RGB* is performed as follows:

$$\begin{pmatrix} X \\ Y \\ Z \end{pmatrix} = \begin{pmatrix} S_r X_r & S_g X_g & S_b X_b \\ S_r Y_r & S_g Y_g & S_b Y_b \\ S_r Z_r & S_g Z_g & S_b Z_b \end{pmatrix} \begin{pmatrix} R \\ G \\ B \end{pmatrix}, \tag{3.41}$$

where $(X, Y, Z)_r$, $(X, Y, Z)_g$, and $(X, Y, Z)_b$ are computed from the color primaries using Equation 3.39. Further, (S_r, S_g, S_b) is computed using

$$\begin{pmatrix} S_r \\ S_g \\ S_b \end{pmatrix} = \begin{pmatrix} X_r & X_g & X_b \\ Y_r & Y_g & Y_b \\ Z_r & Z_g & Z_b \end{pmatrix} \begin{pmatrix} X_w \\ Y_w \\ Z_w \end{pmatrix}, \tag{3.42}$$

where $(X, Y, Z)_w$ is the reference white, often expressed in terms of CIE standard illuminant. Note that the *RGB* values must be linear and in the nominal range [0, 1]. Transformation from *XYZ* to *RGB* is simply the inverse of Equation 3.41:

$$\begin{pmatrix} R \\ G \\ B \end{pmatrix} = \begin{pmatrix} S_r X_r & S_g X_g & S_b X_b \\ S_r Y_r & S_g Y_g & S_b Y_b \\ S_r Z_r & S_g Z_g & S_b Z_b \end{pmatrix}^{-1} \begin{pmatrix} X \\ Y \\ Z \end{pmatrix}. \tag{3.43}$$

Different *RGB* working spaces exist, each resulting in different transformation matrices. Two often used working spaces are specified here. The NTSC is often used in digital cameras and videos, which use the C illuminant as reference white. The color primaries are $xyY_r = (0.6700, 0.3300, 0.2988)^T$, $xyY_g = (0.2100, 0.7100, 0.5868)^T$, and $xyY_b = (0.1400, 0.0800, 0.1144)^T$ and the transformation between RGB_{NTSC} and *XYZ* can be done as follows:

$$\begin{pmatrix} X \\ Y \\ Z \end{pmatrix} = \begin{pmatrix} 0.6069 & 0.1735 & 0.2003 \\ 0.2989 & 0.5866 & 0.1145 \\ 0.0000 & 0.0661 & 1.1162 \end{pmatrix} \begin{pmatrix} R \\ G \\ B \end{pmatrix}, \tag{3.44}$$

and the transformation from XYZ to RGB_{NTSC} is done using

$$\begin{pmatrix} R \\ G \\ B \end{pmatrix} = \begin{pmatrix} 1.9100 & -0.5325 & -0.2882 \\ -0.9847 & 1.9992 & -0.0283 \\ 0.0583 & -0.1184 & 0.8976 \end{pmatrix} \begin{pmatrix} X \\ Y \\ Z \end{pmatrix}. \tag{3.45}$$

Alternatively, the *sRGB* is a standard working space specified for use in monitors, printers, and the internet. This color space uses slightly different color primaries, $xyY_r = (0.6400, 0.3300, 0.2127)^T$, $xyY_g = (0.3000, 0.6000, 0.7152)^T$, and $xyY_b = (0.1500, 0.0600, 0.0722)^T$, and uses D65 as reference white. Transformation between RGB_{sRGB} and XYZ is specified as

$$\begin{pmatrix} X \\ Y \\ Z \end{pmatrix} = \begin{pmatrix} 0.4125 & 0.3576 & 0.1804 \\ 0.2127 & 0.7152 & 0.0722 \\ 0.0193 & 0.1192 & 0.9503 \end{pmatrix} \begin{pmatrix} R \\ G \\ B \end{pmatrix}. \tag{3.46}$$

and the transformation from XYZ to RGB_{sRGB} is done using

$$\begin{pmatrix} R \\ G \\ B \end{pmatrix} = \begin{pmatrix} 3.2405 & -1.5371 & -0.4985 \\ -0.9693 & 1.8760 & -0.0416 \\ 0.0556 & -0.2040 & 1.0572 \end{pmatrix} \begin{pmatrix} X \\ Y \\ Z \end{pmatrix}. \tag{3.47}$$

Various other RGB working spaces exist but are beyond the scope of this book.

3.5.3 Opponent Color Spaces

Once the spectral sensitivities of a camera are known to match the human observer, any *numbering scheme* can be assigned to colors. As long as each possible color value is assigned to a unique number, different schemes describe the same color information. However, a different coding scheme might highlight certain properties of color, as seen in the forthcoming sections. It even can be advantageous to assign similar numbers to different values. For example, by assigning all shades of a color to a single value, one obtains invariance to intensity changes caused by shadow and shading, as discussed in Chapter 4.

One of the properties of RGB is that the values of the three channels are highly correlated (e.g., a high value in one of the three channels usually corresponds to high values in the other two channels as well). Decorrelating the RGB color space leads to an opponent color space. The opponent color theory started at about the year 1500 when Leonardo da Vinci came to the conclusion that colors are produced by the mixture of yellow and blue, green and red, and white and black. Arthur Shopenhauer noted the same opposition of red–green, yellow–blue, and white–black. This opponent color theory has been completed by Edwald Hering concluding that the working of the eye is based on the three kinds of opposite colors. A demonstration of opponent color theory is given by the so-called afterimage: looking for a while at a green sample will cause a red after-image (see Fig. 2.3 also). Focusing on the chromatic channels (i.e., red–green and blue–yellow), they are opponent in two different ways. First, no color seems to be a mixture of both members of any opponent pair (e.g., no color ever seems yellowish-blue, while greenish-blue *is* often encountered). Secondly, each member of an opponent pair exhibits the other, that is, by adding a balanced portion of two opponent colors, gray will be the result. The opponent color theory has been confirmed in 1950 when opponent color signals were detected in the optical connection between eye and brain.

Several models have been proposed to model opponent color theory. One of the simplest models is denoted as *opponent color space* in this book and can be computed by simply rotating the *RGB* color system:

$$O_1 = \frac{R - G}{\sqrt{2}}, \tag{3.48}$$

$$O_2 = \frac{R + G - 2B}{\sqrt{6}}, \tag{3.49}$$

$$O_3 = \frac{R + G + B}{\sqrt{3}}. \tag{3.50}$$

Note that O_1 roughly corresponds to the red–green channel, O_2 corresponds to the yellow–blue channel, and O_3 corresponds to the intensity channel. Besides being intuitive, an additional advantage of this color system is that it largely decorrelates the *RGB* color channels. Further, the opponent color space is device dependent and not perceptually uniform.

3.5.4 Perceptually Uniform Color Spaces

In order to overcome these disadvantages, the CIE proposed two opponent color systems that are designed to be perceptually uniform (i.e., the numerical distance between two colors can be related to perceptual differences) but at the cost of intuitiveness. These two systems are computed from *XYZ* and hence are device independent. The first color system is intended to describe light source colors, and is called CIE $L^*u^*v^*$:

$$L^* = \begin{cases} 116\left(\frac{Y}{Y_w}\right)^{\frac{1}{3}} - 16 & \text{if } \frac{Y}{Y_w} > \epsilon \\ 903.3\left(\frac{Y}{Y_w}\right) & \text{if } \frac{Y}{Y_w} \leq \epsilon \end{cases} \tag{3.51}$$

$$u^* = 13L^*(u' - u'_w), \tag{3.52}$$

$$v^* = 13L^*(v' - v'_w), \tag{3.53}$$

$$u' = \frac{4X}{X + 15Y + 3Z}, \tag{3.54}$$

$$v' = \frac{9Y}{X + 15Y + 3Z}, \tag{3.55}$$

$$u'_w = \frac{4X_w}{X_w + 15Y_w + 3Z_w}, \tag{3.56}$$

$$v'_w = \frac{9Y_w}{X_w + 15Y_w + 3Z_w}, \tag{3.57}$$

where $\epsilon = \frac{216}{24389} = 0.008856$. The color channels u^* and v^* become unstable and meaningless when the intensity is low (i.e., when $(X + 15Y + 3Z)$ is close to zero). The second color system is intended for use with surface colors, and is called CIE $L^*a^*b^*$:

$$L^* = 116f(\frac{Y}{Y_w}) - 16, \tag{3.58}$$

$$a^* = 500\left(f(\frac{X}{X_w}) - f(\frac{Y}{Y_w})\right), \tag{3.59}$$

$$b^* = 200\left(f(\frac{Y}{Y_w}) - f(\frac{Z}{Z_w})\right), \tag{3.60}$$

$$f(t) = \begin{cases} t^{\frac{1}{3}} & \text{if } t > \epsilon \\ \dfrac{903.3t + 16}{116} & \text{if } t \leq \epsilon \end{cases}. \tag{3.61}$$

These color spaces are advantageous in computer vision applications, such as image retrieval or image quality assessment, where the aim is to align to human vision. On the other hand, it does not make much sense to use these color spaces in applications that have no direct link to human vision, such as stereo or motion tracking. In those cases, it is more efficient not to perform the time-consuming nonlinear transformations from RGB to $L^*a^*b^*$.

3.5.5 Intuitive Color Spaces

Besides the opponent color space, color systems discussed until now are not expressed in intuitive terms; the different color channels do not have an intuitive meaning. To this end, different color systems are introduced, which are based on an artist's reasoning. All such color systems express colors in terms of hue, saturation, and intensity. However, many different definitions of these terms exist and none of these definitions are standardized. Further, different definitions are often used for different abbreviations, for example, HSV, HSI, HSL, etc. In this book, subscripts will be used to indicate the corresponding definitions.

One of the common definitions of *hue* is that it is described by the dominant wavelength of a spectral power distribution, that is, hue is described with the words that we normally use to describe any given color: red, blue, orange, yellow, etc. Mathematically, hue can be computed in angular degrees using the following cartesian to polar coordinates transformation:

$$H_{RGB} = \arctan\left(\frac{\sqrt{3}(G-B)}{(R-G)+(R-B)}\right). \tag{3.62}$$

Alternatively, hue is defined as the angle between a reference line (e.g., horizontal axis) and the color point:

$$H_{rgb} = \arctan\left(\frac{r-1/3}{g-1/3}\right). \tag{3.63}$$

A device-independent version of hue can be computed from either CIE $L^*u^*v^*$ or CIE $L^*a^*b^*$:

$$H_{uv} = \arctan\left(\frac{v^*}{u^*}\right), \tag{3.64}$$

$$H_{ab} = \arctan\left(\frac{b^*}{a^*}\right). \tag{3.65}$$

Note that hue is undefined for achromatic colors (i.e., $R = G = B$, $u^* = v^* = 0$, or $a^* = b^* = 0$).

Saturation is usually defined as the purity of a color, which decreases when more achromaticity is mixed into a color. Completely desaturated colors coincide with the gray axis, while fully saturated colors coincide with pure colors. Mathematically, saturation is defined as the distance of a color to the achromatic axis, but different equations can be used to compute this distance; for example,

$$S_{rgb} = \sqrt{(r-1/3)^2 + (g-1/3)^2 + (b-1/3)^2}, \tag{3.66}$$

$$S_{RGB} = 1 - \frac{\min(R,G,B)}{R+G+B}, \tag{3.67}$$

$$S_{HSL} = \max(R,G,B) - \min(R,G,B), \tag{3.68}$$

$$S_{HSV} = 1 - \frac{\min(R,G,B)}{\max(R,G,B)}. \tag{3.69}$$

Note that saturation is undefined for dark pixels (i.e., $R + G + B = 0$). Another color channel, related to saturation, is *chroma*, which can be computed from CIE $L^*u^*v^*$ or CIE $L^*a^*b^*$ color systems as follows:

$$C^*_{uv} = \sqrt{(u^*)^2 + (v^*)^2}, \tag{3.70}$$

$$C^*_{ab} = \sqrt{(a^*)^2 + (b^*)^2}, \tag{3.71}$$

and which, similarly as saturation, describes the purity of the color.

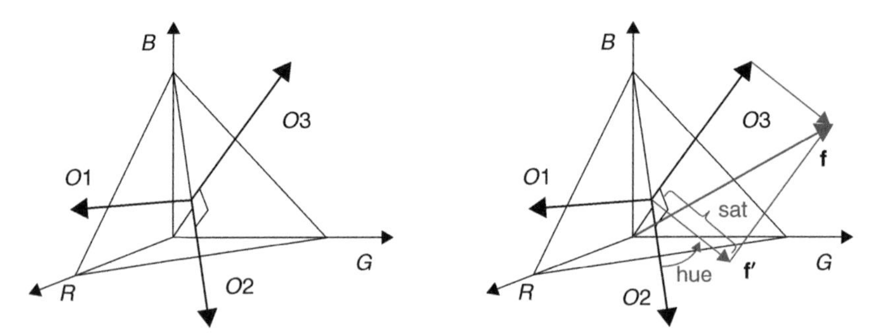

Figure 3.5 (a) The opponent axes are depicted. (b) The relation between the opponent and hue-saturation-intensity coordinate system is shown. Projection of **f** on O3 is the intensity. Let **f**′ be the projection of **f** on the O1-O2 plane; then its length is the *saturation* and the angle in the plane is the *hue*.

Within the context of coordinate transformations, the hue-saturation-intensity transformation is computed by performing a polar coordinate transformation on the opponent color plane formed by O1 and O2, according to

$$\begin{pmatrix} h \\ s \\ i \end{pmatrix} = \begin{pmatrix} \arctan\left(\frac{O1}{O2}\right) \\ \sqrt{O1^2 + O2^2} \\ O3 \end{pmatrix} = \begin{pmatrix} \arctan\left(\frac{\sqrt{3}(R-G)}{R+G-2B}\right) \\ \sqrt{\frac{4}{6}\left(R^2 + G^2 + B^2 - RG - RB - GB\right)} \\ \frac{R+G+B}{\sqrt{3}} \end{pmatrix} \tag{3.72}$$

This transformation is depicted in Figure 3.5. Note that it is more correct to talk about saturation strength in this transformation because the saturation has not been normalized by the intensity as in the equations above.

3.6 Summary

In this chapter, we have introduced the color image formation process. We have discussed several physics-based reflectance models. They provide us with tools to analyze the various physical causes that have led to the color measurement. More precisely, they will allow us to distinguish shadow, specular, and material variations in the image. We will see that this information can be used in many stages of color image understanding, such as improved feature detection, feature description, and meaningful image segmentation.

Additionally, we have enumerated a set of relevant color spaces. Depending on the task at hand, one color space might be preferable over the other. The reflectance models that were introduced in this chapter are used in Chapter 4 to analyze the photometric properties of the color spaces. This will provide further insight into what color space could be best used for each application.

PART II

PHOTOMETRIC INVARIANCE

4 Pixel-Based Photometric Invariance

Computer vision systems have to deal with widely varying imaging conditions. To obtain robust vision systems, an important property is photometric invariance or the so-called color invariance. Color invariance is derived from color spaces that are more or less insensitive to disturbing imaging conditions such as variations in the light source (both intensity and color), camera viewpoint, and object position.

In the previous chapter, it has been shown that from the *RGB* color space, several linear and nonlinear transformations can be applied to obtain new color spaces. In this chapter, an overview is given of the *color invariant* properties of these transformed color spaces. From the dichromatic reflection model, Equation 3.6, it can be derived that the recorded color value at each (pixel) location is highly dependent on the light source characteristics and the object geometry (e.g., the absence/presence of shadows or highlights partly depends on the position of the object with respect to the light source), see Figure 4.1. Many computer vision tasks such as image segmentation and object recognition require stable and repeatable image properties rather than color measurements that are sensitive to imaging conditions. For this purpose, color invariance is needed.

In this chapter, color invariance is obtained by color transformations at a pixel. Both linear and nonlinear color transformations are presented. In general, nonlinear transformations tend to intensify the amount of noise. As a consequence, a small perturbation of *RGB* values will cause a large jump in the transformed values. The way to deal with noise amplification is to analyze how perturbations in *RGB* values propagate through these nonlinear color transformations. The field

Color in Computer Vision: Fundamentals and Applications, First Edition.
Theo Gevers, Arjan Gijsenij, Joost van de Weijer, and Jan-Mark Geusebroek.

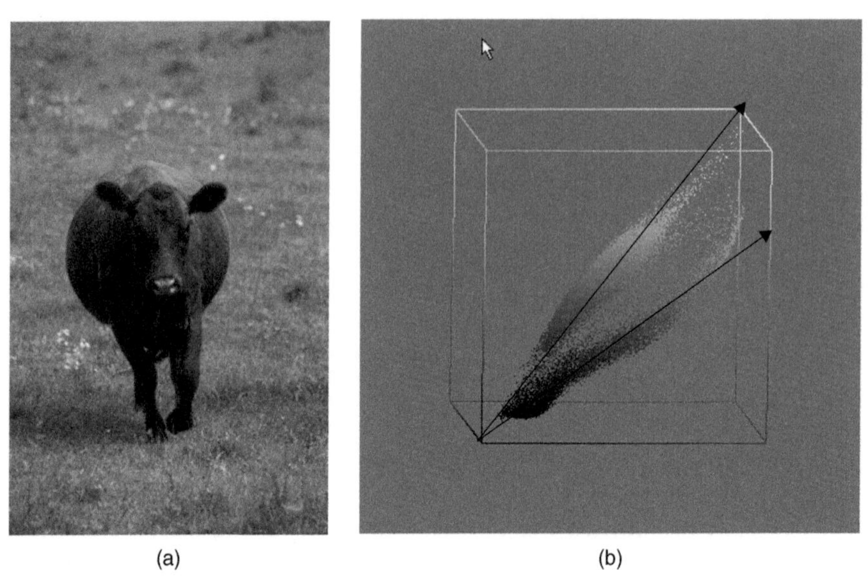

(a) (b)

Figure 4.1 Pixel values are highly dependent on the light source characteristics and the object geometry. Assuming Lambertian reflection and white illumination, *RGB* colors of a homogeneously colored surface will generate elongated streaks in color space. (a) Original image and (b) pixels in *RGB* color space.

of error analysis or error propagation provides such a principled approach and is discussed in detail.

In summary, we first present different color transformations and their invariant characteristics. Then, computational methods are discussed to analyze the influence of noise on these color invariants. Finally, as an application, color invariants are used in object recognition.

4.1 Normalized Color Spaces

In Figure 4.2, it is shown that color pixel values are highly dependent on the light source characteristics and the object geometry, where pixel values of a homogeneously colored surface will generate elongated streaks in *RGB* color space. As these streaks are mainly caused by intensity changes (object geometry and shadows) and not by chromaticity changes, invariant values can be obtained by normalizing *RGB* values by their intensity ($I = R + G + B$) resulting in the *rgb* color system (or normalized *rgb*):

$$r = \frac{R}{R+G+B}, \tag{4.1}$$

$$g = \frac{G}{R+G+B}, \tag{4.2}$$

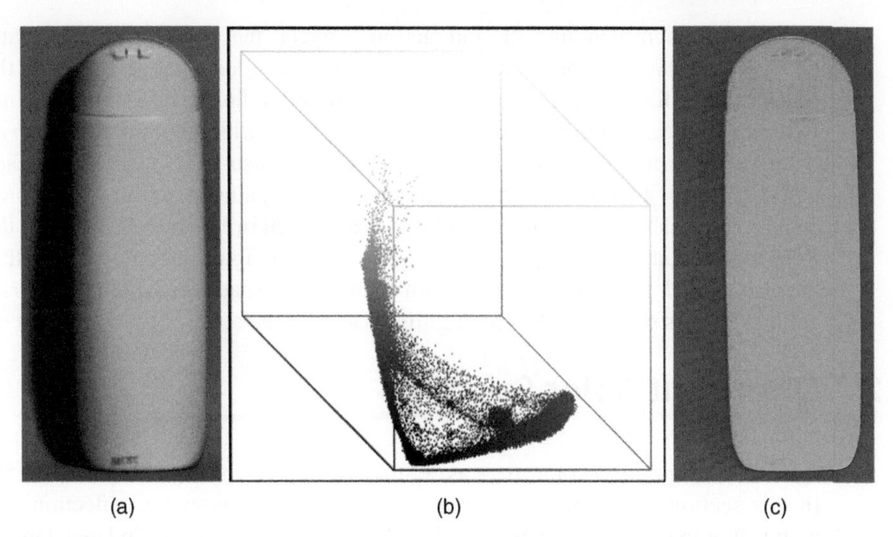

(a) (b) (c)

Figure 4.2 The *rgb* transformation is independent of surface orientation, illumination direction, and illumination intensity. (a) Original image, (b) pixels in *RGB* color space, and (c) normalized color image (*RGB* representation).

$$b = \frac{B}{R+G+B}. \tag{4.3}$$

Because intensity is factored out, this color system has the property that its channels are robust to surface orientation, illumination direction, and illumination intensity when we assume dichromatic reflection and white illumination according to Equation 3.8. In fact, from Equation 3.8, Lambertian reflectance under *white* illumination is given by

$$f^c(\mathbf{x}) = m^b(\mathbf{x}) \int_\omega s(\lambda, \mathbf{x}) \rho^c(\lambda)\, \mathrm{d}\lambda. \tag{4.4}$$

By substituting Equation 4.4 into Equations 4.1–4.3, the invariant properties of the *rgb* color system can be derived:

$$r = \frac{m^b(\mathbf{x})\, k_R}{m^b(\mathbf{x})\,(k_R + k_G + k_B)} = \frac{k_R}{k_R + k_G + k_B}, \tag{4.5}$$

$$g = \frac{m^b(\mathbf{x})\, k_G}{m^b(\mathbf{x})\,(k_R + k_G + k_B)} = \frac{k_G}{k_R + k_G + k_B}, \tag{4.6}$$

$$b = \frac{m^b(\mathbf{x})\, k_B}{m^b(\mathbf{x})\,(k_R + k_G + k_B)} = \frac{k_B}{k_R + k_G + k_B}, \tag{4.7}$$

where $k_c = \int_\omega s(\lambda, \mathbf{x}) \rho^c(\lambda)\, \mathrm{d}\lambda$ and $c \in \{R, G, B\}$. k_c is used in the remainder as a compact representation to simplify the notation.

The coefficient denoting the interaction between the white light source and surface reflectance (represented by $m^b(\mathbf{x})$) is cancelled out, resulting in the independence for the surface orientation, illumination direction, and illumination intensity. Hence, assuming Lambertian reflection and white illumination, the *rgb* color space is only dependent on k_c, which is the sensor $\rho^c(\lambda)$ and the surface albedo $s(\lambda, \mathbf{x})$. In Figure 4.2a, a shampoo bottle against a uniformly colored background is shown. The *RGB* pixel values are given in Figure 4.2b. Further, the *RGB* representation of the *rgb* color image is given in Figure 4.2c. It is shown that the normalized color image is free from shading and shadows. Note that the *rgb* color system is still dependent on highlights and on the color of the light source.

4.2 Opponent Color Spaces

Consider the opponent color space as defined in Chapter 3 by Equations 3.48–3.50. In this section, we focus on O_1 and O_2. Assuming dichromatic reflection and white illumination, the opponent color channels O_1 and O_2 are independent of highlights as follows from substituting Equation 3.8 in Equations 3.48 and 3.49:

$$O_1 = \frac{\left(m^b(\mathbf{x})\, k_R + m^i(\mathbf{x})\right) - \left(m^b(\mathbf{x})\, k_G + m^i(\mathbf{x})\right)}{\sqrt{2}} \tag{4.8}$$

$$= \frac{m^b(\mathbf{x})\, k_R - m^b(\mathbf{x})\, k_G}{\sqrt{2}}, \tag{4.9}$$

$$O_2 = \frac{\left(m^b(\mathbf{x})\, k_R + m^i(\mathbf{x})\right) + \left(m^b(\mathbf{x})\, k_G + m^i(\mathbf{x})\right) - 2\left(m^b(\mathbf{x})\, k_B + m^i(\mathbf{x})\right)}{\sqrt{6}} \tag{4.10}$$

$$= \frac{m^b(\mathbf{x})\, k_R + m^b(\mathbf{x})\, k_G - 2m^b(\mathbf{x})\, k_B}{\sqrt{6}}. \tag{4.11}$$

Note that O_1O_2 is still dependent on $m^b(\mathbf{x}) \int_\omega s(\lambda, \mathbf{x}) \rho^c(\lambda)\, d\lambda$ and consequently sensitive to object geometry, shading, and the intensity of the light source. Note that O_3 corresponds to intensity and contains no invariance at all.

4.3 The HSV Color Space

The hue defined in Chapter 3 by Equation 3.63 is invariant to surface orientation, illumination direction, and illumination intensity under the assumption of dichromatic reflectance and white illumination. This can be derived by substituting Equation 3.8 in Equation 3.62:

$$H_{RGB} = \arctan\left(\frac{\sqrt{3}m^b(\mathbf{x})\left(k_G - k_B\right)}{m^b(\mathbf{x})\left(\left(k_R - k_G\right) + \left(k_R - k_B\right)\right)}\right) \tag{4.12}$$

$$= \arctan\left(\frac{\sqrt{3}(k_G - k_B)}{(k_R - k_G) + (k_R - k_B)}\right), \tag{4.13}$$

and is only dependent on k_c, which is the surface albedo and image sensor. The light source and the surface reflectance are cancelled out, resulting in the independence for the surface orientation, illumination direction, and illumination intensity. Moreover, the hue is invariant to highlights.

Because saturation S_{RGB} corresponds to the radial distance from a pixel color to the main diagonal in the *RGB* color space, S_{RGB} is an invariant for matte, dull surfaces illuminated by white light. This can be derived by substituting Equation 4.4 into Equation 3.67:

$$S_{RGB} = 1 - \frac{\min(m^b(\mathbf{x})\, k_R, m^b(\mathbf{x})\, k_G, m^b(\mathbf{x})\, k_B)}{m^b(\mathbf{x})\, k_R + m^b(\mathbf{x})\, k_G + m^b(\mathbf{x})\, k_B} \tag{4.14}$$

$$= 1 - \frac{m^b(\mathbf{x}) \min(k_R, k_G, k_B)}{m^b(\mathbf{x})\,(k_R + k_G + k_B)} \tag{4.15}$$

$$= 1 - \frac{\min(k_R, k_G, k_B)}{k_R + k_G + k_B}. \tag{4.16}$$

The dependencies on the illumination and surface reflectance are eliminated. The result is only dependent on the image sensors and the surface albedo.

4.4 Composed Color Spaces

Until now, we considered existing color spaces that allow for a meaningful representation of color, either physically or perceptually as described in Chapter 3. Many computer-vision-related tasks, however, do not require such a representation, introducing the possibility of generating new color invariant representations. This section discusses several such color invariant representations. First, a set of color invariants will be discussed under the assumption of Lambertian reflectance and white illumination. Then, the assumption of Lambertian reflectance is relaxed resulting in a slightly more complex set of invariants. More information can be found in Reference 38.

4.4.1 Body Reflectance Invariance

Considering Lambertian reflection and white illumination, Equation 4.4 reveals that the measured color depends on the surface albedo and camera filters (i.e., $k_c = \int_\omega s(\lambda, \mathbf{x}) \rho^c(\lambda)\, d\lambda$), in combination with the local intensity of the illuminant and the roughness and shape of the object $m^b(\mathbf{x})$. The first component mainly determines the chromaticity of the color, while the latter component mainly determines the intensity of the color. In other words, a uniformly colored surface that is curved (i.e., varying surface orientation) may give rise to a broad variance of intensity values (i.e., generating elongated streak as shown in Figs. 4.1 and 4.2). To reduce the effects of the illuminant intensity and the shape of the surface on these measured color values, expressions need to be derived that factor out these dependencies.

To this end, the following basic set of *irreducible color invariants* is provided [38]:

$$\frac{f^i(\mathbf{x})}{f^j(\mathbf{x})} = \frac{f^i}{f^j}, \tag{4.17}$$

where $\mathbf{x}$ can be omitted as this invariant is computed at the same surface location. Substituting Equation 4.4 into Equation 4.17 proves the invariance of this basic set:

$$\frac{f^i}{f^j} = \frac{m^b(\mathbf{x})\, k_i}{m^b(\mathbf{x})\, k_j} = \frac{k_i}{k_j}, \tag{4.18}$$

where $k_i = \int_\omega s(\lambda, \mathbf{x}) \rho^i(\lambda)\, d\lambda$ for any sensor $\rho^i(\lambda)$. The expression is only dependent on the surface albedo and the sensors. Dependencies on the viewpoint, surface orientation, illumination direction, and illumination intensity are cancelled out. From now on, we focus on color images and hence $f^i \in \{f^R, f^G, f^B\}$.

Any linear combination of the basic set of irreducible color invariants will result in a new color invariant. A systematic approach to compute invariants for f^R, f^G, and f^B is given by

$$C_{RGB} = \frac{\sum_i a_i (f^R)_i^p (f^G)_i^q (f^B)_i^r}{\sum_j b_j (f^R)_j^s (f^G)_j^t (f^B)_j^u}, \tag{4.19}$$

where $p+q+r = s+t+u$ and $p,q,r,s,t,u \in \mathbb{R}$. Further, $i,j \geq 1$, and $a_i, b_i \in \mathbb{R}$. Assuming Lambertian reflectance and white illumination, C_{RGB} is independent of the viewpoint, surface orientation, illumination direction, and illumination intensity. By substituting Equation 4.4 into Equation 4.19, we obtain

$$C_{RGB} = \frac{\sum_i a_i (f^R)_i^p (f^G)_i^q (f^B)_i^r}{\sum_j b_j (f^R)_j^s (f^G)_j^t (f^B)_j^u} \tag{4.20}$$

$$= \frac{\sum_i a_i (m^b(\mathbf{x})\, k_R)_i^p (m^b(\mathbf{x})\, k_G)_i^q (m^b(\mathbf{x})\, k_B)_i^r}{\sum_j b_j (m^b(\mathbf{x})\, k_R)_j^s (m^b(\mathbf{x})\, k_G)_j^t (m^b(\mathbf{x})\, k_B)_j^u} \tag{4.21}$$

$$= \frac{\sum_i a_i (m^b(\mathbf{x}))^{p+q+r} ((k_R)_i^p (k_G)_i^q (k_B)_i^r)}{\sum_j b_j (m^b(\mathbf{x}))^{s+t+u} ((k_R)_j^s (k_G)_j^t (k_B)_j^u)}. \tag{4.22}$$

Since $p+q+r = s+t+u$, Equation 4.22 can be further simplified by factoring out the dependencies on the viewpoint, surface orientation, and illumination direction and intensity:

$$C_{RGB} = \frac{\sum_i a_i((k_R)_i^p (k_G)_i^q (k_B)_i^r)}{\sum_j b_j((k_R)_j^s (k_G)_j^t (k_B)_j^u)}. \tag{4.23}$$

Numerous invariants can be obtained. For ease of use, they can be classified into orders, for example, the set of first-order color invariants involves the set where $p+q+r = s+t+u = 1$:

$$\left\{\frac{R}{G}, \frac{R}{B}, \frac{G}{B}, \frac{-B}{R}, \frac{R}{R+G+B}, \frac{R-G}{R+G}, \frac{R+G+B}{2G+3B}, \frac{3(B-G)}{2R+G+3B}, \ldots\right\}, \tag{4.24}$$

and the set of second-order color invariants involves the set, where $p+q+r = s+t+u = 2$:

$$\left\{\frac{RG}{B^2}, \frac{R^2+B^2}{RB}, \frac{2RG-3RB}{R^2+G^2}, \frac{RG+RB+GB}{R^2+G^2+B^2}, \ldots\right\}. \tag{4.25}$$

Each of these expressions is a color invariant for Lambertian reflectance under white illumination. Note that the *rgb* color channels, Equations 4.1–4.3, are instantiations of the first-order color invariants.

4.4.2 Body and Surface Reflectance Invariance

Assuming dichromatic reflection with white illumination (Eq. 3.8), the observed colors of a uniformly colored (but shiny) surface will form a dichromatic plane spanned by the body and surface reflection components, which originates from the main diagonal axis (Fig. 4.3). Therefore, any expression defining colors on this dichromatic plane is a color invariant for the dichromatic reflection model.

To this end, the following basic set of *irreducible color invariants* at location $\mathbf{x}$ is chosen [38]:

$$\frac{f^i(\mathbf{x}) - f^j(\mathbf{x})}{f^k(\mathbf{x}) - f^m(\mathbf{x})} = \frac{f^i - f^j}{f^k - f^m}, \tag{4.26}$$

where $f_k \neq f_m$ and $\mathbf{x}$ can be omitted again, as this invariant is computed at the same surface location. Color invariants can be computed in a systematic way in terms of f^R,f^G, and f^B as follows:

$$L_{RGB} = \frac{\sum_i a_i(f^R - f^G)_i^p (f^R - f^B)_i^q (f^G - f^B)_i^r}{\sum_j b_j(f^R - f^G)_j^s (f^R - f^B)_j^t (f^G - f^B)_j^u}, \tag{4.27}$$

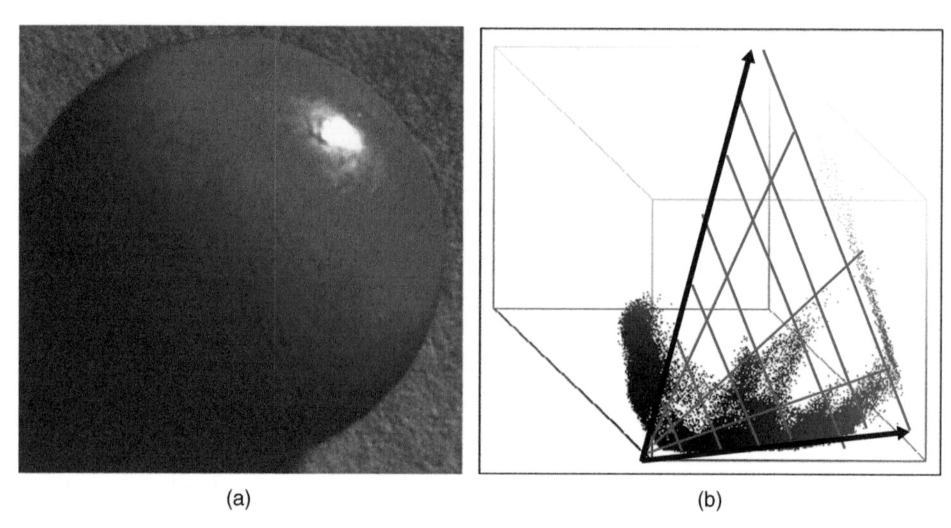

(a) (b)

Figure 4.3 In *RGB* color space, the observed colors of a uniformly colored (shiny) surface will form a dichromatic plane spanned by the body and surface reflection components. Any expression defining colors on this dichromatic plane is a color invariant for the dichromatic reflection model. (a) Original image and (b) pixels in *RGB* color space.

where $p+q+r = s+t+u$ and $p,q,r,s,t,u \in \mathbb{R}$. Further, $i,j \geq 1$, and $a_i, b_i \in \mathbb{R}$. Assuming dichromatic reflectance and white illumination, L_{RGB} is independent of the viewpoint, surface orientation, illumination direction, and illumination intensity and highlights. The subtraction of two terms in Equation 4.27 results in the independence of specular reflection:

$$(f^i - f^j) = (m^b(\mathbf{x})\, k_i + m^i(\mathbf{x})) - (m^b(\mathbf{x})\, k_j + m^i(\mathbf{x})) \tag{4.28}$$

$$= m^b(\mathbf{x})\, k_i - m^b(\mathbf{x})\, k_j \tag{4.29}$$

for $i \neq j, i \in \{R, G, B\}$, which is just the subtraction of the two body reflection components. Then, substituting Equation 3.8 into Equation 4.27 results, for *RGB*, in

$$L_{RGB} = \frac{\sum_i a_i (f^R - f^G)_i^p (f^R - f^B)_i^q (f^G - f^B)_i^r}{\sum_j b_j (f^R - f^G)_j^s (f^R - f^B)_j^t (f^G - f^B)_j^u} \tag{4.30}$$

$$= \frac{\sum_i a_i (m^b(\mathbf{x})(k_R - k_G))_i^p (m^b(\mathbf{x})(k_R - k_B))_i^q (m^b(\mathbf{x})(k_G - k_B))_i^r}{\sum_j b_j (m^b(\mathbf{x})(k_R - k_G))_j^s (m^b(\mathbf{x})(k_R - k_B))_j^t (m^b(\mathbf{x})(k_G - k_B))_j^u} \tag{4.31}$$

$$= \frac{\sum_i a_i (m^b(\mathbf{x}))^{p+q+r} ((k_R - k_G)_i^p (k_R - k_B)_i^q (k_G - k_B)_i^r)}{\sum_j b_j (m^b(\mathbf{x}))^{s+t+u} ((k_R - k_G)_j^s (k_R - k_B)_j^t (k_G - k_B)_j^u)}. \tag{4.32}$$

Since $p+q+r = s+t+u$, Equation 4.32 can be further simplified by

$$L_{RGB} = \frac{\sum_i a_i((k_R - k_G)_i^p (k_R - k_B)_i^q (k_G - k_B)_i^r)}{\sum_j b_j((k_R - k_G)_j^s (k_R - k_B)_j^t (k_G - k_B)_j^u)}. \quad (4.33)$$

Similar to C_{RGB}, the various different color invariants can be classified by their order. The set of first-order color invariants involves the set where $p+q+r = s+t+u = 1$:

$$\left\{ \frac{R-G}{R-B}, \frac{R-B}{G-B}, \frac{G-B}{R-G}, \frac{R-G}{(R-G)+(R-B)}, \frac{(R-B)+3(B-G)}{(R-G)+2(R-G)}, \ldots \right\} \quad (4.34)$$

and the set of second-order color invariants involves the set, where $p+q+r = s+t+u = 2$:

$$\left\{ \frac{(R-G)(R-B)}{(R-B)^2}, \frac{(G-B)(R-B)}{(R-G)(B-R)}, \frac{(R-G)^2+(R-B)(G-B)}{(R-B)^2+2(G-B)^2}, \ldots \right\}. \quad (4.35)$$

Each of these expressions is a color invariant for dichromatic reflectance under white illumination.

As an illustration, consider an instantiation of L_{RGB}, for instance, $\frac{(R-G)^2}{(R-G)^2+(G-B)^2+(R-B)^2}$, which is a different (nonangular) representation of hue. In Figure 4.4a, a pill-shaped object is shown against a uniformly colored background. The RGB representation of the color invariant image is given in

(a) (b)

Figure 4.4 The transformed color space $\frac{(R-G)^2}{(R-G)^2+(G-B)^2+(R-B)^2}$ is independent of varying photometric conditions. (a) Original image and (b) color invariant space (RGB representation).

Figure 4.4b. It is shown that the color invariant image is free from shading, shadow, and highlights. This transformation is indeed independent of varying imaging conditions as illustrated in Figure 4.4. Note that the hue H_{RGB} color channel, Equation 3.62, is one of the instantiations of the first-order color invariants as a function of arctan. For example, hue can be obtained by substituting $a_1 = \sqrt{3}, a_2 = 0$, and $b_1 = b_2 = 1$. Note the instability in the center of the highlight. This is because of the lack of saturation. Hence, it is problematic to determine the color (hue) of a pixel value. In the next section, instabilities of color invariants are addressed.

4.5 Noise Stability and Histogram Construction

The color transformations used to compute the color invariants, as discussed in the previous section, bring with them several drawbacks since these transformations are singular at some *RGB* values and unstable at others. For example, *rgb* is undefined at the black point ($R = G = B = 0$) and hue H at the achromatic axis ($R = G = B$). As a consequence, a small perturbation of sensor values near these *RGB* values may cause a large jump in the transformed values. Traditionally, the effect of noise blowup at unstable color invariant values is simply ignored or suppressed by ad hoc color thresholding. For instance, in object recognition based on color histograms, when constructing histograms, all *RGB* values along and near the achromatic axis are discarded by eliminating all *RGB* values having a saturation and intensity value smaller than 5% of the total range. Another approach is given by Burns and Berns [39], analyzing the error propagation through the CIE $L^*a^*b^*$ color space. Shafarenko et al. [40] use an adaptive filter for noise reduction in the CIE $L^*u^*v^*$ space before 3D color histogram construction. In fact, the filter width is steered based on the covariance matrix of the noise distribution in the CIE $L^*u^*v^*$ space.

In this section, a more principled method is discussed to suppress the effect of noise during histogram construction from color invariants [41]. In fact, variable kernel density estimation is used to construct color invariant histograms. To apply variable kernel density estimation in a proper way, computational methods are used for the propagation of sensor noise through color invariant transformations. As a result, the associated uncertainty is measured for each color invariant value. The associated uncertainty is used to derive the optimal parameterization of the variable kernel used during histogram construction.

4.5.1 Noise Propagation

Assuming that sensor noise is normally distributed, then for an indirect measurement, the true value of a variable u is related to its N arguments, denoted by u_j, as follows:

$$u = q(u_1, u_2, \cdots, u_N). \tag{4.36}$$

Assume that the estimate $\hat{u}$ of the variable u can be obtained by the substitution of $\hat{u}_j$ for u_j. Then, when $\hat{u}_1, \ldots, \hat{u}_N$ are measured with corresponding standard deviations $\sigma_{\hat{u}_1}, \ldots, \sigma_{\hat{u}_N}$, we obtain [42]:

$$\hat{u} = q(\hat{u}_1, \ldots, \hat{u}_N). \tag{4.37}$$

It is known that the approximation of a given function can be written in the form of Taylor series. For $N = 2$, the Taylor series with respect to noise is given by

$$q(\hat{u}_1, \hat{u}_2) = q(u_1, u_2) + \left(\frac{\partial}{\partial u_1} \mathcal{E}_1 + \frac{\partial}{\partial u_2} \mathcal{E}_2 \right) q(u_1, u_2) + \cdots \tag{4.38}$$

$$+ \frac{1}{m!} \left(\frac{\partial}{\partial u_1} \mathcal{E}_1 + \frac{\partial}{\partial u_2} \mathcal{E}_2 \right)^m q(u_1, u_2) + R_{m+1}, \tag{4.39}$$

where $\hat{u}_1 = u_1 + \mathcal{E}_1, \hat{u}_2 = u_2 + \mathcal{E}_2 (\mathcal{E}_1$ and $\mathcal{E}_2$ are the errors of $\hat{u}_1$ and $\hat{u}_2$), and R_{m+1} is the remainder term. Further, $\partial q / \partial \hat{u}_j$ is the partial derivative of q with respect to $\hat{u}_j$.

As the general form of the error of an indirect measurement is

$$E = \hat{u} - u = q(\hat{u}_1, \hat{u}_2) - q(u_1, u_2), \tag{4.40}$$

we obtain in terms of the Taylor series the following:

$$E = \left(\frac{\partial}{\partial u_1} \mathcal{E}_1 + \frac{\partial}{\partial u_2} \mathcal{E}_2 \right) q(u_1, u_2) + \cdots + \frac{1}{m!} \left(\frac{\partial}{\partial u_1} \mathcal{E}_1 + \frac{\partial}{\partial u_2} \mathcal{E}_2 \right)^m q(u_1, u_2) + R_{m+1}. \tag{4.41}$$

In general, only the first linear term is used to compute the error

$$E = \frac{\partial q}{\partial u_1} \mathcal{E}_1 + \frac{\partial q}{\partial u_2} \mathcal{E}_2. \tag{4.42}$$

Then, for N arguments, it follows that if the uncertainties in $\hat{u}_1, \ldots, \hat{u}_N$ are independent, random, and relatively small, the predicted uncertainty in q is given by Taylor [42]:

$$\sigma_q = \sqrt{\sum_{j=1}^{N} \left(\frac{\partial q}{\partial \hat{u}_i} \sigma_{\hat{u}_i} \right)^2}, \tag{4.43}$$

the so-called square-root sum method.

4.5.2 Examples of Noise Propagation through Transformed Colors

As an example, assume that we want to compute the noise for the normalized color system *rgb*. First, the amount of noise should be estimated for each *RGB* color channel. Assuming normally distributed random quantities, the standard way to calculate the standard deviations (noise) σ_R, σ_G, and σ_B is to compute the mean and variance estimates computed from homogeneously colored surface patches in an image. For example, the measured amount of noise for the image shown in Figure 4.5 is $\sigma_R = 4.6$, $\sigma_G = 3.8$, and $\sigma_B = 4.0$ Further, the amount of noise (or uncertainty) for *rgb* is obtained by substituting Equations 4.1–4.3 in Equation 4.43:

$$\sigma_r = \sqrt{\frac{R^2\left(\sigma_B^2 + \sigma_G^2\right) + (B+G)^2\sigma_R^2}{(R+G+B)^4}}, \tag{4.44}$$

$$\sigma_g = \sqrt{\frac{G^2\left(\sigma_B^2 + \sigma_R^2\right) + (B+R)^2\sigma_G^2}{(B+G+R)^4}}, \tag{4.45}$$

$$\sigma_b = \sqrt{\frac{B^2\left(\sigma_R^2 + \sigma_G^2\right) + (R+G)^2\sigma_B^2}{(R+G+B)^4}}. \tag{4.46}$$

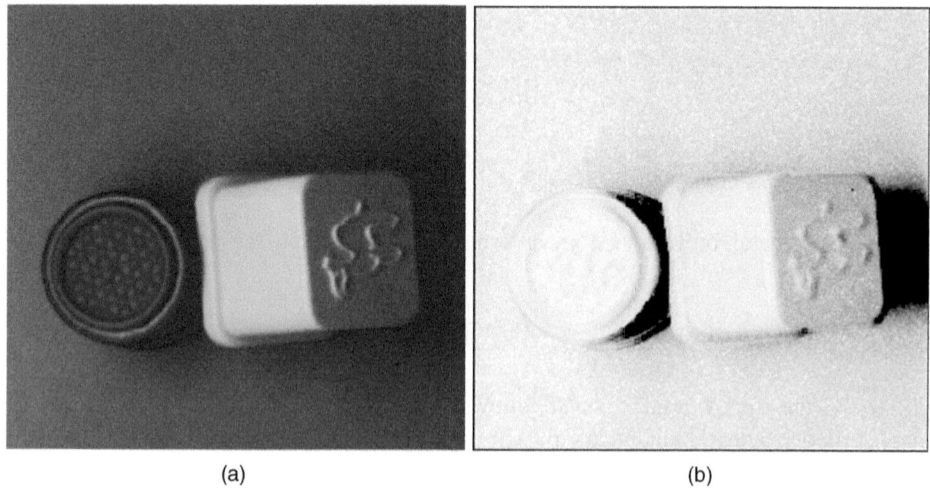

(a) (b)

Figure 4.5 The amount of noise/uncertainty of normalized color is inversely related to the amount of intensity. (a) Original image and (b) the amount of noise, in black, for *r*.

From the analytical study of Equations 4.44–4.46, it can be derived that normalized color becomes unstable around the black point $R = G = B = 0$. If the intensity

$I = R+G+B$ is low, the denominator is relatively small and hence the amount of noise is relatively high. In other words, if the intensity increases, the denominator increases and therefore the amount of noise decreases. In conclusion, the amount of noise/uncertainty of normalized color is inversely related to the amount of intensity (Fig. 4.5).

The noise/uncertainties of O_1 and O_2 are given by

$$\sigma_{O_1} = \frac{\sqrt{\sigma_G^2 + \sigma_R^2}}{\sqrt{2}}, \tag{4.47}$$

$$\sigma_{O_2} = \frac{\sqrt{4\sigma_B^2 + \sigma_G^2 + \sigma_R^2}}{\sqrt{3}}, \tag{4.48}$$

which are the same (stable) at all *RGB* values. Hence, opponent color does not vary with changes in *RGB* values.

Furthermore, the substitution of Equation 3.62 in Equation 4.43 gives the uncertainty for hue:

$$\sigma_\theta = \sqrt{\frac{3}{4}\frac{\sigma_B^2(G-R)^2 + \sigma_G^2(B-R)^2 + \sigma_R^2(B-G)^2}{(B^2B(G+R) + G^2 - GR + R^2)^2}}, \tag{4.49}$$

which is unstable at low saturation (i.e., the gray axis $R = G = B$). This can be interpreted as follows. If the saturation is low, the denominator is relatively small and hence the amount of noise is relatively high. If the saturation increases, the denominator increases and therefore the amount of noise decrease. In conclusion, the amount of noise/uncertainty of the hue is inversely related to the amount of saturation. This means that hue is unstable at gray values.

In conclusion, normalized color is unstable at low intensity. Hue is unstable at low saturation. Opponent color is relatively stable at all *RGB* values.

4.5.3 Histogram Construction by Variable Kernel Density Estimation

A common approach to object recognition is to represent and match images on the basis of histograms derived from color invariants. To suppress the effect of noise for unstable color invariant values, histograms can be computed by variable kernel density estimàtors. The associated uncertainty is used to derive the parameterization of the variable kernel for the purpose of robust histogram construction.

More precisely, a density function f gives a description of the distribution of the measured data. A well-known density estimator is the histogram. The (one-dimensional) histogram is defined as

$$\hat{f}(x) = \frac{1}{nh}\,(\text{number of } X_i \text{ in the same bin as } x), \tag{4.50}$$

where n is the number of pixels with value X_i in the image, h is the bin width, and x is the range of the data. Two choices have to be made when constructing a histogram. First, the bin-width parameter needs to be chosen. Second, the position of the bin edges needs to be established. Both choices affect the resulting estimation. In general, bin widths and edges are chosen in an ad hoc way (e.g., by hand).

In contrast, the kernel density estimator is insensitive to the placement of the bin edges

$$\hat{f}(x) = \frac{1}{nh}\sum_{i=1}^{n} K\left(\frac{x - X_i}{h}\right). \tag{4.51}$$

Here, kernel K is a function satisfying $\int K(x)\mathrm{d}x = 1$. In the *variable* kernel density estimator, the single h is replaced by n values $\alpha(X_i), i = 1, \ldots, n$. This estimator is of the form

$$\hat{f}(x) = \frac{1}{n}\sum_{i=1}^{n} \frac{1}{\alpha(X_i)} K\left(\frac{x - X_i}{\alpha(X_i)}\right). \tag{4.52}$$

The kernel centered around X_i has associated with it its own scale parameter $\alpha(X_i)$, thus allowing different degrees of smoothing. To use variable kernel density estimators for color images, we let the scale parameter be a function of the *RGB* values and the color space transform. Assuming normally distributed noise, the distribution is given by the Gaussian distribution [42]:

$$K(x) = \frac{1}{\sqrt{2\pi}} \exp^{-x^2/2}. \tag{4.53}$$

Then, the variable kernel method estimating the density of color channel C is as follows:

$$\hat{f}(C) = \frac{1}{n}\sum_{i=1}^{n} \sigma_{C_i}^{-1} K\left(\frac{(C - C_i)}{\sigma_{C_i}}\right), \tag{4.54}$$

where σ_C is the amount of noise for color channel C.

For example, the variable kernel method for the bivariate normalized *rg* kernel is given by

$$\hat{f}(r, g) = \frac{1}{n}\sum_{i=1}^{n} \sigma_{r_i}^{-1} K\left(\frac{r - r_i}{\sigma_{r_i}}\right) \sigma_{g_i}^{-1} K\left(\frac{g - g_i}{\sigma_{g_i}}\right), \tag{4.55}$$

where σ_r and σ_g are defined by Equations 4.44 and 4.45, respectively. In Figure 4.6, kernel density estimation is illustrated. For pixel values that generate color transformation instabilities, a smoother kernel is obtained. For stable color invariant values, narrow kernel sizes are used. In this way, kernel sizes are steered by the amount of uncertainty of the color invariant values.

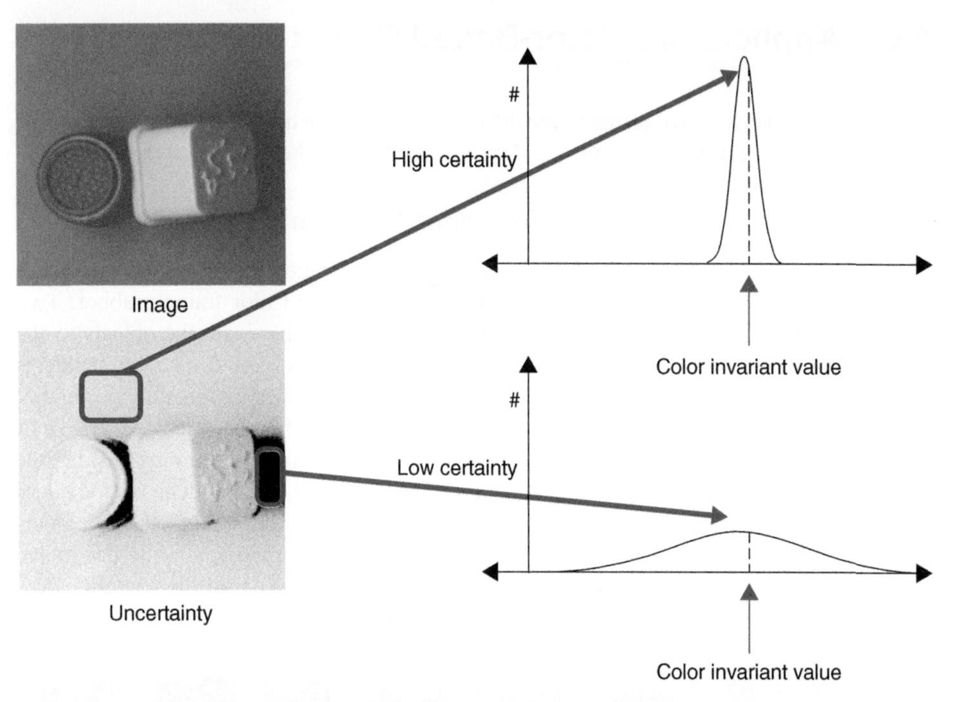

Figure 4.6 The uncertainty is used to derive the parameterization of the variable kernel for the purpose of robust histogram construction. For pixel values that generate color transformation instabilities, a smoother kernel is obtained. For stable color invariant values, narrow kernel sizes are provided. In this way, kernel sizes are steered by the amount of uncertainty of the color invariant values.

The variable kernel method estimating the directional hue density is given by

$$\hat{f}(\theta) = \frac{1}{n}\sum_{i=1}^{n}\sigma_{\theta_i}^{-1}K\left(\frac{(\theta-\theta_i)\bmod(\pi)}{\sigma_{\theta_i}}\right), \tag{4.56}$$

where σ_θ is defined by Equation 4.49.

Finally, the variable kernel method for the bivariate normalized O_1O_2 kernel is given by

$$\hat{f}(O_1, O_2) = \frac{1}{n}\sum_{i=1}^{n}\sigma_{O_{1_i}}^{-1}K\left(\frac{O_1 - O_{1_i}}{\sigma_{O_{1_i}}}\right)\sigma_{O_{2_i}}^{-1}K\left(\frac{O_2 - O_{2_i}}{\sigma_{O_{2_i}}}\right), \tag{4.57}$$

where σ_{O_1} and σ_{O_2} are defined by Equations 4.47 and 4.48.

In conclusion, to reduce the effect of sensor noise during density estimation, variable kernels are used where the normal distribution defines the shape of the kernel. Further, kernel sizes are steered by the amount of uncertainty of the color invariant values.

4.6 Application: Color-Based Object Recognition

In this section, we compare the different ways to construct color histograms in the context of object recognition. More information can be found in Reference [41].

4.6.1 Dataset and Performance Measure

In Figure 4.7, various images are shown. These images are recorded by a SONY XC-003P CCD color camera and the Matrox Magic Color frame grabber. Two light sources of average daylight color are used to illuminate the objects in the scene. The database consists of $N_1 = 500$ target images taken from colored objects such as tools, toys, food cans, and art. Objects are recorded in isolation (one per image), that is, 500 images are recorded from 500 different objects. The size of the images is 256×256 with 8 bits per color. The images show a considerable amount of shadows, shading, and highlights. A second, independent set (the query set) of $N_2 = 70$ query or test recordings is made of randomly chosen objects already in the object database. These objects are recorded again one per image with a new, arbitrary position and orientation with respect to the camera, some recorded upside down, some rotated, and some at different distances.

Figure 4.7 Various images that are included in the image database of 500 images. The images are representative of the images in the database. Objects were recorded in isolation (1 per image).

Then, for each image, traditional histograms (Eq. 4.50) and histograms based on variable density estimation are constructed on the basis of the rg (Eq. 4.55) and the hue space (Eq. 4.56). For the traditional (raw) histograms, the appropriate bin size is determined by varying the number of bins on the axes over $q \in \{2, 4, 8, 16, 32, 64, 128, 256\}$. The results show (not presented here) that the number of bins was of little influence on the recognition accuracy when the number of bins ranges from $q = 32$ and up. Therefore, the color histogram bin size for each axis used during histogram formation is $q = 32$.

For a measure of match quality, let rank r^{Q_i} denote the position of the correct match for test image $Q_i, i = 1, \ldots, N_2$, in the ordered list of N_1 match values. The rank r^{Q_i} ranges from $r = 1$ from a perfect match to $r = N_1$ for the worst possible match.

Then, for one experiment, the average ranking percentile is defined by

$$\bar{r} = \left(\frac{1}{N_2} \sum_{i=1}^{N_2} \frac{N_1 - r^{Q_i}}{N_1 - 1} \right) 100\%. \tag{4.58}$$

In the remaining sections, we use 70 test images and 500 target images. Matching is based on histogram intersection [43].

4.6.2 Robustness Against Noise: Simulated Data

The effect of noise is produced by adding independent zero-mean additive Gaussian noise with $\sigma \in \{2, 4, 8, 16, 32, 64\}$ to the query images. In Figure 4.8, two objects are shown generating together 10 images by adding noise with $\sigma \in \{8, 16, 32, 64, 128\}$.

Figure 4.8 Two images generating together 10 images by adding noise with $\sigma \in \{8, 16, 32, 64, 128\}$.

We concentrate on the quality of the recognition rate with respect to different noise levels. To compare histogram matching, we constructed four different histograms:

1. No thresholding is performed. This histogram construction scheme does not cope with unstable color invariant values. Hence, all color invariant values are equally weighted in the histogram as used by Swain and Ballard [43]. The color histogram without thresholding is denoted by $\mathcal{H}_{\theta_1}$ based on the hue θ color model and $\mathcal{H}_{rg_1}$ for the rg color model.
2. rg and θ values are discarded when the intensity is below 5% of the total range. For this histogram construction scheme, we denote $\mathcal{H}_{\theta_2}$ based on θ and $\mathcal{H}_{rg_2}$ derived from rg.
3. rg and hue values are discarded during histogram construction when the intensity and saturation are within the range of 4σ centered at the origin of the RGB space yielding $\mathcal{H}_{\theta_3}$ and $\mathcal{H}_{rg_3}$.
4. The proposed variable kernel density estimator is given by $\mathcal{H}_{\theta_4}$ and $\mathcal{H}_{rg_4}$.

The influence of noise differentiated by the various histogram construction schemes, shown in Figure 4.9 based on the hue color model, reveals that kernel density estimation outperforms the ad hoc thresholding schemes. In fact, the histogram intersection based on kernel density estimation gives good results up to considerable amounts of noise ($\sigma = 64$). Further, the thresholded

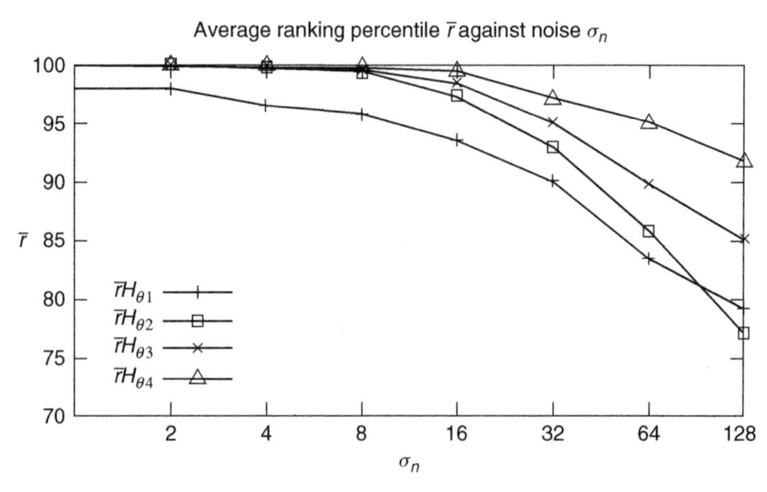

Figure 4.9 The discriminative power of the matching process differentiated for the various histogram construction schemes based on θ with respect to noise. The average percentile $\bar{r}$ for histogram $\mathcal{H}_{\theta_1}$, $\mathcal{H}_{\theta_2}$, $\mathcal{H}_{\theta_3}$, and $\mathcal{H}_{\theta_4}$ is given by $\bar{r}_{\mathcal{H}_{\theta_1}}$, $\bar{r}_{\mathcal{H}_{\theta_2}}$, $\bar{r}_{\mathcal{H}_{\theta_3}}$, and $\bar{r}_{\mathcal{H}_{\theta_4}}$, respectively.

histogram construction schemes always give higher recognition accuracy than no thresholding at all.

Further, on the basis of the *rg* color model, the impact of noise differentiated by the various histogram construction schemes is shown in Figure 4.10. Again the kernel density estimator provides higher recognition accuracy than the other schemes.

4.6.2.1 Robustness Against Noise: Realistic Data To measure the sensitivity of different histogram construction schemes with respect to varying SNR (signal-to-noise ratio), 10 objects were randomly chosen from the image dataset. Then, each object was recorded again under a global change in illumination intensity (i.e., dimming the light source) generating images with SNR $\in \{24, 12, 6, 3\}$ (Fig. 4.11).

These low intensity images can be seen as images of snap shot quality, a good representation of views from everyday life, as it appears in home video, the news, and consumer digital photography in general.

Matching based on the tradition histogram construction scheme, computed for *rg* is denoted by $\mathcal{H}_{rg_T}$, and for θ, we obtain $\mathcal{H}_{\theta_T}$. Thresholding has been applied on the images (not on the query image) and consequently *rg* and θ values are discarded when the intensity is below 5% of the total range. The kernel density estimation, based on *rg*, is denoted by $\mathcal{H}_{rg_K}$, and for θ, we have $\mathcal{H}_{\theta_K}$. The discriminative power of the histogram matching process based on *rg* and θ differentiated for the different histogram construction methods plotted against the amount of SNR is shown in Figure 4.12.

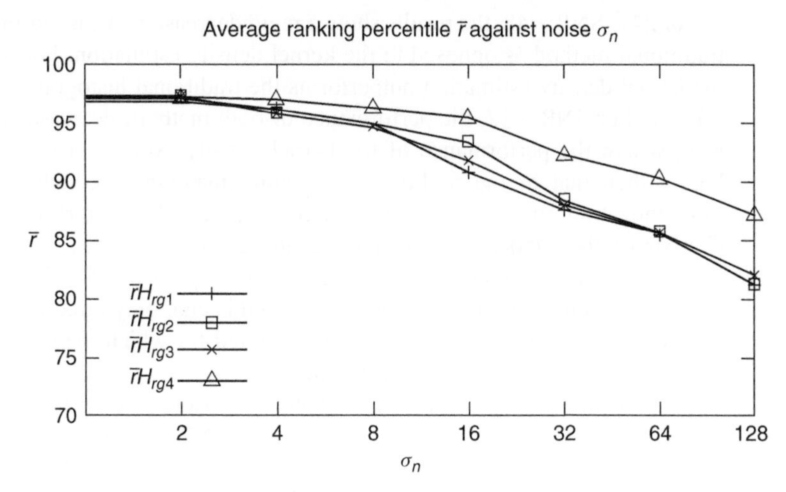

Figure 4.10 The discriminative power of the matching process differentiated for the various histogram construction schemes based on *rg* with respect to noise. The average percentile $\bar{r}$ for histogram $\mathcal{H}_{rg_1}, \mathcal{H}_{rg_2}, \mathcal{H}_{rg_3}$, and $\mathcal{H}_{rg_4}$ is given by $\bar{r}_{\mathcal{H}rg_1}, \bar{r}_{\mathcal{H}rg_2}, \bar{r}_{\mathcal{H}rg_3}$, and $\bar{r}_{\mathcal{H}rg_4}$, respectively.

Figure 4.11 Two objects under varying illumination intensity each generating four images with SNR $\in \{24, 12, 6, 3\}$.

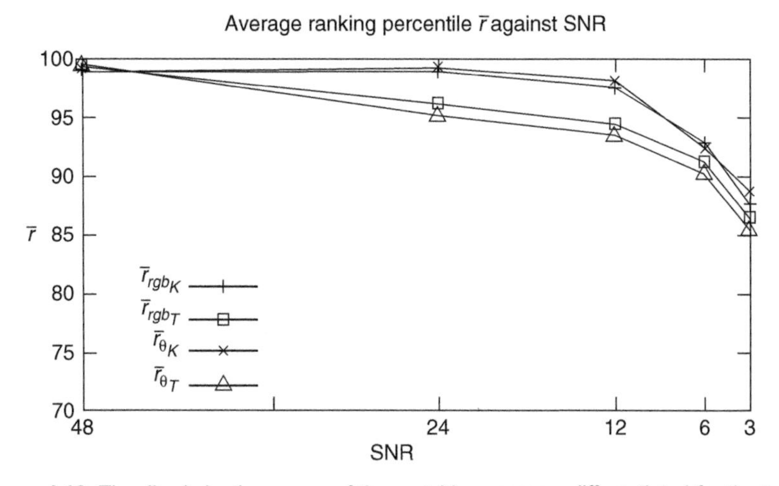

Figure 4.12 The discriminative power of the matching process, differentiated for the traditional histogram and kernel density estimation scheme, based on *rg* and θ with respect to SNR.

For $24 < \text{SNR} < 48$, the results show a rapid decrease in the performance of the traditional method as opposed to the kernel density estimation. For these SNRs, the kernel density estimation outperforms the traditional histogram construction scheme. For $\text{SNR} < 12$, the performance of both methods decreases in the same way, where the performance of the kernel density estimation remains slightly higher than that of the traditional histogram matching methods. This is due to quantization errors for very low intensity pixels that disturb the underlying Gaussian noise model. In fact, quantization errors are caused by reducing the image intensity and consequently limiting the range of *RGB* color values from which the color invariants are computed. To this end, only a reduced number of different color invariant values can be generated for which the assumption of a Gaussian noise model is not valid anymore.

In conclusion, the kernel density estimator outperforms the traditional histogram method up to considerable amounts of noise ($\text{SNR} = 12$). However, for very low intensity images ($\text{SNR} < 12$), due to quantization errors, the kernel density estimation behaves the same as traditional histogram methods.

4.7 Summary

Sets of color invariant models have been discussed, which are independent of the viewpoint, geometry of the object, and illumination conditions. These sets of color models are derived from the dichromatic reflection model. Different color transformations and their invariant characteristics are presented. Computational methods have been discussed to measure the effect of noise through these color invariants. As an application, color invariants are used for object recognition.

5 Photometric Invariance from Color Ratios

With contributions by Cordelia Schmid

In the previous chapter, several color channels were proved to be invariant to photometric changes under the assumption of white illumination. However, in real-world images, the light source can have different spectral power distributions. Although for outdoor images an often assumed illuminant is D65 (which is approximately white), the actual variety in outdoor light sources is much larger. Moreover, for indoor images, the color of the light source can take on even larger variations. The observed light reflected by the objects in the world is the product of the spectral power distribution of the illuminant and the object reflectance. Approaches that aim to describe scene reflectances invariant with respect to illuminant changes can be divided into two groups. The first group explicitly computes the illuminant and subsequently corrects the input image. These methods are discussed in detail in Part III (color constancy) of this book. The second group, which is discussed in detail in this chapter, does not aim to explicitly estimate the illuminant first to correct the image. Rather, the so-called color ratios combine measurements into dimensionless numbers that are invariant with respect to the color of the light source.

Fascinated by the human ability to observe object reflectance independent of the light source, Land and McCann [13] conducted a series of ingenious experiments

Color in Computer Vision: Fundamentals and Applications, First Edition.
Theo Gevers, Arjan Gijsenij, Joost van de Weijer, and Jan-Mark Geusebroek.

aimed to unravel the underlying mechanisms. Observers were presented with planar scenes of colored patches, called *Mondrians* in a reference to the paintings of the Dutch artist (Fig. 5.1). For these images, observers correctly reported the patch reflectances as being red, green, yellow, etc., independent of the light source illuminating the scene. In their analysis of how humans manage to ignore the undesired influence of the illuminant color, they observed that taking ratios of adjacent points in images leads to an edge detection, which is invariant with respect to the illumination. The main underlying assumption of their theory, known as *retinex theory*, is that the illuminant of two adjacent (or nearby) points in a scene is the same. On the other hand, the reflectance changes are assumed to be abrupt. After the initial research of Land and McCann, the theory of color ratios was further developed by Nayar and Bolle [45]. They show that the reflectance ratios also hold for curved surfaces in a 3D world under the assumption of locally smooth surfaces. In other words, illuminant changes are assumed to be spatially low frequent, whereas reflectance changes are high frequent. Note that this assumption does allow for multiple illuminants as long as the changes in chromaticity and intensity are low frequent. This assumption of *locally constant illumination* is at the basis of all the color ratios presented in this chapter.

An important application area for color ratios is the field of color indexing for image retrieval. To successfully index objects, image representations should be robust to scene incidental events, such as variations in viewpoint, shadow, shading, and illuminant color, which is exactly the strength of color ratios. Color indexing was first proposed by Ballard and Swain [43] and applied to object recognition. Their method recognizes objects by using *RGB* color histograms. Funt and Finlayson [46] pointed out that this method lacks robustness with

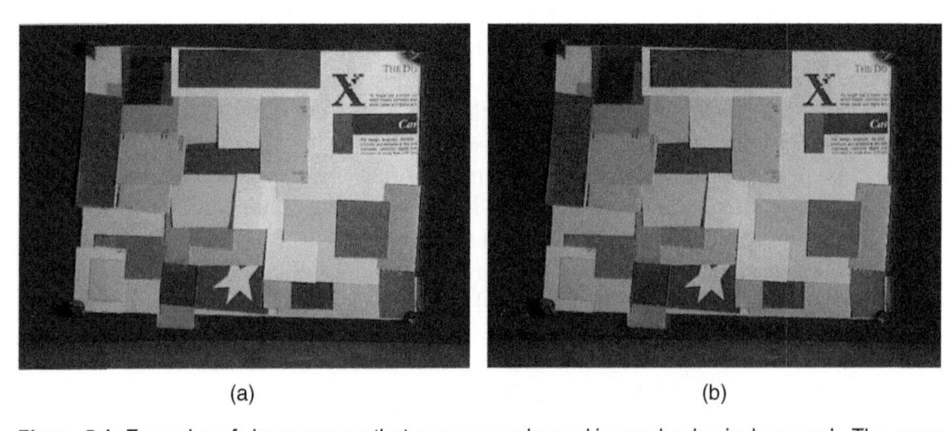

Figure 5.1 Examples of planar scenes that are commonly used in psychophysical research. The same scene is illuminated by two different light sources. Land and McCann [13] found that humans are able to describe the reflectance color of the patches in such scenes independent of the illuminant color. The images are taken from Reference 44.

respect to changes in the illuminant's color. They showed that using color ratios ensures robustness with respect to illuminant changes. In addition, they show how to compute color ratios based on image derivatives. However, these descriptors are still dependent on the lighting geometry. Hence, abrupt changes due to object orientation or camera viewpoint alter the object's description. A solution to this problem was proposed by Gevers and Smeulders [47]. They introduced an invariant that is both robust to variations of illuminant color and lighting geometry. Finally, Van de Weijer and Schmid observed that color ratios based on image derivatives are dependent on the smoothness of the edge [48, 49]. To overcome this problem, they proposed a set of color ratios that is robust to blur.

In this chapter, we present various color ratios. We will show that they can be computed either as ratios of pixels (different spatial locations) or as image derivatives. As a matter of fact, they could have equally well been classified as illuminant invariant image derivatives. Because of this double role in between pixel-based invariants of Chapter 4 and the photometric invariant derivatives discussed in Chapter 6, we now introduce the color ratios.

5.1 Illuminant Invariant Color Ratios

To derive the color ratios, we start with the reflectance model introduced in Chapter 3. Under the assumption of Lambertian reflectance, Equation 5.3 can be written as

$$f^c(\mathbf{x}) = m^b(\mathbf{x}) \int_\lambda e(\lambda, \mathbf{x}) s(\lambda, \mathbf{x}) \rho^c(\lambda) \mathrm{d}\lambda, \tag{5.1}$$

recalling that for three channels $c \in \{R, G, B\}, m^b(\mathbf{x})$ contains changes due to illuminant intensity changes, object geometry, and lighting geometry. e is the illuminant and $s(\lambda, \mathbf{x})$ denotes object reflectance. Further, assuming narrow-band sensor sensitivities, such that the spectral response can be approximated by delta functions $\rho^c(\lambda) = \delta(\lambda - \lambda_c)$, Equaton 5.1 can be simplified as

$$f^c(\mathbf{x}) = m^b(\mathbf{x}) e^c(\mathbf{x}) s^c(\mathbf{x}), \tag{5.2}$$

where $e^c(\mathbf{x})$ is short for $e(\lambda_c, \mathbf{x})$ and $s^c(\mathbf{x})$ for $s(\lambda_c, \mathbf{x})$. Narrow-band sensors imply that only light of a particular wavelength is passed. If the color camera provides narrow-band sensors then the measured values that are available from the camera can be used. This approximation is found to be acceptable for most narrow-band color sensors [50].

Equation 5.2 is at the basis of all the color ratios. The equation states that the measured values are a multiplication of the scene and light geometry, the light source color, and the object color. Land and McCann [13] made the observation that in most real-world scenes, the light source is locally constant, meaning that for two adjacent points x_1 and x_2 the following holds: $e^c(\mathbf{x_1}) = e^c(\mathbf{x_2})$. On the basis of this observation, Funt and Finlayson [46] (in a similar derivation as Nayar and Bolle [45]) propose to use color ratios for the purpose of object recognition:

$$F(f^c_{\mathbf{x}_1}, f^c_{\mathbf{x}_2}) = \frac{f^c_{\mathbf{x}_1}}{f^c_{\mathbf{x}_2}}. \tag{5.3}$$

In fact, for $c \in \{R, G, B\}$, the color ratios are computed from colors at two neighboring image locations, $\mathbf{x}_1$ and $\mathbf{x}_2$, and are given by

$$F_1 = \frac{R_{\mathbf{x}_1}}{R_{\mathbf{x}_2}}, \tag{5.4}$$

$$F_2 = \frac{G_{\mathbf{x}_1}}{G_{\mathbf{x}_2}}, \tag{5.5}$$

$$F_3 = \frac{B_{\mathbf{x}_1}}{B_{\mathbf{x}_2}}. \tag{5.6}$$

Substituting Equation 5.2 into Equation 5.3, we obtain

$$F = \frac{m^b(\mathbf{x}_1)e^c(\mathbf{x}_1)s^c(\mathbf{x}_1)}{m^b(\mathbf{x}_2)e^c(\mathbf{x}_2)s^c(\mathbf{x}_2)}. \tag{5.7}$$

The effect of the illuminant e is factored out under the assumption of locally constant illumination $e^c(\mathbf{x}_1) = e^c(\mathbf{x}_2)$. Note that this assumption still allows for varying illumination across the scene, for example, multiple light sources, but only requires that the illuminant shows no sudden local changes. Further, under the assumption that neighboring points have the same surface orientation ($m^b(\mathbf{x}_1) = m^b(\mathbf{x}_2)$), for example, locally smooth surfaces, the body reflectance term factors out, leaving only the surface albedo of the two neighboring points:

$$F = \frac{s^c(\mathbf{x}_1)}{s^c(\mathbf{x}_2)}. \tag{5.8}$$

Hence, computing the ratio between two neighboring points results in a color invariant that is insensitive to object geometry, illumination direction, intensity, and color under the assumption of smooth continuous surfaces.

As can be seen from Equation 5.3, F is unbounded. If the color signal of the second location is small then F can take on huge values: $f^c_{\mathbf{x}_2} \to 0 \Rightarrow F \to \infty$. In order to turn F into a well-behaved function, Nayar and Bolle [45] propose a slightly different ratio (also called the *Michelson contrast*):

$$N(f^c_{\mathbf{x}_1}, f^c_{\mathbf{x}_2}) = \frac{f^c_{\mathbf{x}_1} - f^c_{\mathbf{x}_2}}{f^c_{\mathbf{x}_1} + f^c_{\mathbf{x}_2}}. \tag{5.9}$$

In this case, $-1 \leq N \leq 1$ if the two neighboring points are not both black.

The underlying assumption of the color ratio F is that the neighboring points have the same surface normal (and therefore $m^b(\mathbf{x}_1) = m^b(\mathbf{x}_2)$). This restriction excludes many real-world objects that have abrupt geometry changes, such as the transitions between sides of a cube for which $m^b(\mathbf{x}_1) \neq m^b(\mathbf{x}_2)$. To overcome this, Gevers and Smeulders [47] proposed a color ratio that is not only invariant to the color of the light source but also discounts the object's geometry:

$$M(f_{\mathbf{x}_1}^{c_1}, f_{\mathbf{x}_1}^{c_2}, f_{\mathbf{x}_2}^{c_1}, f_{\mathbf{x}_2}^{c_2}) = \frac{f_{\mathbf{x}_1}^{c_1} f_{\mathbf{x}_2}^{c_2}}{f_{\mathbf{x}_1}^{c_2} f_{\mathbf{x}_2}^{c_1}}, \tag{5.10}$$

where f^{c_1} and f^{c_2} are two different color channels. For an RGB image, three different color channels can be derived as follows:

$$M_1 = \frac{R_{\mathbf{x}_1} G_{\mathbf{x}_2}}{R_{\mathbf{x}_2} G_{\mathbf{x}_1}}, \tag{5.11}$$

$$M_2 = \frac{R_{\mathbf{x}_1} B_{\mathbf{x}_2}}{R_{\mathbf{x}_2} B_{\mathbf{x}_1}}, \tag{5.12}$$

$$M_3 = \frac{G_{\mathbf{x}_1} B_{\mathbf{x}_2}}{G_{\mathbf{x}_2} B_{\mathbf{x}_1}}. \tag{5.13}$$

Note that the third is dependent on the first two according to $M_3 = M_2/M_1$. It can be shown that these ratios are invariant to the color of the light source (under the assumption of locally uniform illumination), intensity of the light source, viewpoint, and surface geometry:

$$M = \frac{(m^b(\mathbf{x}_1) e^{c_1}(\mathbf{x}_1) s^{c_1}(\mathbf{x}_1))(m^b(\mathbf{x}_2) e^{c_2}(\mathbf{x}_2) s^{c_2}(\mathbf{x}_2))}{(m^b(\mathbf{x}_1) e^{c_2}(\mathbf{x}_1) s^{c_2}(\mathbf{x}_1))(m^b(\mathbf{x}_2) e^{c_1}(\mathbf{x}_2) s^{c_1}(\mathbf{x}_2))} \tag{5.14}$$

$$= \frac{s^{c_1}(\mathbf{x}_1) s^{c_2}(\mathbf{x}_2)}{s^{c_2}(\mathbf{x}_1) s^{c_1}(\mathbf{x}_2)}, \tag{5.15}$$

as $e^{c_i}(\mathbf{x}_1) = e^{c_i}(\mathbf{x}_2)$ under the assumption of locally constant illumination.

5.2 Illuminant Invariant Edge Detection

In the previous section, the color ratios have been computed by taking the ratio of spatially varying points in the image. In this section, we show that the color ratios can also be written as image derivatives, an observation that was first made by Funt and Finlayson [46]. They noted that if the color ratio F is invariant for illuminant changes then so is $\ln(F)$. Rewriting $\ln(F)$ shows that computing the color ratios F is equal to taking the derivative of the logarithm of the channels:

$$\ln(F_1) = \ln\left(\frac{R_{\mathbf{x}_1}}{R_{\mathbf{x}_2}}\right) = \ln(R_{\mathbf{x}_1}) - \ln(R_{\mathbf{x}_2}) = \tfrac{\partial}{\partial \mathbf{x}} \ln(R(\mathbf{x})). \tag{5.16}$$

Hence, the derivative of the logarithm of an image is invariant to illuminant changes under the assumption of locally constant surface normals. Using the fact that $\frac{\partial}{\partial x}\ln(f(x)) = \frac{f_x(x)}{f(x)}$, the three ratios can also be computed by

$$\{F_1, F_2, F_3\} = \left\{\frac{R_{\mathbf{x}}}{R}, \frac{G_{\mathbf{x}}}{G}, \frac{B_{\mathbf{x}}}{B}\right\}, \tag{5.17}$$

where the subscript $\boldsymbol{x}$ indicates the spatial derivative.

A similar derivation holds for the color ratios proposed by Gevers and Smeulders [47]. Starting from the logarithm of the color ratio M, this can be rewritten as

$$\ln\left(M_1\right) = \ln\left(\frac{R^{\mathbf{x}_1}G^{\mathbf{x}_2}}{R^{\mathbf{x}_2}G^{\mathbf{x}_1}}\right) = \ln\left(\frac{R^{\mathbf{x}_1}}{G^{\mathbf{x}_1}}\right) - \ln\left(\frac{R^{\mathbf{x}_2}}{G^{\mathbf{x}_2}}\right) = \frac{\partial}{\partial \mathbf{x}}\ln\left(\frac{R(\mathbf{x})}{G(\mathbf{x})}\right). \tag{5.18}$$

Hence the derivative of the logarithm of the division of two different channels is independent of the illuminant color. The two color ratios can be computed according to

$$\{M_1, M_2\} = \left\{\frac{R_{\mathbf{x}}G - G_{\mathbf{x}}R}{RG}, \frac{G_{\mathbf{x}}B - B_{\mathbf{x}}G}{GB}\right\}. \tag{5.19}$$

In Figure 5.2, an illustration of the photometric invariant color ratios is provided.

5.3 Blur-Robust and Color Constant Image Description

Apart from the previously discussed photometric variations, blur changes are another frequently encountered phenomenon. They can be caused, among others, by out-of-focus, relative motion between the camera and the object, and aberrations in the optical system [51]. For zero-order descriptions (e.g., normalized *RGB*), variations in blur have little influence. However, a change in blur will drastically change edge-based descriptions. Edge-based color methods measure two intertwined phenomena: the color change between two regions and the edge sharpness of the transition between the regions. A change in blur will have little influence on the color change, but it will influence the edge sharpness of the transition. Therefore, representations based on derivatives have the undesirable effect that they vary under image blur. We now discuss the influence of blur on the color constant ratios discussed previously. We further discuss a method to reduce the sensitivity of color ratios to image blur.

Let us assume that the illuminant invariant derivatives discussed previously are computed by derivation with a Gaussian derivative at scale σ_d. As a consequence the ratios have a certain scale, for example, $F_1^{\sigma_d} = R_{\mathbf{x}}^{\sigma_d}/R^{\sigma_d}$. We model blur by

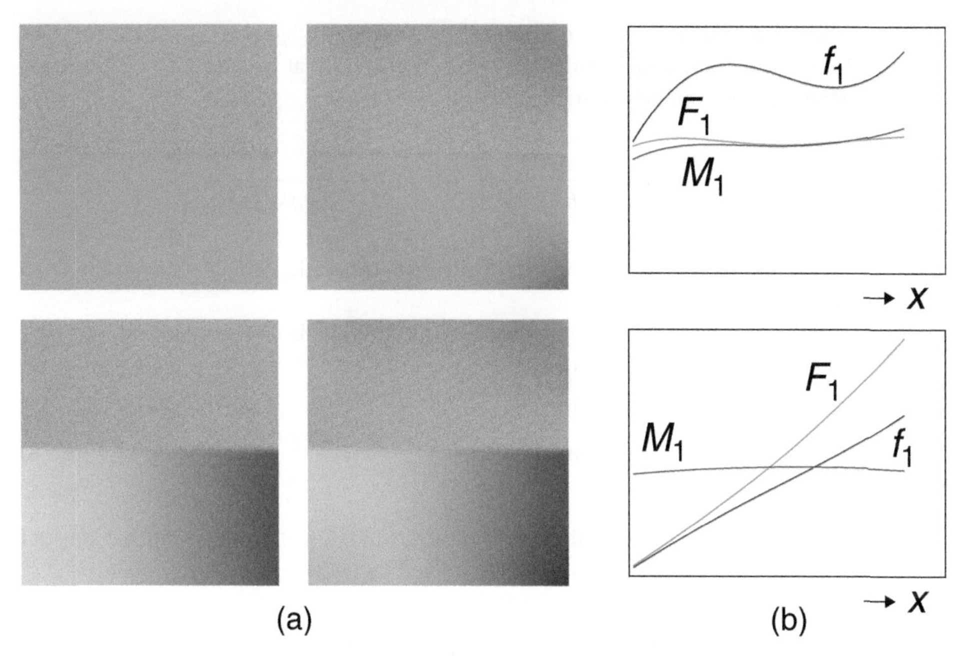

Figure 5.2 Illuminant invariant edge detection: (a) two color transitions are shown. The left-bottom figure contains an additional fall in intensity from left to right in the blue part. This models variations that are common due to changes in scene geometry. In the middle, the two color transitions on the left are illuminated by a locally smooth varying illuminant. (b) The two graphs depict the edge response computed on the images in the middle along the line separating the two regions. The responses are given for normal image derivative f_1 and the color ratios F_1 and M_1. For the top image both color ratios remain fairly stable, whereas the normal image derivative response f_1 varies significantly due to the illuminant changes. As predicted by theory, when we also vary the scene geometry, as in the bottom row, the response of F_1 varies, whereas only M_1 remains stable under a combination of geometrical and illuminant variations.

a convolution with a Gaussian kernel with σ_b. Then blurring will have a similar effect as computing the ratio at a different scale $\sigma = \sqrt{\sigma_d^2 + \sigma_b^2}$, since

$$F_1^\sigma = \frac{(R \otimes G^{\sigma_b}) \otimes \frac{\partial}{\partial \mathbf{x}} G^{\sigma_d}}{R \otimes G^{\sigma_b} \otimes G^{\sigma_d}} = \frac{R \otimes \frac{\partial}{\partial \mathbf{x}} G^{\sqrt{\sigma_b^2+\sigma_d^2}}}{R \otimes G^{\sqrt{\sigma_b^2+\sigma_d^2}}}, \tag{5.20}$$

and hence robustness with respect to blur is equal to robustness with respect to changing the scale of the ratios.

Next we analyze the influence of scale on the ratios. Assume that an edge can be modeled by a step edge $R(x) = \alpha u(x) + \beta$. Then,

$$F_1^\sigma = \frac{\frac{\partial}{\partial \mathbf{x}} (\alpha u(x) + \beta) \otimes G^\sigma}{(\alpha u(x) + \beta) \otimes G^\sigma} = \frac{\alpha \delta(x) \otimes G^\sigma}{(\alpha u(x) + \beta) \otimes G^\sigma}, \tag{5.21}$$

where we used the fact that the derivative of the step edge $u(x)$ is the delta function $\delta(x)$. Let us now consider the ratio response exactly at the edge, $x = 0$. Here the denominator remains constant, and

$$F_1^\sigma = \frac{\alpha}{\beta + \frac{1}{2}\alpha} G^\sigma(0) = \frac{\alpha}{\beta + \frac{1}{2}\alpha} \frac{1}{\sigma\sqrt{2\pi}}. \tag{5.22}$$

This response is clearly not independent of the scale, which proves that color ratios vary with blur.

To obtain robustness with respect to blur, Van de Weijer and Schmid [48] proposed the following color angles $\varphi_F = \{\varphi_F^1, \varphi_F^2\}$:

$$\varphi_F^1 = \arctan\left(\frac{F_1}{F_2}\right), \varphi_F^2 = \arctan\left(\frac{F_2}{F_3}\right). \tag{5.23}$$

The dependence on blur is factored out by the division of the color ratios. Consider the edge of the green channel to be modeled by $G(x) = \lambda u(x) + \gamma$, then

$$\varphi_F^1 = \arctan\left(\frac{\alpha\left(\gamma + \frac{1}{2}\lambda\right)}{\left(\beta + \frac{1}{2}\alpha\right)\lambda}\right), \tag{5.24}$$

which is independent of the scale σ, and therefore robust to variation of blur. Moreover, φ_F^1 is invariant for illuminant color changes since both F_1 and F_2 are invariant. Note that the usage of the arctan is not necessary to obtain the invariance. However, the arctan maps the output to the domain of $[-\pi, \pi]$, which can be better represented in a histogram.

A similar derivation of dependence to blur can be given for the color constant and lighting geometry invariant ratios, M_1 and M_2. To obtain robustness the following color angle can be computed:

$$\varphi_M = \arctan\left(\frac{M_1}{M_2}\right). \tag{5.25}$$

When using the color angles proposed in Equations 5.23 and 5.25, one should take the reliability into account [41]. Application of error analysis to any of the color angles yields the following results:

$$\left(\partial \arctan\left(\frac{a}{b}\right)\right)^2 = \frac{(\partial\epsilon)^2}{\sqrt{a^2 + b^2}}, \tag{5.26}$$

where we assume $\partial a = \partial b = \partial \varepsilon$. This equation informs us that color angles, for which $\sqrt{a^2 + b^2}$ is small, are less reliable.

5.4 Application: Image Retrieval Based on Color Ratios

As an illustration of how color ratios can be used, we apply them to an image retrieval task. The task is designed to test the image descriptions with respect to changes in illuminant color variations. The performance of the retrieval is assessed by the rank results of the correct matches, where the rank indicates after how many images the correct image was retrieved. We also analyze the normalized average rank (NAR) that is defined for a single query as

$$\mathrm{NAR} = \frac{1}{NN_R}\left(\sum_{i=1}^{N_R} R_i - \frac{N_R\left(N_R+1\right)}{2}\right), \tag{5.27}$$

where N is the number of images in the database, N_R is the number of relevant images to the query, and R_i is the rank at which the ith relevant image is retrieved. An NAR of zero indicates perfect results, and an NAR $= 0.5$ is equal to random retrieval. We will give the average NAR results over all queries, indicated by ANAR.

Histograms of the color ratios and color angles are constructed to represent the image. We have used 16 bins in each color dimension (there are three dimensions for $\boldsymbol{F}$, two dimensions for φ_F and $\boldsymbol{M}$, and one dimension for φ_M). To robustify the construction of the histograms of the color angles, we use Equation 5.26. For example, for φ_M we update the histogram with $\sqrt{M_1^2 + M_2^2}$. The retrieval is based on the Euclidean distance between the histograms and the derivatives are computed with Gaussian derivative filters with a standard deviation of $\sigma = 2$. The first two experiments are performed on a set of 20 colorful objects, all taken under 10 different light sources with varying object orientations [44], of which examples are given in Figure 5.3.

5.4.1 Robustness to Illuminant Color

First, we test the image descriptions with respect to robustness to illuminant color variations. For each of the 20 objects, we pick 1 single image as a query. For each query, there exist 10 relevant images of the same object taken under different light sources and in various object orientations. The results are summarized in Table 5.1a. These images were all taken at a similar distance and hence the edges are equally sharp in most images. Therefore, robustness with respect to blur is not required and the two color ratios, F and M, obtain good results. The added robustness with respect to blur for color angles results in lower discriminative power; however, for φ_F the drop in performance is minimal. For the 16-bin representation of φ_M, the performance drop due to loss of discriminative power is higher.

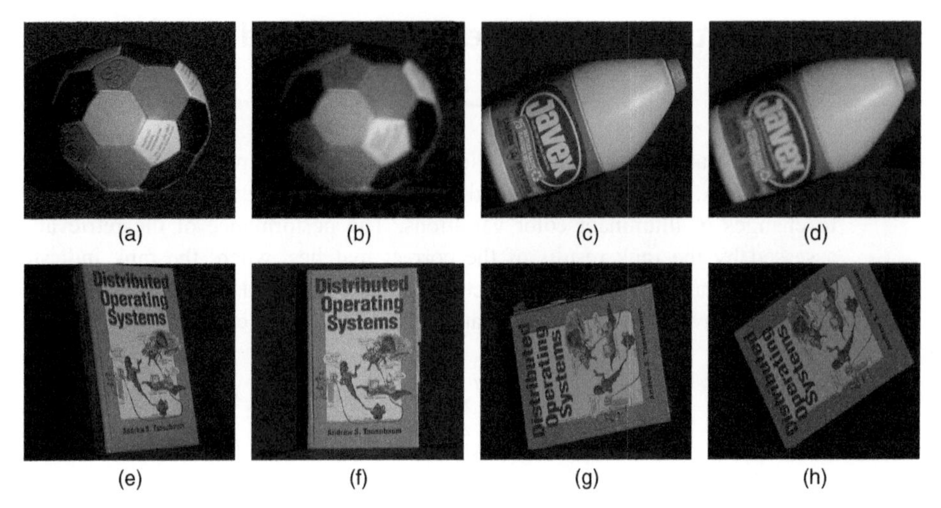

Figure 5.3 Examples of object images from the Simon Fraser data set (637 × 468 pixels). (a–d) Images of two objects and their smoothed versions used to test robustness with respect to Gaussian blur. (e–h) Four instantiations of a single object under four different illuminants and with varying object orientation. These are used to test the image description with respect to illuminant color and illuminant geometry changes.

5.4.2 Robustness to Gaussian Blur

Next we test the image descriptions with respect to changes in blur. To this end, we take a single image of all 20 objects taken under the same illuminant. Next, Gaussian smoothing with standard deviation of $\sigma = 2$ is applied to the images, which leads only to a slight visual change in the images (Fig. 5.3). We used the nonsmoothed image as a query to find its smoothed counterpart in the set of 20 smoothed images. The retrieval results of this experiment are given in Table 5.1. The sensitivity of the color ratios $\boldsymbol{F}$ and $\boldsymbol{M}$ under blur is apparent: only for a few of the queries the relevant image was found with rank 1. The two color angles, which were designed to be robust with respect to blur, obtain good results. For φ_F, only for a single image the relevant image was *not* the first image to be retrieved. In conclusion, color angles provide a robust image description under image blur.

5.4.3 Robustness to Real-World Blurring Effects

This experiment is performed on a set of 20 pairs of images. Each pair consists of two images of the same scene; however, the images vary in blur. The blur is caused by changing the acquisition parameters such as shutter time and aperture, and due to relative movement between the camera and the object (for examples,

---

This data is available on http://www.colorincomputervision.com.

see Fig. 5.4). Table 5.1 provides the results. The variations in blur cause the color ratios, F and M, to perform poorly. Although not all real-world blurring effects can be modeled by a Gaussian [51], the proposed blur-robust color angles obtain good results: for φ_F only a single image is not retrieved within the first two images.

Table 5.1 Rank and ANAR for the retrieval experiment.

(a) Robustness to illuminant color

Rank	1–10	11–20	>20	ANAR
F	180	5	15	0.010
φ_F	169	17	14	0.012
M	155	22	23	0.024
φ_M	115	23	65	0.049

(b) Robustness to Gaussian blur

Rank	1	2	>2	ANAR
F	5	0	15	0.218
φ_F	19	1	0	0.003
M	1	3	16	0.258
φ_M	15	3	2	0.023

(c) Robustness to real-world effects

Rank	1	2	>2	ANAR
F	7	2	11	0.365
φ_F	16	3	1	0.018
M	6	2	12	0.303
φ_M	13	1	6	0.053

(a) (b) (c)

Figure 5.4 Examples database: (a) out-of-focus blur, (b) change in focus from foreground to background, and (c) motion blur.

5.5 Summary

Assuming white illumination, as we saw in earlier chapters, might be a too restrictive assumption for real-world applications. The assumption that illuminants in a scene can be colored and vary spatially throughout the scene is more realistic. In this chapter, we discussed the color ratios that are invariant for such illuminant changes. The underlying insight is that illuminants are locally constant, and are therefore equal on both sites of an edge. Dividing observations on both sites of the edge will then factor out the illuminant color.

Several extensions on the basic principle have been discussed. We have shown how to obtain invariance with respect to geometry and blur variations. Furthermore, we have derived how to compute the color ratios as image derivatives. Finally, we have a shown several experiments to illustrate the descriptive power of the color ratios.

6 Derivative-Based Photometric Invariance

With contributions by Rein van den Boomgaard and Arnold W. M. Smeulders

Image derivatives are essential to describe the local structure in images. First-order derivatives reveal information about the location of edges in images or the speed of objects in videos. Second-order derivatives of images allow us to identify corners in images and object acceleration in videos. Being essential operations, image derivatives are applied in the vast majority of computer vision applications, including basic operations such as edge detection, feature extraction, and optical flow and more complex applications such as shape from shading, image segmentation, and object detection.

A problem in classical derivative-based computer vision, which is based only on luminance or *RGB*, is that derivatives describe both scene incidental edges such as shadow and specularity transitions, as well as relevant material transitions. For example, luminance-based optical flow estimation is flawed by moving shadows or RGB-based object segmentations fails in the presence of specularities. These problems can be solved by extending the photometric invariance theory described in Chapter 4 to the computation of image derivatives. The differential structure of images could then be split up into separate parts according to their invariance.

Color in Computer Vision: Fundamentals and Applications, First Edition.
Theo Gevers, Arjan Gijsenij, Joost van de Weijer, and Jan-Mark Geusebroek.

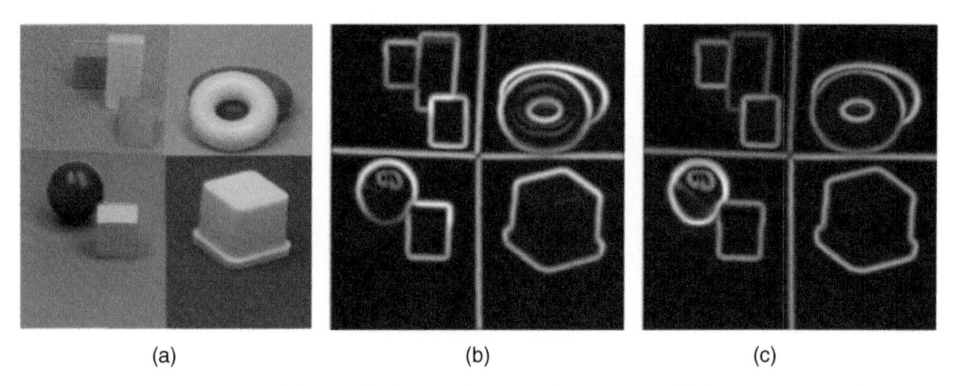

(a) (b) (c)

Figure 6.1 (a) Input image. (b) The shadow and shading full invariant. (c) The shadow and shading quasi-invariant. Both invariants do not have a response on edges, which are caused by shadows and shading.

For example, image derivatives could be derived to be invariant to shading, specularities, and illuminant changes. In Figure 6.1, an example of photometric invariant edge detection is provided. The image shows two examples of invariants that do not respond to shadow and shading edges. This can, for example, be seen from the sharp shading edges on the green cube, which are ignored by both edge detectors.

Before exploring how to extend photometric invariance theory to image derivatives, we shortly introduce the difference between feature detection and feature description (both are considered part of feature extraction), which will help us explain the different approaches to photometric invariant edge detection. Feature detection is the task of locating features in images, and it includes edge, corner, and t-junction detection [52–54]. On the other hand, feature description aims to describe these local features in images. Examples of often used feature descriptors are the SIFT descriptor [55] and the shape-context descriptor [56]. In many computer vision applications, first a detector is applied to locate the features, after which a descriptor is applied to describe the local features [57]. The difference between feature detection and description is important to understand the difference between two different photometric invariance theories that we describe in this chapter.

The two approaches to compute photometric invariant features, which are handled in this chapter, are called *full invariants* and *quasi-invariants* [58]. Section 6.1 derives full photometric invariants, which can be used for both feature detection and feature description. Their properties are such that the resulting invariant values are truly independent of the physical events they are designed to ignore. Hence, these features can be used for recognition of an object under different imaging conditions, as will be illustrated in Section 6.1. The alternative approach is that of quasi-invariant discussed in Section 6.2, which concentrates on the detection of features independent of certain photometric events. Similar to full invariants, quasi-invariants do not respond to changes in the images, which are purely caused by the event, which they are designed to ignore (e.g., a pure shadow edge). However, the strength of these invariants varies at places where there is a mixture of events (such as a shadow variation on a material edge). As a result, quasi-invariants are restricted to be used for feature detection, but cannot be applied for feature description. For the task of feature detection, we will see that quasi-invariants have improved discriminative power and reduced edge displacement with respect to full invariants.

To illustrate the subtleties between full invariants and quasi-invariants, we provide a practical example. In Figure 6.2, the edge response of the standard color gradient and the full and quasi-invariants along the red dotted line are provided. The red dotted line crosses three edges, first, a material transition from purple to green, then, a shading edge on the green cube, and finally, a material transition from green to purple. As expected, the color gradient yields a response for all three edges. Both invariants do not have a response on the shading edge and only respond to the material transitions. The figure also illustrates the difference between the full and quasi-invariants. The full invariant has exactly the same response for both green–purple transitions. However, the response of the quasi-invariant varies between the two green–purple transitions. The response changes due to variations in intensity of the object and background.

In conclusion, color edge detection has the added advantage that the color information allows us to separate different causes of edges into shadow, shading, or specular transitions. In this chapter, we discuss two approaches to obtain photometric invariant edge derivatives. These derivatives can be used in all computer vision applications that are based on image derivatives. As such, they

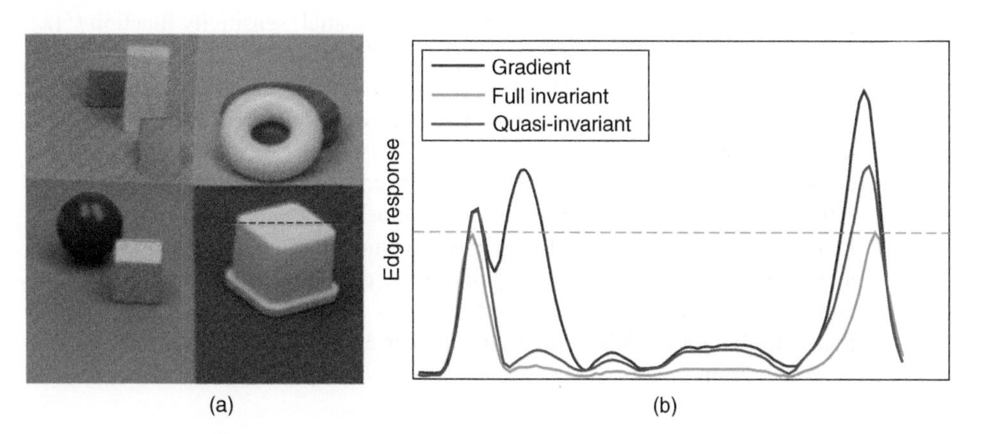

Figure 6.2 (a) Input image. (b) Edge response along the dotted red line in the input image for the color gradient, the full invariant, and the quasi-invariant. Note that the shading variations on the green cube, which results in the second peak in the gradient response, do not lead to any response for both invariants.

help to obtain robustness to scene accidental changes such as shadows, shading effects, and specularities.

6.1 Full Photometric Invariants

The measurement of invariance involves a balance between constancy of the measurement regardless of the disturbing influence of the unwanted transform on the one hand and retained discriminating power between truly different states of the objects on the other. As a rule of thumb, features that are more invariant have less discriminative power. Hence, both invariance and discriminating power of a method should be investigated simultaneously. Only this allows for assessing the practical performance of the proposed method. The emphasis in this section is on invariance, with an experimental assessment of discriminative power. In Section 6.2, full invariance is sacrificed to improve discriminating power which results in improved sensitivity and accuracy for feature detection.

In this section, we consider the introduction of wavelength in the scale-space paradigm, as suggested by Koenderink [59]. This leads to a spatiospectral family of Gaussian aperture functions, which smooth and differentiate the data. Hence, the mathematical framework is that of differential calculus and differential invariants. The general idea is that derivatives lead to orthogonalization of the influence of the various parameters on the measurement, with the assumption that small fluctuations can be accounted to a linear decomposition of the variation over the effective parameters. This leads to the well-known chain rule, the fundamental rule for differential calculus. The idea developed in this chapter is that one can measure the spatial *and* spectral derivatives of the energy density $E(\lambda, \mathbf{x})$ in Equation 3.30. The integral in that equation acts as the effective differentiator operator by ''choosing'' the right spectral (and spatial) sensitivity function $f^c(\lambda)$. The next section illustrates how to ''choose'' the sensitivity function by an appropriate linear transformation of the camera *RGB* sensitivities, such that the result mimics the Gaussian smoothing and derivative operator up to second order (as there are three spectral sensitivities). This model is termed the *Gaussian color model* by Koenderink and is also referred to as the *local color model* [59, Section 5.6]. Spatial derivatives are operationally defined by convolving the image with the derivatives of the Gaussian smoothing operator, a by now basic technique in computer vision. Exploiting the orthogonality of these derivatives and the chain rule of differentiation, one can apply differential calculus to the image formation models outlined in Chapter 3 and truly obtain the differential invariants from images.

6.1.1 The Gaussian Color Model

Physical measurements imply integration over the spectral and spatial (and time) dimensions. The integration reduces the infinitely dimensional Hilbert space of spectra at infinitesimally small spatial neighborhood to a limited amount of measurements. The Gaussian color model is not an essentially new color model, but rather a theory of color measurement. The Gaussian color model may be considered an extension of the Gaussian derivative framework into the spatiospectral domain. As such, the model extents to the spatiospectral scale-space and allows the measurement of combined photometric and geometric differential invariants.

From scale-space theory we know how to probe a function at a certain scale; the probe should have a Gaussian shape in order to prevent the creation of extra details into the function when observed at a higher scale (lower resolution) [60]. We consider the Gaussian as a general probe for the measurement of spatiospectral differential quotients. We follow [61] for the Gaussian color model. Let $E(\lambda)$ be the energy distribution of the incident light, where λ denotes wavelength and let $G(\lambda_0; \sigma_\lambda)$ be the Gaussian at spectral scale σ_λ positioned at λ_0. The spectral energy distribution may be approximated by a Taylor expansion at λ_0:

$$E(\lambda) = E^{\lambda_0} + \lambda E_\lambda^{\lambda_0} + \frac{1}{2}\lambda^2 E_{\lambda\lambda}^{\lambda_0} + \cdots. \tag{6.1}$$

Measurement of the spectral energy distribution with a Gaussian aperture yields a weighted integration over the spectrum. The observed energy in the Gaussian color model $E(\lambda)$, at infinitely small spatial resolution and spectral scale σ_λ, is in second order equal to [59]

$$E^{\sigma_\lambda} = E^{\lambda_0,\sigma_\lambda} + \lambda E_\lambda^{\lambda_0,\sigma_\lambda} + \frac{1}{2}\lambda^2 E_{\lambda\lambda}^{\lambda_0,\sigma_\lambda} + \mathcal{O}(\lambda^3), \tag{6.2}$$

where $E^{\lambda_0,\sigma_\lambda} = \int E(\lambda)G(\lambda; \lambda_0, \sigma_\lambda)d\lambda$ measures the spectral intensity. Then, differentiation $E_\lambda^{\lambda_0,\sigma_\lambda} = \int E(\lambda)G_\lambda(\lambda; \lambda_0, \sigma_\lambda)d\lambda$ gives the first-order spectral derivative and $E_{\lambda\lambda}^{\lambda_0,\sigma_\lambda} = \int E(\lambda)G_{\lambda\lambda}(\lambda; \lambda_0, \sigma_\lambda)d\lambda$ measures the second-order spectral derivative. The aperture functions G, G_λ, and $G_{\lambda\lambda}$ denote derivatives of the Gaussian with respect to λ, the sensitivities shown in Figure 6.3. The Gaussian model of spectral measurement allows probing the differential structure of the spectrum. The measurements are obtained by integrating over the incoming spectrum, weighted by derived Gaussian sensitivity functions. Hence, the Gaussian color model measures the coefficients $E^{\lambda_0,\sigma_\lambda}$, $E_\lambda^{\lambda_0,\sigma_\lambda}$, and $E_{\lambda\lambda}^{\lambda_0,\sigma_\lambda}$ of the Taylor expansion of the Gaussian-weighted spectral energy distribution at λ_0 and scale σ_λ.

Introduction of spatial extent in the Gaussian color model yields a local Taylor expansion at wavelength λ_0 and position $\vec{x}_0$. Each measurement of a spatiospectral energy distribution has a spatial as well as a spectral resolution. The measurement is obtained by probing an energy density volume in a three-dimensional spatiospectral space (Fig. 6.4). The size of the probe is determined by the observation scale σ_λ and $\sigma_{\vec{x}}$.

$$E(\lambda, \vec{x}) = E + \begin{pmatrix} \vec{x} \\ \lambda \end{pmatrix}^T \begin{bmatrix} E_{\vec{x}} \\ E_\lambda \end{bmatrix} + \frac{1}{2}\begin{pmatrix} \vec{x} \\ \lambda \end{pmatrix}^T \begin{bmatrix} E_{\vec{x}\vec{x}} & E_{\vec{x}\lambda} \\ E_{\lambda\vec{x}} & E_{\lambda\lambda} \end{bmatrix}\begin{pmatrix} \vec{x} \\ \lambda \end{pmatrix} + \cdots. \tag{6.3}$$

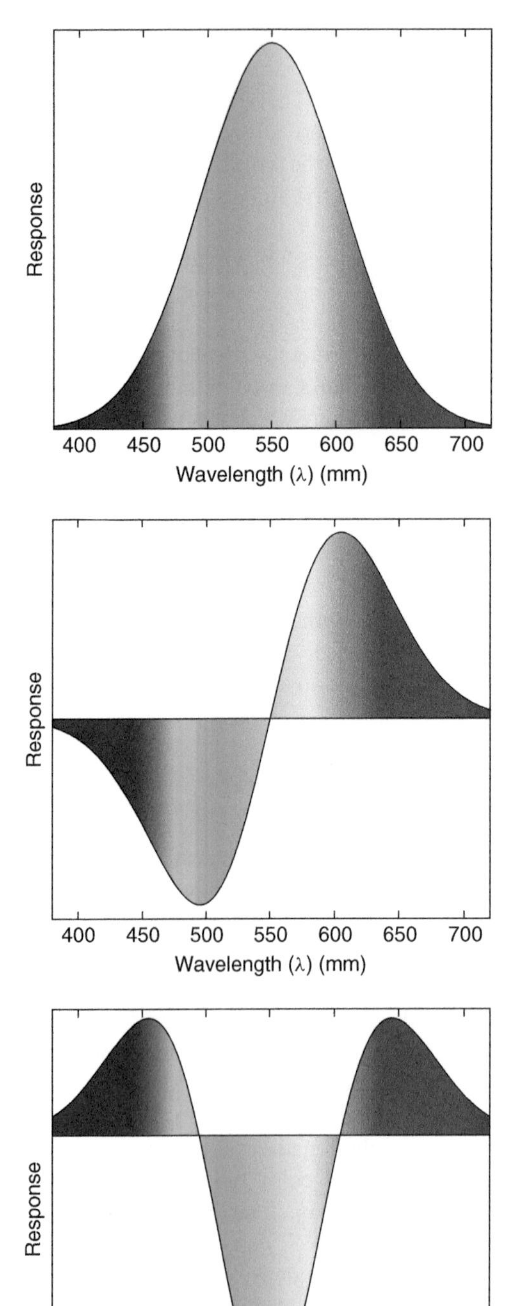

Figure 6.3 The Gaussian sensitivity functions over the wavelengths. The incoming spectrum $E(\lambda)$ is weighted and integrated over the three sensitivity curves $\{G(\lambda; \lambda_0, \sigma_\lambda), G_\lambda(\lambda; \lambda_0, \sigma_\lambda),$ and $G_{\lambda\lambda}(\lambda; \lambda_0, \sigma_\lambda)\}$, yielding the three spectral measurements E, E_λ, and $E_{\lambda\lambda}$. Gaussian central wavelength $\lambda_0 = 520$ nm and scale $\sigma_\lambda = 55$ nm are chosen such that compatibility with human vision is achieved.

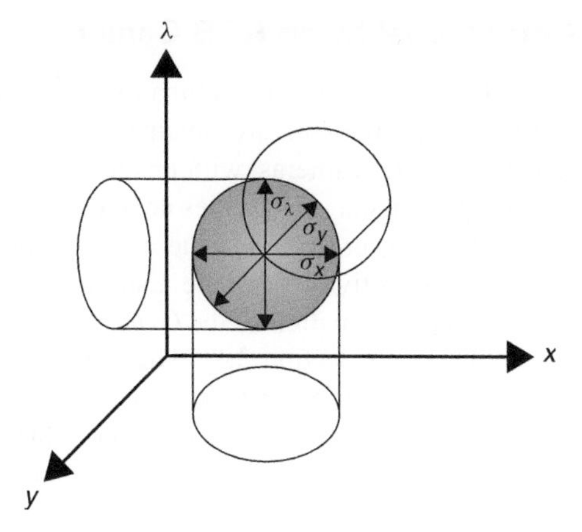

Figure 6.4 Probing the spatiospectral energy density boils down to integrating with a Gaussian sensitivity function over the spatial and spectral dimensions.

The mth differentiation with respect to λ and the nth differentiation with respect to $\vec{x}$ may be transported using Gaussian derivative filters in the well-known N-jet [62]

$$E_{\lambda^m \vec{x}^n}(\lambda, \vec{x}) = E(\lambda, \vec{x}) * G_{\lambda^m \vec{x}^n}(\lambda, \vec{x}; \sigma_\lambda, \sigma_{\vec{x}}). \tag{6.4}$$

Here, $G_{\lambda^m \vec{x}^n}(\lambda, \vec{x}; \sigma_\lambda, \sigma_{\vec{x}})$ are the Gaussian-shaped spatiospectral probes, or color receptive fields Figure 6.5. The coefficients of the Taylor expansion of $E(\lambda, \vec{x})$ together form a complete representation of the local image structure. Truncation of the Taylor expansion results in an approximate representation, optimal in a least squares sense.

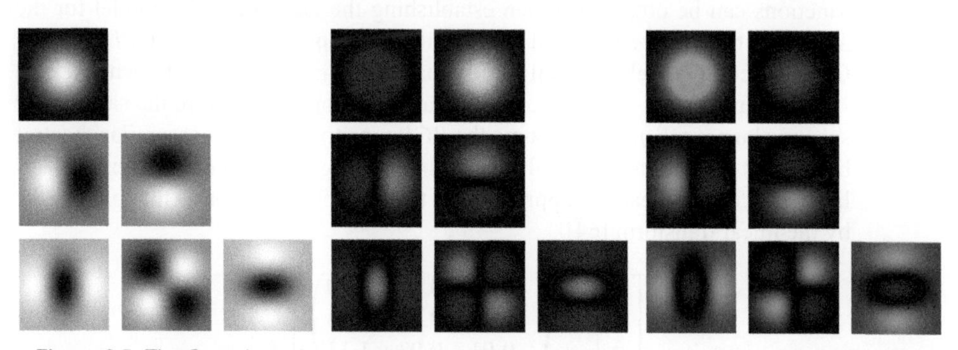

Figure 6.5 The Gaussian color smoothing and derivative filters up to second order in (x, y) and λ.

It appears that the above Gaussian color model approximates the Hering basis [63] for human color vision when truncated at second order and taking the parameters $\lambda_0 \simeq 520$ nm and $\sigma_\lambda \simeq 55$ nm [61]. We follow this case and denote spectral differential quotients by E, E_λ, and $E_{\lambda\lambda}$, and spatial differential quotients by E_x, $E_{\lambda x}$, and $E_{\lambda\lambda x}$.

6.1.2 The Gaussian Color Model by an RGB Camera

Spectral differential quotients are obtained by a linear combination of given (RGB) sensitivities, whereas spatial differential quotients are obtained by convolution with Gaussian derivative filters. For general cameras, with unknown characteristics, it is reasonably safe to assume the camera sensitivities are close to Gaussian functions, centered somewhere at the red, green, and blue areas of the visible spectrum in such a way that the RGB sensitivities capture a similar color space as human perception. In that case, an approximation of the Gaussian color model is given by a simple opponent color space (Section 3.5). The intensity channel $I = R+G+B$ represents the Gaussian-weighted spectral response, the yellow–blue channel $YB = R+G-2B$ the first-order derivative comparing one (blue) half of the spectrum with the other half (yellow), and the reddish–green channel $RG = R-2G+B$ the second-order derivative comparing the center of the spectrum with the outsides. Hence,

$$\begin{bmatrix} \hat{E} \\ \hat{E}_\lambda \\ \hat{E}_{\lambda\lambda} \end{bmatrix} = \frac{1}{3} \begin{pmatrix} 1 & 1 & 1 \\ 1 & 1 & -2 \\ 1 & -2 & 1 \end{pmatrix} \begin{bmatrix} R \\ G \\ B \end{bmatrix}. \tag{6.5}$$

Note that we try to achieve derivative filters in the spectral domain by transforming the spectral responses as given by the RGB filters. The transformed filters may be imperfect, but are likely to offer accurate estimates of differential measurements. When the spectral responses of the RGB filters are known, a better transform can be obtained.

When a camera is calibrated and the XYZ responses are known, a more elaborate and precise alignment of the camera sensitivities to the Gaussian basis functions can be obtained. When establishing the Gaussian color model for the XYZ sensitivities, we note that the first three components E, E_λ, and $E_{\lambda\lambda}$ of the Gaussian color model very well approximate the CIE 1964 XYZ basis when taking $\lambda_0 = 520$ nm and $\sigma_\lambda = 55$ nm. A camera is developed to capture the same color space as humans, hence we assume the RGB sensitivities to span a similar spectral bandwidth and to have a similar central wavelength. When camera response is linearized, an RGB camera approximates the CIE 1964 XYZ basis for colorimetry by the linear transform [64]

$$\begin{bmatrix} \hat{X} \\ \hat{Y} \\ \hat{Z} \end{bmatrix} = \begin{pmatrix} 0.62 & 0.11 & 0.19 \\ 0.3 & 0.56 & 0.05 \\ -0.01 & 0.03 & 1.11 \end{pmatrix} \begin{bmatrix} R \\ G \\ B \end{bmatrix}. \tag{6.6}$$

The best linear transform from XYZ values to the Gaussian color model is given by [61]

$$\begin{bmatrix} \hat{E} \\ \hat{E}_\lambda \\ \hat{E}_{\lambda\lambda} \end{bmatrix} = \begin{pmatrix} -0.48 & 1.2 & 0.28 \\ 0.48 & 0 & -0.4 \\ 1.18 & -1.3 & 0 \end{pmatrix} \begin{bmatrix} \hat{X} \\ \hat{Y} \\ \hat{Z} \end{bmatrix}. \tag{6.7}$$

Figure 6.6 An example image and its color components E, E_λ, and $E_{\lambda\lambda}$. For the latter two images, negative values are indicated by dark intensities and positive values by bright intensities.

The product of Equations 6.6 and 6.7 gives the desired implementation of the Gaussian color model in *RGB* terms.

$$\begin{bmatrix} \hat{E} \\ \hat{E}_\lambda \\ \hat{E}_{\lambda\lambda} \end{bmatrix} = \begin{pmatrix} 0.06 & 0.63 & 0.27 \\ 0.3 & 0.04 & -0.35 \\ 0.34 & -0.6 & 0.17 \end{pmatrix} \begin{bmatrix} R \\ G \\ B \end{bmatrix}. \tag{6.8}$$

The resulting conversion from an RGB image to the opponent color space is illustrated in Figure 6.6.

6.1.3 Derivatives in the Gaussian Color Model

Image derivatives for gray-value images are obtained by smoothing and differentiation in both the x- and y-direction. In our notation, the intensity gradient is denoted by $\nabla \mathbf{E} = (E_x, E_y)$, with its magnitude $E_w = |\nabla \mathbf{E}| = \sqrt{E_x^2 + E_y^2}$. Each of the gradient components are obtained by convolving the intensity channel of the image E with the Gaussian derivative functions $G_x(x,y)$ and $G_y(x,y)$. The above outlined Gaussian color model extends this framework to spectral derivatives. Here, the spectral parameters of the Gaussian functions are fixed—the Gaussian is centered at a fixed wavelength and has a fixed standard deviation (spectral bandwidth), and there are three spectral derivatives available: a zero-order derivative E being the intensity, a first-order derivative E_λ comparing the yellow to the blue part of the spectrum, and a second-order derivative $E_{\lambda\lambda}$ comparing the green middle part to the two outer regions of the spectrum. These parameters are implemented in the camera device, in accordance with the properties of human color vision. Of course, there is still the freedom to choose the spatial position and scale by convolving the spectral Gaussian measurements E, E_λ, and $E_{\lambda\lambda}$ with the Gaussian derivative kernel. As such, we are able to take spatiospectral derivatives of a color image by simply combining the appropriate channel of the Gaussian color model with the appropriate spatial derivative operator. For example, the total edge strength due to the color information in an image may be obtained by

$$E_w = \sqrt{E_x^2 + E_y^2 + E_{\lambda x}^2 + E_{\lambda y}^2 + E_{\lambda\lambda x}^2 + E_{\lambda\lambda y}^2}. \tag{6.9}$$

The image processing operators involved to obtain the result E_w include (i) obtaining three x-derivative images by filtering the Gaussian color channels E, E_λ, and $E_{\lambda\lambda}$ with a Gaussian x-derivative filter G_x; (ii) obtaining three y-derivative images by filtering the Gaussian color channels E, E_λ, and $E_{\lambda\lambda}$ with a Gaussian y-derivative filter G_y; (iii) pixelwise squaring of each of the six derivate response image; (iv) pixelwise addition of the six squared images together into a single image; and (v) returning its pixelwise square root as the final edge strength image.

The Gaussian color model and Gaussian derivative operator smooth *and* differentiate the spatial and spectral data in, say, one go. However, as both operations are linear and follow linear systems theory, we may consider the smoothing and differentiation as two independent steps of an operation. Although in practice we cannot separate the two, we can use them separately in our theoretical derivations. This observation is fundamental in the derivation of the differential invariants that follow. Let us make this more explicit by considering the properties of linear system theory.

$$\int G_{x^n}(x;\sigma)f(x)\mathrm{d}x = \int \left\{\frac{\partial^n}{\partial x^n}G(x;\sigma)\right\}f(x)\,\mathrm{d}x \tag{6.10}$$

$$= \int \frac{\partial^n}{\partial x^n}\{G(x;\sigma)f(x)\}\,\mathrm{d}x = \int G(x;\sigma)\left\{\frac{\partial^n}{\partial x^n}f(x)\right\}\mathrm{d}x. \tag{6.11}$$

This implies that, independent of the fact that at which exact step we apply the derivative operator, we may regard it as actually differentiating the underlying function and measuring (by integrating over the Gaussian kernel) its response. This of course under the assumption that responses are linear (up to an arbitrary scaling), which does hold for many cameras that do not compress their data. Even under compression, the artifacts may be considered noise and are smoothed out by the Gaussian smoothing operator. Hence, the Gaussian color model allows the assessment of derivatives of the spatiospectral energy function *just before observation* by the camera system. This is the light field in front of the camera as is modeled by the reflection models outlined in Chapter 3.

6.1.4 Differential Invariants for the Lambertian Reflection Model

Using the Gaussian color model, we may apply differential calculus to establish photometric invariant properties, as is the goal of this chapter. Therefore, we may consider any reflectance model, like the well-known Lambertian model, the dichromatic reflection model, or the Kubelka–Munk model. Before going into depth, we first give an illustration using a simplified Lambertian model to sketch the steps involved. Consider the simplified Lambertian reflectance model:

$$E(\lambda, x) = m^b(x)s(\lambda, x), \tag{6.12}$$

(see Equation 3.1, where we assume a white light source, that is, $e(\lambda) = c$). Here, m^b represents the intensity term due to the object geometry, that is, the "cosine rule" of Lambertian reflection. Furthermore, s represents the object albedo function. Now, taking the spatial derivative results in

$$E_x(\lambda, x) = \frac{\partial}{\partial x}\left\{m^b(x)s(\lambda, x)\right\} \tag{6.13}$$

$$= s(\lambda, x)\frac{\partial m^b(x)}{\partial x} + m^b(x)\frac{\partial s(\lambda, x)}{\partial x}. \tag{6.14}$$

The right-hand side of this equation expresses the properties of photometry, with the spatial derivative operator "propagated" through the underlying physics. Here, the chain rule has been applied to explicate the influence of fluctuations in each of the components $s(.)$ and $m^b(.)$ on the total fluctuation. The left-hand side expresses what will be measured by the camera, the measurement obtained by smoothing with the Gaussian kernel. Hence, the equation expresses that the response of a Gaussian derivative filter on an intensity image (the *left-hand* side) yields something (the *right-hand* side) that depends on both the geometry term (m^b) and the material reflectance term (s) in the Lambertian model. This is of course a well-known fact made explicit by our framework.

Let us now consider the case of a spectral derivative of the simplified Lambertian model:

$$E_\lambda(\lambda, x) = \frac{\partial}{\partial \lambda}\left\{m^b(x)s(\lambda, x)\right\}, \tag{6.15}$$

$$= m^b(x)\frac{\partial s(\lambda, x)}{\partial \lambda}. \tag{6.16}$$

Here, the geometrical term $m^b(x)$ does not depend on the wavelength, and hence is considered a constant in the partial derivative. Again, the right-hand side expresses the photometry, now with the spectral derivative operator propagated through the underlying physics. The left-hand side expresses the yellow–blue component of the Gaussian color model. As can be observed from the equation, the measured yellow–blue color channel E_λ does linearly depend on the intensity term m^b and hence on the geometry of the imaged Lambertian object, again a well-known fact. However, normalizing by the measured intensity yields

$$\frac{E_\lambda(\lambda, x)}{E(\lambda, x)} = \frac{m^b(x)}{m^b(x)s(\lambda, x)}\frac{\partial s(\lambda, x)}{\partial \lambda}, \tag{6.17}$$

$$= \frac{1}{s(\lambda, x)}\frac{\partial s(\lambda, x)}{\partial \lambda}. \tag{6.18}$$

Now, the right-hand side expresses that the photometry of E_λ normalized by E only depends on the surface albedo s, hence on the body reflectance, as the geometry term m^b is canceled out in the expression. The left-hand side expresses the measurements that need to be combined to yield the invariant, being the first-order spectral derivative E_λ, the yellow–blue opponent color channel, divided by the intensity E to yield the invariant. For pixel-based invariance, these properties are evaluated per pixel. For the Gaussian color model, these properties can be evaluated after smoothing of the channels E and E_λ (and $E_{\lambda\lambda}$), as the opponent color channels are decorrelated, as a result of which the Gaussian smoothing does not introduce color blending artifacts. We denote the resulting invariant by C_λ

$$C_\lambda = \frac{E_\lambda}{E}, \tag{6.19}$$

which is calculated by pixelwise division of the yellow–blue smoothed opponent channel E_λ by the smoothed intensity channel E.

The second-order spectral derivative of the simplified Lambertian model yields a similar expression as Equation 6.15,

$$E_{\lambda\lambda}(\lambda, x) = \frac{\partial^2}{\partial\lambda^2}\left\{m^b(x)s(\lambda, x)\right\}, \tag{6.20}$$

$$= m^b(x)\frac{\partial^2 s(\lambda, x)}{\partial\lambda^2}. \tag{6.21}$$

Hence, for the reddish–green opponent channel, representing the second-order derivative of wavelength, we can derive a similar normalization expression,

$$\frac{E_{\lambda\lambda}(\lambda, x)}{E(\lambda, x)} = \frac{m^b(x)}{m^b(x)s(\lambda, x)}\frac{\partial^2 s(\lambda, x)}{\partial\lambda^2} \tag{6.22}$$

$$= \frac{1}{s(\lambda, x)}\frac{\partial^2 s(\lambda, x)}{\partial\lambda^2}. \tag{6.23}$$

Again, the geometry term m^b is canceled out. Hence, under our simplified Lambertian reflectance model, the normalized reddish–green channel depends only on the surface reflectance function, resulting in

$$C_{\lambda\lambda} = \frac{E_{\lambda\lambda}}{E}, \tag{6.24}$$

So far we introduced a framework of spatiospectral derivatives and showed their applicability by deriving an invariant expression for the simplified Lambertian color model of Equation 6.12. Now that we understand the underlying principles, we can elaborate on our simplified Lambertian model. Extending the reflection model to include local intensity variation

$$E(\lambda, x) = i(x)m^b(x)s(\lambda, x), \tag{6.25}$$

the local intensity term $i(x)$ models a nonuniform illumination intensity and intensity variations due to, for example, shadows. As the model does not essentially differ from the original model, the intensity fluctuation i can be absorbed in the local geometry term m^b without any consequences for the derived invariant expression C_λ and $C_{\lambda\lambda}$. Hence, the C invariant disregards any changes in intensity, for example, due to shadow, shading, local illumination variation, or any other fluctuations affecting the intensity.

Once we have a lowest order invariant, any higher order derivative of that expression inherits the same invariant properties. Hence, by taking higher order derivatives, we can produce a hierarchy of differential invariants of arbitrarily large order n. The lowest order invariant is referred to as the fundamental invariant. For example, spatial derivatives of C_λ and $C_{\lambda\lambda}$ are given by

$$C_{\lambda x} = \frac{E_{\lambda x}E - E_\lambda E_x}{E^2}, \tag{6.26}$$

$$C_{\lambda\lambda x} = \frac{E_{\lambda\lambda x}E - E_{\lambda\lambda}E_x}{E^2}. \tag{6.27}$$

Now, the invariant edge strength C_w^2 can be obtained by taking the squared sum of the invariant derivatives $C_{\lambda x}$, $C_{\lambda y}$, $C_{\lambda\lambda x}$, and $C_{\lambda\lambda y}$

$$C_w = \sqrt{C_{\lambda x}^2 + C_{\lambda y}^2 + C_{\lambda\lambda x}^2 + C_{\lambda\lambda y}^2}, \tag{6.28}$$

which yields the gradient magnitude independent of local intensity changes. Hence, this edge detector is invariant to shadow and shading. Figure 6.7 illustrates the C invariant.

A further extension of the simplified Lambertian reflection model is to include a colored illuminant. Originally, in Equation 6.12, we assumed white light by the strict assumption $e(\lambda, x) = c$. We relaxed the assumption by allowing spatial intensity fluctuations and introduced a spatial fluctuation term $i(x)$. Now, we further relax our model by allowing a colored light source, leading to a term

 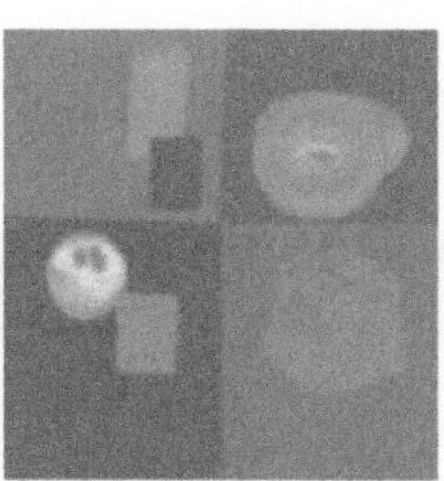 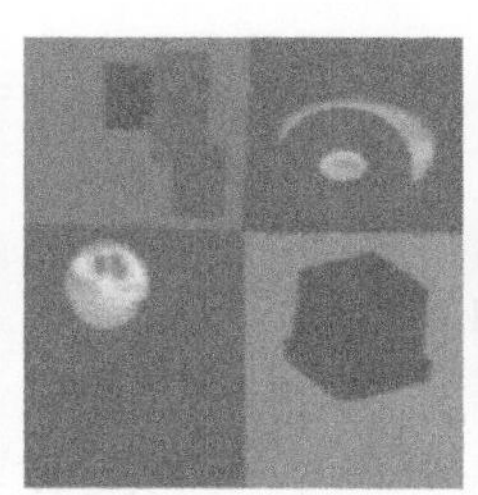 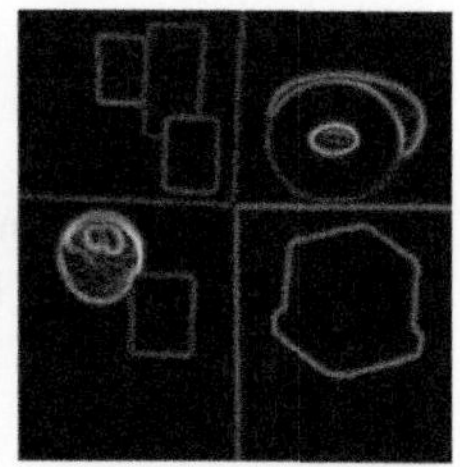

Figure 6.7 Examples of the normalized colors C_λ denoting the first spectral derivative, $C_{\lambda\lambda}$ denoting the second spectral derivative, and the gradient magnitudes C_w. Note that intensity edges are being suppressed, whereas highlights are still present.

$e(\lambda)i(x)$, which models both the spectrum of the light source and the spatial fluctuation over the intensity

$$E(\lambda, x) = e(\lambda)i(x)s(\lambda, x), \tag{6.29}$$

where, for simplicity of notation, we absorbed the geometry term $m^b(x)$ into the spatial fluctuation $i(x)$. The extended Lambertian reflection model allows us to derive invariants independent of the illumination. Note that a spectral or a spatial derivative will act on either e (the spectral derivative) or i (the spatial derivative) while regarding the other as constant. Both derivatives will act on s. Hence, by taking a second-order derivative, we might be able to cancel the variation by light source and geometry.

Let us start with the first-order spectral derivative of the extended Lambertian reflection model, yielding

$$E_\lambda(\lambda, x) = \frac{\partial}{\partial \lambda}\{e(\lambda)i(x)s(\lambda, x)\}, \tag{6.30}$$

$$= i(x)\frac{\partial e(\lambda)s(\lambda, x)}{\partial \lambda}, \tag{6.31}$$

$$= i(x)s(\lambda, x)\frac{\partial e(\lambda)}{\partial \lambda} + i(x)e(\lambda)\frac{\partial s(\lambda, x)}{\partial \lambda}. \tag{6.32}$$

Here we see the full power of derivative-based invariance, as the chain rule of differentiation separates the influence of each individual factor in our model. Hence, normalization of the spectral derivative E_λ to the intensity E gives

$$\frac{E_\lambda(\lambda, x)}{E(\lambda, x)} = \frac{1}{e(\lambda)}\frac{\partial e(\lambda)}{\partial \lambda} + \frac{1}{s(\lambda, x)}\frac{\partial s(\lambda, x)}{\partial \lambda}. \tag{6.33}$$

Now, the results are composed of a term depending on λ only, being the influence of the spectrum of the light source, and a term depending on both λ and x, being related to the body reflectance. After partial differentiation to x,

$$\frac{\partial}{\partial x}\left\{\frac{E_\lambda(\lambda, x)}{E(\lambda, x)}\right\} = \frac{\partial}{\partial x}\left\{\frac{1}{s(\lambda, x)}\frac{\partial s(\lambda, x)}{\partial \lambda}\right\}, \tag{6.34}$$

only the term depending on λ has vanished, and the result depends on the object albedo only. Hence,

$$N_{\lambda x} = \frac{\partial}{\partial x}\left\{\frac{E_\lambda(\lambda, x)}{E(\lambda, x}\right\}, \tag{6.35}$$

$$= \frac{E_{\lambda x}E - E_\lambda E_x}{E^2}, \tag{6.36}$$

is invariant for local intensity changes and for the color of the light source under Lambertian reflection. The second-order spectral derivative in this case is obtained by further differentiation to λ,

$$N_{\lambda\lambda x} = \frac{E_{\lambda\lambda x}E^2 - E_{\lambda\lambda}E_x E - 2E_{\lambda x}E_\lambda E + 2E_\lambda^2 E_x}{E^3}. \tag{6.37}$$

One observation here is that color constancy apparently can only be achieved by considering a comparison between local pixel values, here implemented by edge detection through the Gaussian derivative filter. If one compares the derivatives of the invariants C with the above-derived expressions for N, indeed, $C_{\lambda x}$ is identical to $N_{\lambda x}$. Hence, one can expect to already achieve some independence of the light source spectrum when using the C invariant edge detectors. The invariant edge magnitude N_w can be obtained by

$$N_w = \sqrt{N_{\lambda x}^2 + N_{\lambda y}^2 + N_{\lambda\lambda x}^2 + N_{\lambda\lambda y}^2}, \tag{6.38}$$

which yields the gradient magnitude independent of local intensity changes and the color of the light source. Hence, this edge detector is color constant and invariant to shadow and shading.

Note that the above-derived expressions for reflection of light for the Lambertian model are also valid for the case of light transmission under the Beer–Lambert law. Recalling the law in Equation 3.24, and including our assumption above of a colored light source, we obtain

$$E(\lambda, x) = e(\lambda)i(x)\exp\left(-d(x)c(x)\alpha(\lambda, x)\right), \tag{6.39}$$

$$= e(\lambda)i(x)t(\lambda, x), \tag{6.40}$$

where t represents the total extinction coefficient. Note the resemblance with Equation 6.29. Hence, the expressions derived above are also valid for transparent materials under the Beer–Lambert law. The N invariant, in that case, is robust for a change in local illumination intensity and for illumination color. The property is illustrated in Figure 6.8.

6.1.5 Differential Invariants for the Dichromatic Reflection Model

We continue our analysis of differential invariants and now turn to the dichromatic reflection model of Section 3.2. Consider the simplified dichromatic reflection model,

$$E(\lambda, x) = m^b(x)c^b(\lambda, x) + m^i(x), \tag{6.41}$$

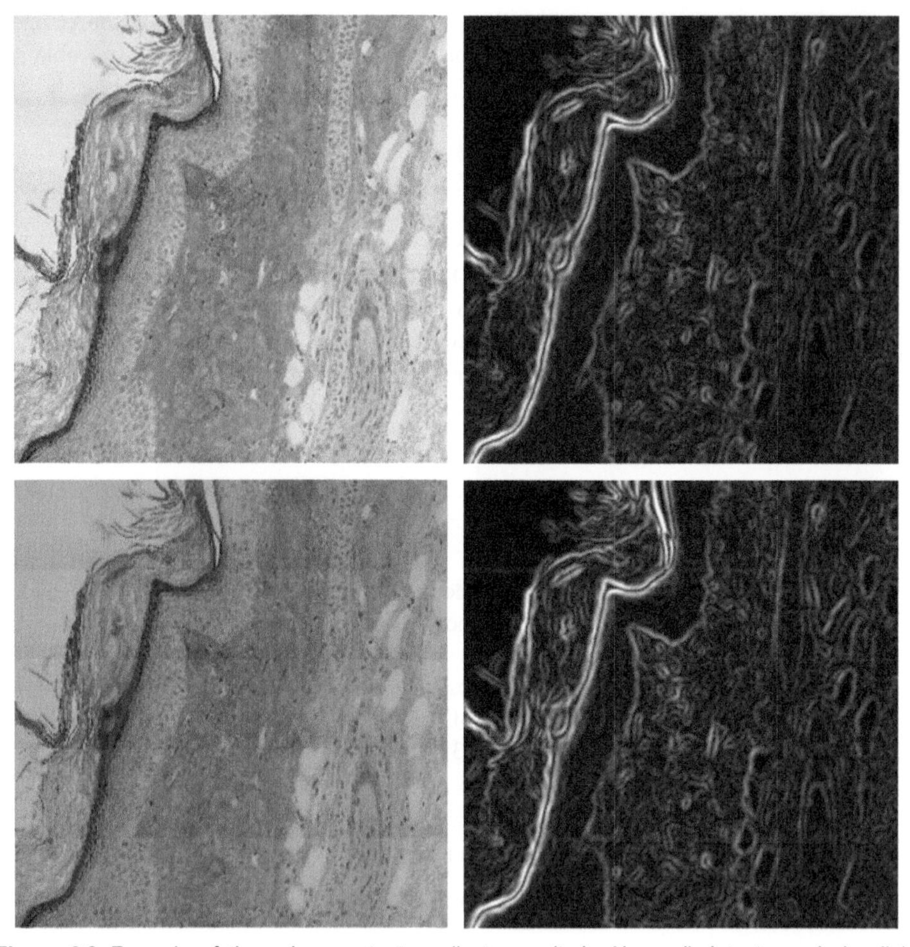

Figure 6.8 Example of the color constant gradient magnitude N_w applied to transmissive light microscopy with an epithelial tissue section under a halogen light source operating at 3400 K and 2500 K. The resulting edges are independent of the color differences induced by the change in illumination color.

where we assume a neutral interface causing specular reflections to directly reflect the spectrum of the light source reduced by a scalar factor m^i. Hence, the model describes colored surface reflection through the term c^b, shading through the term m^b, highlights through the term m^i, and shadow and (local) intensity changes hidden in both geometrical terms m^b and m^i. Note that the geometrical terms are independent of wavelength. Hence, partial differentiation with respect to wavelength yields

$$E_\lambda(\lambda, x) = \frac{\partial}{\partial \lambda} \left\{ m^b(x) c^b(\lambda, x) \right\}, \tag{6.42}$$

$$= m^b(x)\frac{\partial c^b(\lambda,x)}{\partial \lambda}. \tag{6.43}$$

The result depends on the geometry term m^b and the Lambertian reflectance component of the object c^b. Invariance can now be obtained by differentiating again,

$$E_{\lambda\lambda}(\lambda,x) = \frac{\partial^2}{\partial \lambda^2}\left\{m^b(x)c^b(\lambda,x)\right\}, \tag{6.44}$$

$$= m^b(x)\frac{\partial^2 c^b(\lambda,x)}{\partial \lambda^2}, \tag{6.45}$$

after which we normalize them,

$$\frac{E_{\lambda\lambda}(\lambda,x)}{E_{\lambda}(\lambda,x)} = \frac{\dfrac{\partial c^b(\lambda,x)}{\partial \lambda}}{\dfrac{\partial^2 c^b(\lambda,x)}{\partial \lambda^2}}. \tag{6.46}$$

Hence, the ratio of E_λ and $E_{\lambda\lambda}$ is invariant to highlights, shading, and shadows and yields the invariant H

$$H = \arctan\left\{\frac{E_\lambda}{E_{\lambda\lambda}}\right\}, \tag{6.47}$$

where the arctan is introduced because of the interpretation below and for numerical stability.

To interpret H, consider the local Taylor expansion at λ_0 truncated at second order,

$$E(\lambda_0 + \Delta\lambda) \approx E(\lambda_0) + \Delta\lambda E_\lambda(\lambda_0) + \frac{1}{2}\Delta\lambda^2 E_{\lambda\lambda}(\lambda_0). \tag{6.48}$$

The function extremum of $E_\lambda(\lambda_0 + \Delta\lambda)$ is at $\Delta\lambda$ for which the first-order derivative is zero

$$\frac{d}{d\lambda}\left\{E(\lambda_0 + \Delta\lambda)\right\} = E_\lambda(\lambda_0) + \Delta\lambda E_{\lambda\lambda}(\lambda_0) = 0. \tag{6.49}$$

Hence, for $\Delta\lambda$ near the origin λ_0

$$\Delta\lambda_{\max} = -\frac{E_\lambda(\lambda_0)}{E_{\lambda\lambda}(\lambda_0)}. \tag{6.50}$$

In conclusion, the property H is related to the hue (i.e., $\arctan(\lambda_{max})$) of the material. For $E_{\lambda\lambda}(\lambda_0) < 0$ the result is at a maximum and describes a band-pass (prism) color, whereas for $E_{\lambda\lambda}(\lambda_0) > 0$ the result is at a minimum and indicates a band-stop (slit) color.

The gradient magnitude of the H invariant is obtained by spatial differentiation

$$H_x = \sqrt{\frac{E_{\lambda\lambda}E_{\lambda x} - E_{\lambda}E_{\lambda\lambda x}}{E_{\lambda}^2 + E_{\lambda\lambda}^2}}, \tag{6.51}$$

which results in the gradient magnitude

$$H_w = \sqrt{H_x^2 + H_y^2}. \tag{6.52}$$

Beside the hue-related H, an expression for saturation S can be derived:

$$S = \frac{1}{E}\sqrt{E_{\lambda}^2 + E_{\lambda\lambda}^2}. \tag{6.53}$$

Note that saturation S is independent of local intensity changes, shadow, and shading, but not for highlights. This by now can be easily derived by the reader. The hue and saturation invariants are illustrated in Figure 6.9.

Figure 6.9 Example of the invariants associated with H. (a) Example image, (b) H, (c) the derived expression S, and (d) gradient magnitude H_w. Intensity changes and highlights are suppressed in the H and H_w image. The S image shows a low purity at color borders, due to mixing of colors on two sides of the border. For all pictures, $\sigma_{\tilde{x}} = 1$ pixel and the image size is 256×256.

Common expressions for hue are known to be noise sensitive. In the Gaussian derivative framework, the Gaussian smoothing offers a trade-off between noise and detail sensitivity. The influence of noise on hue gradient magnitude H_w for various $\sigma_{\tilde{x}}$ is shown in Figure 6.10. The influence of noise on hue edge detection is drastically reduced for larger observational scale $\sigma_{\tilde{x}}$.

6.1.6 Summary of Full Color Invariants

Various sets of invariants have been derived in this chapter. The invariant sets may be ordered by broadness of invariance, where broader sets allow ignorance of a larger set of disturbing factors than tighter sets (Fig. 6.11). The set of disturbing factors is summarized in Table 6.1.

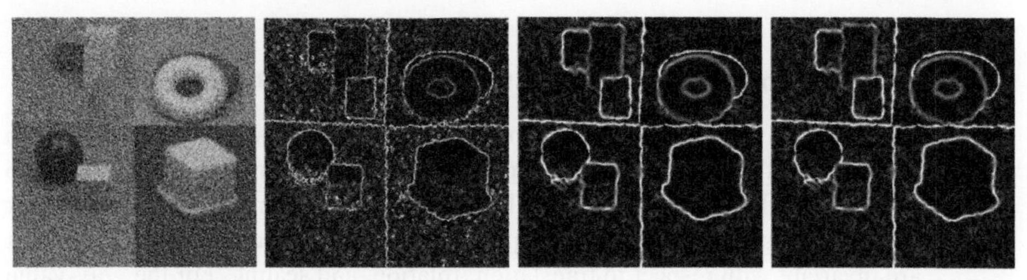

Figure 6.10 The influence of white additive noise on gradient magnitude H_w. Independent Gaussian zero-mean noise is added to each of the *RGB* channels, SNR = 5, and H_w is determined for $\sigma_{\tilde{x}} = 1$, $\sigma_{\tilde{x}} = 2$, and $\sigma_{\tilde{x}} = 4$ pixels. Note the noise robustness of the hue gradient H_w for larger $\sigma_{\tilde{x}}$.

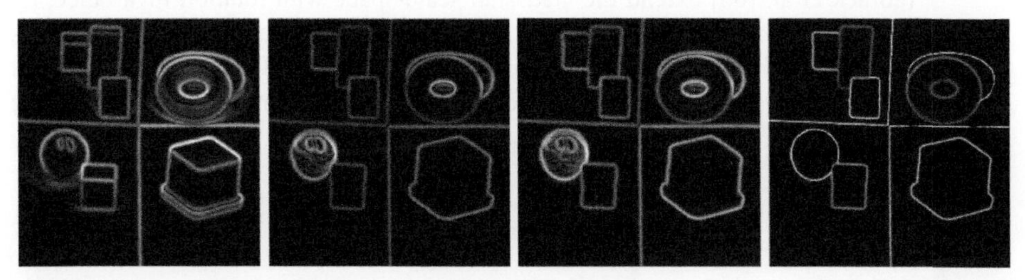

Figure 6.11 Examples of the total color edge strength measures. E_w is shown, which is not invariant. Note that this image shows intensity, color, and highlight boundaries. Further, C_w and N_w invariant for shading, and finally, H_w invariant for shading and highlights are shown. The effect of intensity and highlights on the different invariants are in accordance with Table 6.1.

Table 6.1 Summary of the various color invariant sets and their invariance to specific imaging conditions.

	Viewing direction	surface orientation	Highlights	Illumination direction	Illumination intensity	Illumination color
H	+	+	+	+	+	−
N	+	+	−	+	+	+
C	+	+	−	+	+	−
E	−	−	−	−	−	−

Invariance is denoted by +, whereas sensitivity to the imaging condition is indicated by −. Note that the reflected spectral energy distribution *E* is sensitive to all the conditions cited.

The table offers the solution of using the narrowest set of invariants for known imaging conditions, since $H \subset N \subset C \subset E$. In the case that recording circumstances are unknown, the table offers a broad to narrow hierarchy. Hence, an incremental strategy of invariant feature extraction may be applied. Combination of invariants open up the way to edge type classification as suggested in Reference 65. The vanishing of edges for certain invariants indicate if their cause is shading, specular reflectance, or material boundaries.

6.1.7 Geometrical Color Invariants in Two Dimensions

So far, we have established color invariant descriptors, based on differentials in the spectral and the spatial domain in one spatial dimension. When applied in two dimensions, the result depends on the orientation of the image content. In order to obtain meaningful image descriptions it is crucial to derive descriptors that are invariant with respect to translation, rotation, and scaling. For the gray-value luminance L, geometrical invariants are well established [66]. Translation and scale invariance is obtained by examining the (Gaussian) scale-space, which is a natural representation for investigating the scaling behavior of image features [60]. Florack et al. [66] extend the Gaussian scale-space with rotation invariance, by considering in a systematic manner local gauge coordinates. The coordinate axis w and v are aligned to the gradient and isophote tangents directions, respectively. Hence, the first-order gradient gauge invariant is the magnitude of the luminance gradient

$$L_w = \sqrt{L_x^2 + L_y^2}. \tag{6.54}$$

Note that the first-order isophote gauge invariant is zero by definition. The second-order invariants are given by

$$L_{vv} = \frac{L_x^2 L_{yy} - 2L_x L_y L_{xy} + L_y^2 L_{xx}}{L_w^2}, \tag{6.55}$$

related to isophote curvature,

$$L_{vw} = \frac{L_x L_y \left(L_{yy} - L_{xx}\right) - \left(L_x^2 - L_y^2\right) L_{xy}}{L_w^2}, \tag{6.56}$$

to flow-line curvature, and

$$L_{ww} = \frac{L_x^2 L_{xx} + 2L_x L_y L_{xy} + L_y^2 L_{yy}}{L_w^2}, \tag{6.57}$$

to isophote density. These spatial results may be combined with the color invariants established before, by straightforward substitution of the spatial derivatives of L above by the derivatives of the respective color invariants. See Reference 67 for further details.

6.2 Quasi-Invariants

A straightforward extension of photometric invariance theory to the differential structure of images would be to transfer the image into photometric invariant representations (Chapter 4), such as normalized RGB or the hue representation, and subsequently apply derivatives to these representations to obtain photometric invariant derivatives. This is only successful to a limited degree. The main problem is caused by the nonlinear transformations, which are applied to compute the photometric invariant representations. The nonlinearities result in instabilities of the photometric invariant representation (typically in the dark, or achromatic regions). As a consequence, derivatives based on this representation inherit the instabilities. This observation is the starting point of this chapter where we derive a set of photometric invariant image derivatives called *quasi-invariants*. This set of derivatives is based on linear operations only and therefore does not suffer from the aforementioned problems. The theory of quasi-invariants is closely related to the theory of color subspaces of Zickler et al. [68].

A drawback of the quasi-invariants is that, as discussed in the introduction of this chapter, they are only appropriate for feature detection but not for feature description. However, an advantage of the quasi-invariants is that they are more stable and have a higher discriminative power than detectors based on full photometric variants. In the case of edge detection this means that the quasi-invariant edge detector obtains better edge localization and is able to detect more color transitions.

6.2.1 Edges in the Dichromatic Reflection Model

Recall from Section 3.2 that the dichromatic reflection model divides the reflection in the body (object color) and surface reflection (specularities or highlights) component for optically inhomogeneous materials. If we assume that shadows are not significantly colored, a known illuminant, $\mathbf{c}^i = (\alpha, \beta, \gamma)^T$, and neutral interface reflection, the RGB vector, $\mathbf{f} = (R, G, B)^T$, can be seen as a weighted summation of two vectors (Eq. 3.9):

$$\mathbf{f}(\mathbf{x}) = e(\mathbf{x}) \left(m^b(\mathbf{x}) \mathbf{c}^b(\mathbf{x}) + m^i(\mathbf{x}) \mathbf{c}^i(\mathbf{x}) \right), \tag{6.58}$$

in which $\mathbf{c}^b$ is the color of the body reflectance, $\mathbf{c}^i$ the color of the surface reflectance, and m^b and m^i are scalars representing the corresponding magnitudes of body and surface reflection. Here we introduce the parameter $e(\mathbf{x})$ to describe the variations of the intensity of the light source as a function of the spatial coordinate $\mathbf{x}$.

From the dichromatic reflection model, photometric invariants can be derived (e.g., normalized RGB, hue). These invariants have the disadvantage that they are unstable; normalized RGB is unstable near zero intensity and hue is undefined on the black–white axis (Section 4.5).

The instabilities can be avoided by analyzing the *RGB* values in the *RGB* histogram [28, 29]. Instead of looking at the zeroth-order structure (the *RGB* values), we focus here on the first-order structure of the image (the edges of the image). A straightforward extension of the photometric invariance theory to first-order filters can be obtained by taking the derivative of invariant zero-order representations (e.g., hue). However, these filters would inherit the undesired instabilities of the photometric invariants. Quasi-invariant derivatives are an alternative way to arrive at photometric invariant derivatives.

The spatial derivative of the dichromatic reflection model (Eq. 6.58) yields the following equation for the photometric derivative structure of images:

$$\mathbf{f}_x = em^b\mathbf{c}_x^b + \left(e_x m^b + em_x^b\right)\mathbf{c}^b + \left(em_x^i + e_x m^i\right)\mathbf{c}^i. \qquad (6.59)$$

Here, the subscript indicates spatial differentiation. Since we assume a known illuminant and neutral interface reflection, $\mathbf{c}^i$ is independent of x. The derivative in Equation 6.59 is a summation of three weighted vectors, successively caused by body reflectance, shadow-shading, and specular change.

It is interesting to investigate Equation 6.59 in more detail. The color derivative of an image is formed by three parts. It turns out that we can actually predict the direction of the three parts of $\hat{\mathbf{f}}_\mathbf{x}$ if we know the *RGB* value ($\hat{\mathbf{f}}$). To do so, we first have a look at the second part, which causes the shadow-shading changes. We see that (in the absence of interface reflection) the direction is given by $\mathbf{c}^b$, which coincides with the direction of $\hat{\mathbf{f}} = \frac{1}{\sqrt{R^2+G^2+B^2}}(R, G, B)^T$. The hat is used to denote unit vectors. We call this direction the shadow-shading direction because all shadow-shading variations are in this direction. A slice of the vector field of shadow-shading directions in the *RGB* cube is given in Figure 6.12a. The shadow-shading itself consists of two scalars representing two different physical phenomena. First, $e_x m^b$ indicates a change in intensity, which corresponds to a shadow edge. And em_x^b is a change in the geometry coefficient, which represents a shading edge.

Next, we consider the third part of Equation 6.59, which is the specular direction $\mathbf{c}^i$ in which changes of the specular geometry coefficient m_x^i occur. In Figure 6.12b,

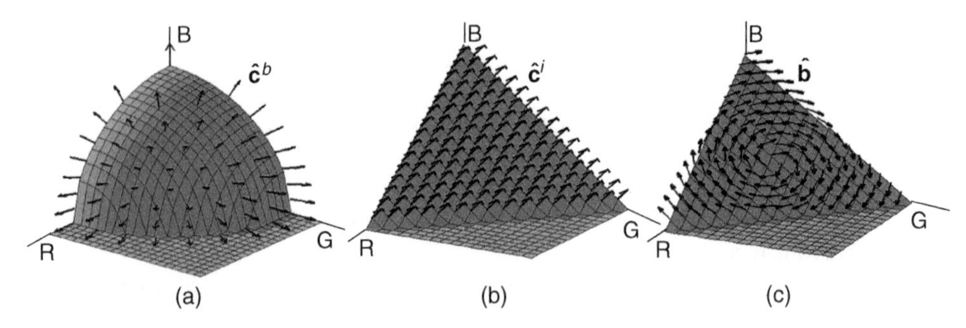

Figure 6.12 (a) Shadow-shading direction $\hat{\mathbf{c}}^b$, (b) specular direction $\hat{\mathbf{c}}^i$, and (c) hue direction $\hat{\mathbf{b}}$. *Source:* Reprinted with permission, © 2005 IEEE.

$\mathbf{c}^i$ is depicted for the case of a white light source for which $\hat{\mathbf{c}}^i = \frac{1}{\sqrt{3}}(1,1,1)^T$. The specular direction is multiplied by two factors. First, em_x^i is a change of the geometric coefficient caused by changes in the angles between viewpoint, object, and light source. Second, the term $e_x m^i$ represents a shadow edge on top of a specular reflection.

The direction of the first part of Equation 6.59, which describes material transitions, is unknown. However, having the direction of two of the causes of an edge, we are able to construct a third direction, which is perpendicular to these two vectors (Fig. 6.12c). This direction, named the hue direction $\hat{\mathbf{b}}$, is computed by the outer product

$$\hat{\mathbf{b}} = \frac{\hat{\mathbf{f}} \times \hat{\mathbf{c}}^i}{\left|\hat{\mathbf{f}} \times \hat{\mathbf{c}}^i\right|}. \tag{6.60}$$

If $\hat{\mathbf{f}}$ and $\hat{\mathbf{c}}^i$ are parallel, we define $\hat{\mathbf{b}}$ to be the zero vector. Note that the hue direction is *not* equal to the direction in which changes of the body reflectance occur, $\hat{\mathbf{c}}_x^b$. However, because it is perpendicular to the two other causes of an edge, we know that changes in the hue direction can only be attributed to a body reflectance change.

In conclusion, there are three causes for an edge in an image: a hue, a shadow-shading, and a specular change. We indicated three directions: the shadow-shading direction, the specular direction, and the hue direction. The quasi-invariant derivatives can now be constructed by projecting the image derivative $\mathbf{f}_\mathbf{x}$ onto these directions.

6.2.2 Photometric Variants and Quasi-Invariants

To construct the quasi-invariants the derivative of an image, $\mathbf{f}_x = (R_x, G_x, B_x)^T$, is projected on the three directions found in the previous section. We call these projections *variants* in case they vary with a physical cause, and we call them *quasi-invariants* in case they do not respond to the physical cause, for example, the projection of the derivative on the shadow-shading direction results in the shadow-shading variant. The projection of the derivative on the hue direction results in a shadow-shading-specular quasi-invariant because the hue direction is perpendicular to changes of these events. Summing the variant and quasi-invariant results in the image derivative $\mathbf{f}_x$. Hence, a variant can be computed by subtracting the quasi-invariant from the image derivative, and vice versa.

The projection of the derivative on the shadow-shading direction is called the shadow-shading variant and is defined as

$$\mathbf{S}_x = \left(\mathbf{f}_x \cdot \hat{\mathbf{f}}\right)\hat{\mathbf{f}}. \tag{6.61}$$

The dot indicates the vector inner product. The second $\hat{\mathbf{f}}$ indicates the direction of the variant. The shadow-shading variant is the part of the derivative that

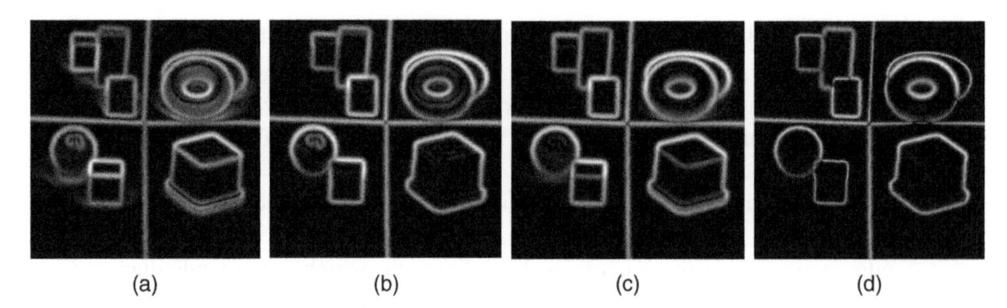

(a) (b) (c) (d)

Figure 6.13 Various derivatives applied to Figure 6.15a: (a) color gradient ($\mathbf{f}_x$), (b) shadow-shading quasi-invariant ($\mathbf{S}_x^c$), (c) the specular quasi-invariant ($\mathbf{O}_x^c$), and (d) the specular-shadow-shading quasi-invariant ($\mathbf{H}_x^c$).

could be caused by shadow or shading. Owing to correlation of the hue and specular direction with the shadow-shading direction, part of $\mathbf{S}_x$ might be caused by changes in hue or specular reflection.

What remains after subtraction of the variant is called the *shadow-shading quasi-invariant*, indicated by superscript c:

$$\mathbf{S}_x^c = \mathbf{f}_x - \mathbf{S}_x. \tag{6.62}$$

The quasi-invariant $\mathbf{S}_x^c$ consists of that part of the derivative which is *not* caused by shadow-shading edges (Fig. 6.13b), hence it only contains specular and hue edges.

The same reasoning can be applied to the specular direction and results in the specular variant and the specular quasi-invariant:

$$\begin{aligned} \mathbf{O}_x &= \left(\mathbf{f}_x \cdot \hat{\mathbf{c}}^i\right) \hat{\mathbf{c}}^i , \\ \mathbf{O}_x^c &= \mathbf{f}_x - \mathbf{O}_x. \end{aligned} \tag{6.63}$$

The specular quasi-invariant is insensitive to highlight edges (Fig. 6.13c).

Finally, we can construct the shadow-shading-specular variant and quasi-invariant by projecting the derivative on the hue direction:

$$\begin{aligned} \mathbf{H}_x^c &= \left(\mathbf{f}_x \cdot \hat{\mathbf{b}}\right) \hat{\mathbf{b}} , \\ \mathbf{H}_x &= \mathbf{f}_x - \mathbf{H}_x^c. \end{aligned} \tag{6.64}$$

$\mathbf{H}_x^c$ does not contain specular or shadow-shading edges (Fig. 6.13d).

6.2.3 Relations of Quasi-Invariants with Full Invariants

This section investigates the relation between quasi-invariants and full invariants. It turns out that there exists a geometrical relation in *RGB* space between the two. This relationship sheds light on the stability and noise robustness of quasi- and full invariants.

An orthogonal transformation that has the shadow-shading direction as one of its components is the spherical coordinate transformation. Transforming the *RGB* color space results in the spherical color space or $r\theta\varphi$-color space. The transformations are

$$\begin{aligned} r &= \sqrt{R^2 + G^2 + B^2} = |\mathbf{f}|, \\ \theta &= \arctan(\tfrac{G}{R}), \\ \varphi &= \arcsin\left(\frac{\sqrt{R^2 + G^2}}{\sqrt{R^2 + G^2 + B^2}}\right). \end{aligned} \tag{6.65}$$

Since r is pointing in the shadow-shading direction, its derivative corresponds to $\mathbf{S}_x$:

$$r_x = \frac{RR_x + GG_x + BB_x}{\sqrt{R^2 + G^2 + B^2}} = \mathbf{f}_x \cdot \hat{\mathbf{f}} = |\mathbf{S}_x|. \tag{6.66}$$

The quasi-invariant $\mathbf{S}_x^c$ is the derivative energy in the plane perpendicular to the shadow-shading direction. The derivative in the $\theta\varphi$-plane is given by

$$\begin{aligned} |\mathbf{S}_x^c| &= \sqrt{(r\varphi_x)^2 + (r\sin\varphi\theta_x)^2}, \\ &= r\sqrt{(\varphi_x)^2 + (\sin\varphi\theta_x)^2}. \end{aligned} \tag{6.67}$$

To conserve the metric of *RGB* space the angular derivatives are multiplied by their corresponding scale factors, which follow from the spherical transformation. For matte surfaces both θ and φ are independent of m^b (substitution of Eq. 6.58 in Eq. 6.65). Hence, the part under the root is a shadow-shading invariant.

By means of the spherical coordinate transformation a relation between the quasi-invariant and the full invariant is found. The difference between the quasi-invariant $|\mathbf{S}_x^c|$ and full invariant $s_x = \sqrt{(\varphi_x)^2 + (\sin\varphi\theta_x)^2}$ is the multiplication with r, which is the $L2$ norm for the intensity (Eq. 6.65). In geometrical terms, the derivative vector that remains after subtraction of the part in the shadow-shading direction is *not* projected on the sphere to produce an invariant. This projection introduces the instability of the full shadow-shading invariants for low intensities

$$\begin{aligned} &\lim_{r\to 0} s_x \quad \text{does not exist} \\ &\lim_{r\to 0} |\mathbf{S}_x^c| = 0. \end{aligned} \tag{6.68}$$

The first limit follows from the nonexistence of the limit for both φ_x and θ_x at zero. The second limit can be concluded from $\lim_{r\to 0} r\varphi_x = 0$ and $\lim_{r\to 0} r\theta_x = 0$. Concluding, the multiplication of the full invariant with $|\mathbf{f}|$ resolves the instability.

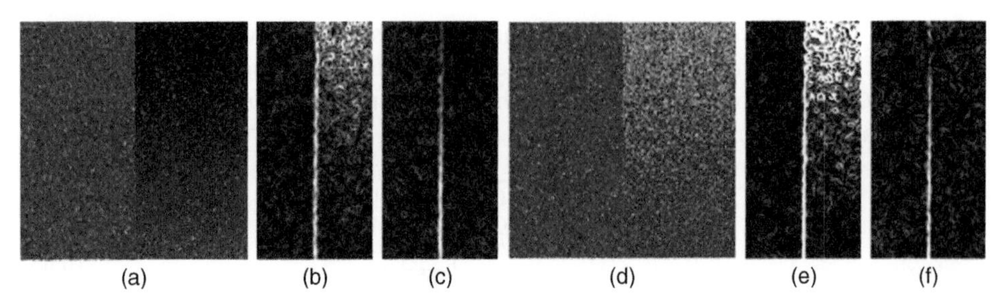

Figure 6.14 (a) Red–blue edge with a decreasing intensity of the blue patch going in the upward direction. Responses of (b) normalized *RGB* derivative and (c) shadow-shading quasi-invariant ($\mathbf{S}_x^c$). (d) Red–blue edge with decreasing saturation going in the upward direction. Responses of (e) hue derivative (h_x) and (f) specular-shadow-shading quasi-invariant ($\mathbf{H}_x^c$).

Hence, the quasi-invariant remains stable in low intensity regions, whereas the full invariant is unstable in these regions.

An example of the responses for the shadow-shading invariant and quasi-invariant is given in Figure 6.14. In Figure 6.14a, synthetic image of a red–blue edge is depicted. The blue intensity decreases along the y-axis. Gaussian uncorrelated noise is added to the *RGB* channels. In Figure 6.14b, the normalized *RGB* response is depicted and the instability of low intensities is clearly visible. For the shadow-shading quasi-invariant (Fig. 6.14c), no instability occurs and the response just diminishes at low intensities. Note that the instable region is particularly inconvenient because shadow-shading edges tend to produce low intensity areas.

The orthonormal transformation which accompanies the specular variant is known as the opponent color space. For a known illuminant $\mathbf{c}^i = (\alpha, \beta, \gamma)^T$ it is given by

$$o1 = \frac{\beta R - \alpha G}{\sqrt{\alpha^2 + \beta^2}},$$
$$o2 = \frac{\alpha\gamma R + \beta\gamma G - (\alpha^2 + \beta^2)B}{\sqrt{\left(\alpha^2 + \beta^2 + \gamma^2\right)\left(\alpha^2 + \beta^2\right)}}, \tag{6.69}$$
$$o3 = \frac{\alpha R + \beta G + \gamma B}{\sqrt{\alpha^2 + \beta^2 + \gamma^2}}.$$

The relations with the variant and its complement are $\left|\mathrm{O}_{\mathrm{x}}\right| = o3_x$ and $\left|\mathrm{O}_{\mathrm{x}}^{\mathrm{c}}\right| = \sqrt{o1_x^2 + o2_x^2}$.

As discussed in Section 6.2.2 the shadow-shading-specular quasi-invariant is both perpendicular to the shadow-shading direction and the specular direction. An orthogonal transformation that satisfies this constraint is the hue saturation intensity transformation. It is actually a polar transformation on the opponent color axis $o1$ and $o2$.

$$h = \arctan\left(\frac{o1}{o2}\right),$$
$$s = \sqrt{o1^2 + o2^2}, \tag{6.70}$$
$$i = o3.$$

The changes of h occur in the hue direction, hence the derivative in the hue direction is equal to the shadow-shading-specular quasi-invariant,

$$\left|\mathbf{H}_x^c\right| = s \cdot h_x. \tag{6.71}$$

The multiplication with the scale factor s follows from the fact that for polar transformations, the angular derivative is multiplied by the radius.

The hue, h, is a well-known full shadow-shading-specular invariant. Equation 6.71 provides a link between the derivative of the full invariant, h_x and the quasi-invariant $\left|\mathbf{H}_x^c\right|$. A drawback of hue is that it is undefined for points on the black–white axis, that is, for small s. Therefore, the derivative of hue is unbounded. In Section 6.2.2, we derived the quasi-invariant as a linear projection of the spatial derivative. For these projections, it holds that $0 < \left|\mathbf{H}_x^c\right| < \left|\mathbf{f}_x\right|$, and hence the shadow-shading specular quasi-invariant is bounded. It should be mentioned that small changes round the gray axis result in large changes of the direction or 'color' of the derivative, for example, from blue to red, in both the quasi-invariant and the full invariant. However, the advantage of the quasi-invariant is that the norm remains bounded for these cases. For example, in Figure 6.14d a red–blue edge is depicted. The blue patch becomes more achromatic along the y-axis. The instability for gray values is clearly visible in Figure 6.14e, whereas in Figure 6.14f, the response of the quasi-invariant remains stable.

Full invariants possess *strong photometric invariance* meaning that they are invariant with respect to a physical photometric parameter, for instance, the geometric term m^b in the case of normalized RGB. Hence, the first-order derivative response of such invariants does not contain any shadow-shading variation. Our approach determines the direction in the RGB cube in which shadow-shading edges exhibit themselves. This direction is then used to compute the quasi-derivative, which shares with full invariants the property that shadow-shading edges are ignored. However, the quasi-invariants are not invariant with respect to m_b. For the shadow-shading quasi-invariant, subtraction from Equation 6.59 of the part in the shadow-shading direction $\mathbf{c}^b$ results in

$$\mathbf{f}_x = em^b\left(\mathbf{c}_x^b - \mathbf{c}_x^b \cdot \hat{\mathbf{c}}^b\right), \tag{6.72}$$

which is clearly not invariant for m^b and e. Therefore, quasi-invariants are said to only possess *weak photometric invariance*. In a similar way the specular-shadow-shading quasi-invariant can also be proved to be dependent on m^b and e.

The fact that quasi-invariant only possess weak photometric invariance means that they are dependent on m^b and e, which limits their applicability. They cannot be used for feature description where edge responses are compared under different circumstances, such as content-based image retrieval. However, they can be used for applications that are based on feature detection, such as shadow-edge-insensitive image segmentation, shadow-shading-specular independent corner detection, and edge classification.

A major advantage of the quasi-invariants is that their response to noise is independent of the signal. In the case of additive uniform noise, the noise in the quasi-invariants is also additive and uniform since it is a linear projection of the derivative of the image. This means that the noise distortion is constant over the image. We have seen that full invariants differ from the quasi-invariants by scaling with a signal-dependent factor (the intensity or saturation), and hence their noise response is also signal dependent. Typically, the shadow-shading full invariant exhibits high noise distortion around low intensities, while the shadow-shading-specular full invariant has high noise dependency for points around the achromatic axis. This is shown in Figure 6.14. The uneven levels of noise throughout an image hinder further processing for the full invariants.

A second advantage of photometric variants and quasi-invariants is that they are expressed in the same units (i.e., being projections of the derivative, they are expressed in *RGB* value per pixel). This allows for a quantitative comparison of their responses. An example is given in Figure 6.15. Responses along two lines in the image are enlarged in Figure 6.15c and Figure 6.15d. The line in Figure 6.15c crosses two object edges and several specular edges. It shows nicely that the specular variant almost perfectly follows the total derivative energy for the specular edges in the middle of the line. In Figure 6.15d a line is depicted that crosses two object edges and three shadow-shading edges. Again, the shadow-shading variant follows the gradient for the three shading edges. A simple classification scheme results in Figure 6.15b. Note that full invariants cannot be compared quantitatively because they have different units.

6.2.4 Localization and Discriminative Power of Full and Quasi-Invariants

In this subsection, we compare the performance of quasi-invariants with that of full invariants on the task of edge detection. We compare both approaches based on edge displacement and discriminative power.

In order to investigate the discriminative power and edge displacement of the proposed invariants, edge detection between 1012 different colors of the

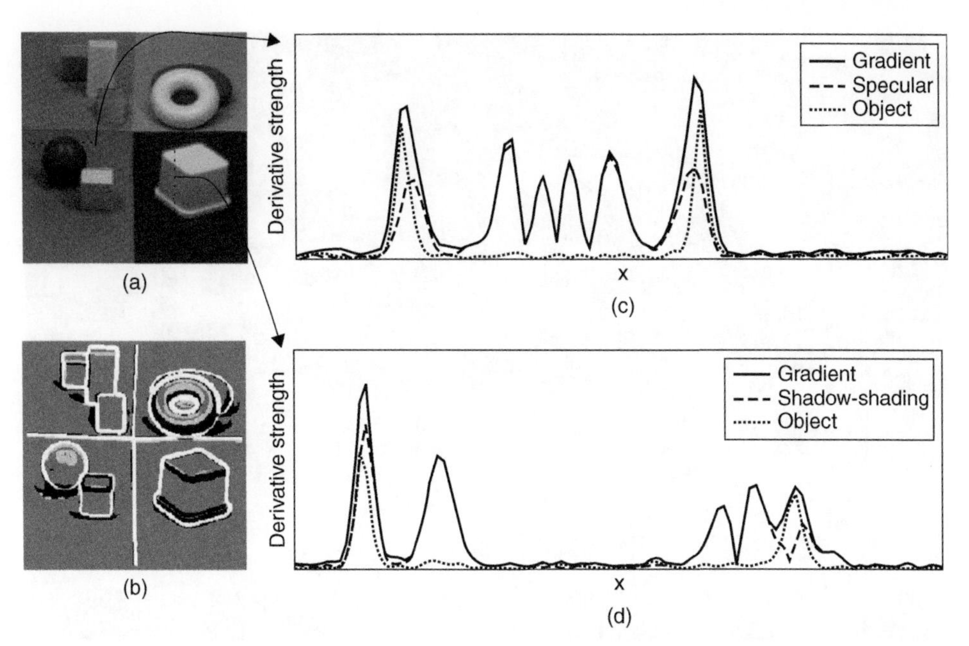

Figure 6.15 (a) Input image with two superimposed dotted lines, which are plotted in (c) (d). (b) Edge classification result, with white object edges, black shadow edges, and light gray specular edges. (c), (d) The derivative strength along lines indicated in (a). *Source:* Reprinted with permission, © 2005 IEEE.

PANTONE color system is examined (Fig. 6.16). The PANTONE colors span a convex, nontriangular set in chromaticity space, hence they may be considered as a mixture of various inks. The set is representative for natural surface reflection spectra, since most reflection functions may be modeled by a linear five- to seven-parameter model [69]. The colors are uniformly distributed in color space. The 1012 PANTONE colors are recorded by an *RGB* camera (Sony DXC-930P), under a 5200 K daylight simulator (Little Light, Grigull, Jungingen, Germany).

In this experiment, we combine two PANTONE patches into a single image and perform edge detection. We evaluate the quality of the edge detection by measuring edge displacement and discriminative power. The discriminative power is the number of edges that can be differed; with increasing photometric invariance this is expected to drop. Edge detection is performed between the 1012 different colors from the PANTONE [70] color system. The patches from PANTONE are reduced to one *RGB* value by a large Gaussian averaging operation. Every one of the 1012 different *RGB* values is combined with all other *RGB* values, resulting in a total of $N = 1012 * 1011/2 = 511566$ edges of $M = 32$ pixels length. The

PANTONE is a trademark of Pantone, Inc.

We use the PANTONE edition 1992-1993, Groupe BASF, Paris, France.

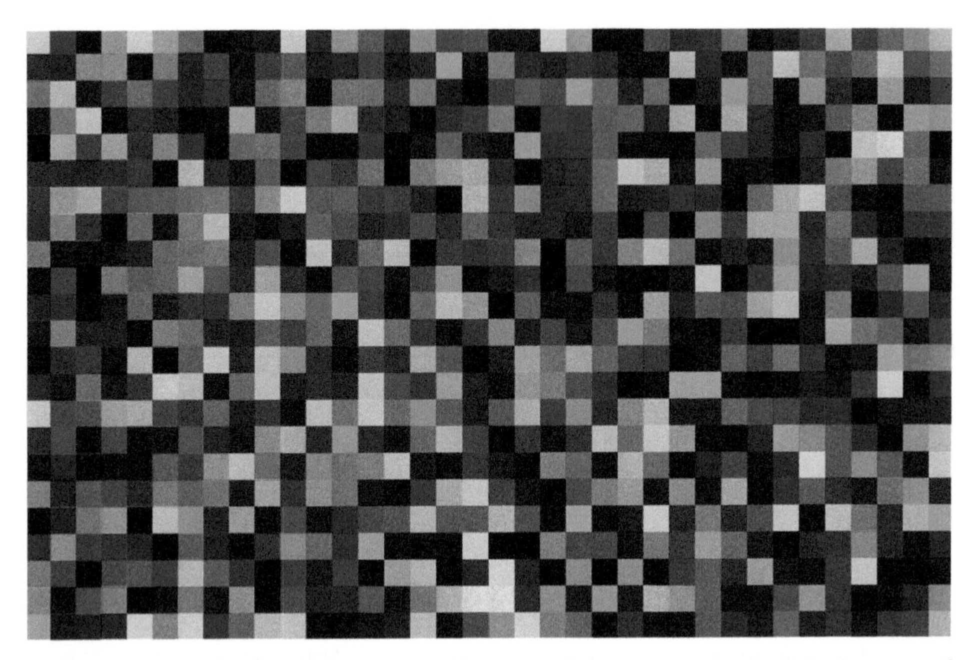

Figure 6.16 Examples of PANTONE colors, which are applied to compute the discriminative power of the various photometric invariants.

edge position is determined by computing the maximum response path of the derivative energy in a region of 20 pixels around the actual edge. This results in an edge estimation, which is compared with the actual edge. We define two error measures. First is the average displacement Δ:

$$\Delta = \frac{\sum_{\{x_{i,j};\,|x_{i,j}-x_0|>0.5\}} |x_{i,j} - x_0|}{N \cdot M}, \tag{6.73}$$

in which $x_{i,j}$ is the jth edge pixel of the ith edge. Because the actual edge is located between two pixels, displacements equal to half a pixel are considered as a perfect match. The second error measure is the percentage of missed edges, ε. An edge is classified missed when the variation over one edge:

$$\text{var}(i) = \frac{1}{M} \sum_{j=1}^{M} \left| x_{i,j} - \frac{1}{M} \sum_{k} x_{i,k} \right|, \tag{6.74}$$

is larger than 1 pixel. For the Gaussian derivative, a scale $\sigma = 1$ is chosen. The experiments are performed with uncorrelated Gaussian noise of standard deviation 5, and 20.

The results are shown in Table 6.2. The standard image derivative $\mathbf{f}_x$ shows very small edge displacement and is able to discriminate most color patches. This is not surprising since the PANTONE colors are made to look different. Looking at the results of the invariants we see two observations that could be expected

Table 6.2 The displacement, Δ, and the percentage of missed edges, ε, for six different edge detectors.

Noise →		5		20	
Detector ↓	Invariance	Δ	ε	Δ	ε
$\mathbf{f}_x$	—	0.003	0.1%	0.07	2.4%
S_x^c	s	0.04	1.1%	0.38	11.7%
C	S	0.20	2.5%	1.34	25.9%
H_x^c	s & h	0.29	6.5%	0.86	25.2%
H	S & H	0.77	11.5%	2.03	39.2%
N	I	0.23	2.7%	1.50	28.9%

Gaussian noise of standard deviation 5, and 20, was added. For each edge detector the photometric invariance is given: *s* indicates shadow-shading invariance, *h* indicates specular invariance, and *i* indicates illuminant invariance. Capitals are used to indicate full invariance.

theoretically. First, we can see that adding invariance lowers the discriminative power and increases the edge displacement. For example, a photometric shadow-shading invariant edge detector cannot distinguish between patches that could be caused by the same material reflectance but with a different intensity. Second, quasi-invariants outperform full invariants on edge detection. In comparison with full invariants, they have about half the edge displacement and miss around half the number of edges. This is also expected, because quasi-invariants do not have the nonlinear instabilities of full invariants. Therefore, in case of feature detection it is advisable to apply quasi-invariants. However, feature description requires full photometric invariance as we also show in more detail in Chapter 14.

6.3 Summary

In this chapter we have discussed how to extend photometric invariance theory to the differential structure of images. We have outlined two approaches based on full and quasi-invariance.

The full invariants are derived from differential calculus, aiming to cancel unwanted factors in the photometric model. By applying the Gaussian measurement framework, the derived theoretical invariants are converted into image features. These features are obtained by the appropriate combinations of Gaussian smoothing and derivative filters, which are applied to the opponent color channels. We have given examples and derivations of full invariants for the Lambertian and dichromatic reflectance model.

The quasi-invariants are derived from the dichromatic reflection model and have been proved to differ from full photometric invariants by a local normalization factor. These quasi-invariants do not have the inherent instabilities of full photometric invariants, and from theoretical and experimental results it is shown that quasi-invariants have better noise characteristics and discriminative power and introduce less edge displacement than full photometric invariants for the task of feature detection. The lack of full photometric invariance limits the applicability of quasi-invariants, and therefore quasi-invariants cannot be used for feature extraction.

7 Photometric Invariance by Machine Learning

With contributions by José M. Àlvarez and Antonio M. López

As shown in the previous chapters, the choice of a color model is of great importance for many computer vision algorithms, as the chosen color model induces the equivalence classes to the actual algorithms. As there are many color models available, the inherent difficulty is how to automatically select a single color model or, alternatively, a weighted subset of color models producing the best result for a particular task. The subsequent hurdle is how to obtain a proper fusion scheme for the algorithms so that the results are combined in a proper setting. In the previous chapters, physical reflection models (e.g., Lambertian or dichromatic reflectance) are used to derive color invariant models. However, this approach may be too restricted to model real-world scenes in which different reflectance mechanisms can hold simultaneously. Instead of modeling the world by a single reflection model, we now focus on how color invariance can be obtained by machine learning.

The learning process is based on the selection of positive examples (e.g., colored image patches of a certain object to be recognized) to obtain color invariant ensembles. Of course, the training examples should include a broad range of pixel values capturing all possible imaging conditions under which the object can be captured. Using these training samples, the aim is to arrive at color

Color in Computer Vision: Fundamentals and Applications, First Edition.
Theo Gevers, Arjan Gijsenij, Joost van de Weijer, and Jan-Mark Geusebroek.

ensembles that yield a proper balance between invariance (repeatability) and discriminative power (distinctiveness).

In this chapter, a learning approach that minimizes the estimation error is presented. The method is also suited to deal with sequential data. A weighting scheme is used to incorporate the dynamics of observations over time. The ensemble is periodically updated considering the new data and its temporal order. The chapter is organized as follows. First, the learning-based fusion scheme is discussed. Then, the evolution of data over time is included. Finally, the approach is applied to two applications: facial skin detection and road detection (RD). More information can be found in Reference 71.

7.1 Learning from Diversified Ensembles

In machine learning, combining multiple classifiers that consider the differences between their components is a powerful technique to improve the performance of single classifiers [72–74]. The measure of disagreement between components is referred to as *diversity*. A promising subset of combining strategies are those using diversity in the process of generating the ensemble [74]. For instance, Melville and Mooney [75] consider diversity as being the disagreement of an ensemble member with the ensemble's prediction. Jacobs [76] proposes a minimum variance estimator where the estimated aggregate has a variance at most as large as the variance of any of the input features. Stokman and Gevers [77] use the Markowitz diversification criterion [78] in the process of defining the ensemble. The method assumes that each descriptor can be characterized by an unimodal distribution and computes the best combination that provides maximal feature discrimination. However, in practice, the distribution of the training data is often not unimodal, leading to estimation errors that are maximized by the quadratic optimization technique used to compute the ensemble [79].

For a given combination strategy, proper selection of its components is important to improve the performance of the strategy. The ideal situation would be a set of classifiers with uncorrelated errors. Then, these classifiers could be combined to minimize the effect of these failures. In fact, the combination of a set of similar classifiers will not outperform the individual members [74]. The improvement that can be obtained by selecting appropriate classifiers can even be larger when the method uses a learning step to adapt to the specific classification problem (e.g., boosting, bagging and random forests). To facilitate the learning procedure, systems use training data corresponding to the object to be recognized (i.e., positive examples) and, for instance, background (i.e., negative examples). Systems using only positive data within the training step are more desirable since obtaining a comprehensive representation of negatives or unknown universe is often unfeasible. In addition, if negative data is not chosen properly, this may lead to a lower classification accuracy [80].

Therefore, the learning phase is based only on positive examples. The aim is to model a homogeneously colored image region (object) recorded under varying imaging conditions (views) by combining different color models (observations). At each view, the image region contains multiple pixels (samples of an observation). Then, the color of the region is modeled using a single value (expected color $E[\xi_O]$) with small deviations from this value (σ_O) by

$$O = E[\xi_O] \pm \sigma_O. \tag{7.1}$$

Definitions are given in Table 7.1 and the modeling process is illustrated in Figure 7.1.

Table 7.1 Definitions and correspondence between symbols and color-related terms.

Definitions	
Abstract Terms	**Color Terms**
Object O	Homogeneously colored image region.
View	Image region recorded under a different imaging condition (i.e., illumination, shading)
Object representation ξ_i	Expected value using the *i*th color model *independent* of the imaging conditions
Observation $\tilde{\xi}_{ij}$	Expected value using the *i*th color model for the *j*th view
Samples of observation ξ_{ijl}	Pixel values used to estimate the *i*th color model data distribution for the *j*th view

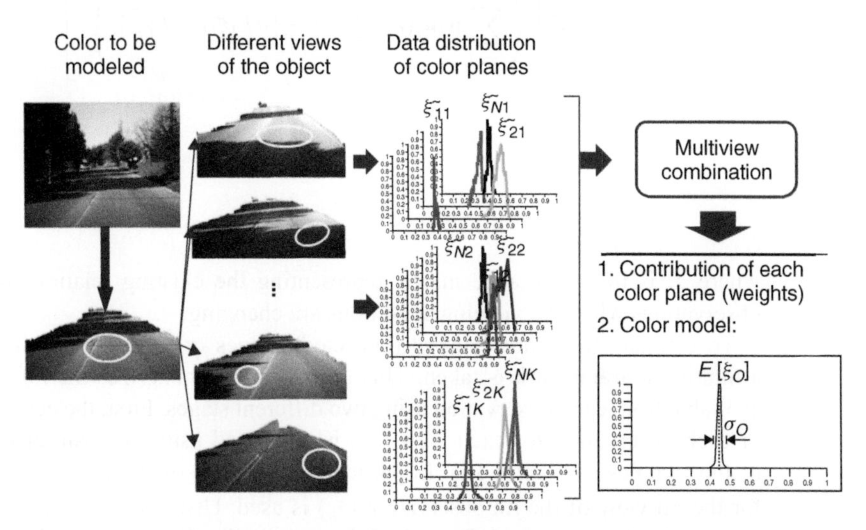

Figure 7.1 The color of an image region is modeled combining the information of different color models from different views.

To build the model, it is assumed that L different samples of N different observations for K views of the object are available ($\xi_{ijl}, i \in [1, 2, \ldots, N], j \in [1, 2, \ldots, K], l \in [1, 2, \ldots, L]$). These samples correspond, for example, to the same object imaged under varying imaging conditions (e.g., shading, highlights, and illumination) generating variations of the observations other than device-dependent recording noise. Multiple samples of each observation are provided to reduce the influence of noise. A set of N orthogonal and nonredundant representations of the object is estimated ($\xi_1, \ldots, \xi_N$) and the object is modeled by a weighted linear combination of the representative (expected) values of each observation $E[\xi_i]$:

$$E[\xi_O] = \sum_{i=1}^{N} w_i E[\xi_i], \tag{7.2}$$

where $\mathbf{w} = [w_1, \ldots, w_N]$ is the contribution of each observation to the final combination. Further, the standard deviation of the object patch is given by

$$\begin{aligned}
\sigma_O^2 &= E[(E[\xi_O] - \overline{E[\xi_O]})^2] \\
&= E\left[\left(\sum_{i=1}^{N} w_i E[\xi_i] - \sum_{i=1}^{N} w_i \overline{E[\xi_i]}\right)^2\right] \\
&= E\left[\left(\sum_{i=1}^{N} w_i (E[\xi_i] - \overline{E[\xi_i]})\right)^2\right] \\
&= E\left[\left(\sum_{i,j=1}^{N} w_i w_j (E[\xi_i] - \overline{E[\xi_i]})(E[\xi_j] - \overline{E[\xi_j]})\right)^2\right] \\
&= \sum_{i,j=1}^{N} w_i w_j \sigma_{ij} \\
&= w \Sigma w^T,
\end{aligned} \tag{7.3}$$

where Σ is the covariance matrix representing the existing relations between observations when the viewing conditions are changing.

To estimate the representative (expected) values of each observation $E[\xi_i]$, a multiview framework is taken. This framework characterizes the information available from each observation using two different stages. First, the central value of each observation for each view (ξ_{ij}) is computed using the data distribution of the samples available (ξ_{ijl}). In particular, the mode of the samples available for the jth view of the ith observation ($\tilde{\xi}_{ij}$) is used. Using this central value, the algorithm minimizes the influence of skewed distributions, thus minimizing the estimation error. Second, the expected value of an observation given the values of different views $E[\xi_i]$ is estimated assuming that each available view has the same probability of appearing. In particular, the mean value of central values for each view is considered as follows:

$$E[\xi_i] = \frac{1}{K}\sum_{j=1}^{K}\tilde{\xi}_{ij}. \tag{7.4}$$

What remains is the estimation of w_i. A proper combination of observations leads to a model for which the expected value of an object ($E[\xi_O]$) is close to a reference value ($E[\xi_{O_R}]$) and for which its variance is minimized. In this way, the combination reduces the deviations from the expected value due to varying viewing conditions. This reference value is, for example, the value which is obtained when ideal acquisition conditions are obtained. Hence, computing w_i can be posed as an optimization problem formulated as follows:

$$\text{minimize} \quad \sum_{i,j=1}^{N} w_i w_j \sigma_{ij} \tag{7.5}$$

$$\text{subject to} \quad E[\xi_O] \geq E[\xi_{O_R}],$$

$$\sum_{i=1}^{N} w_i = 1, \tag{7.6}$$

with the constraint that the total contribution of observations must sum up to 1.

Quadratic optimization techniques [81] can be applied to solve Equation 7.5 and provide a set of optimum solutions (efficient ensembles) called the *efficient frontier* [79]. That is, the efficient frontier contains different values of $E[\xi_O]$ and associated weights that minimize the corresponding σ_O. However, quadratic optimization techniques tend to select components with attractive characteristics. In this way, components with lesser appealing features are not selected. This is the case when the estimation error is likely to be maximal [79, 82]. Therefore, in order to deal with the estimation error and improve the diversity of the ensemble, a resampling technique is adopted. This resampling technique uses a Monte Carlo simulation to obtain a set of efficient ensembles called *resampled frontier* [83]. Ensembles lying on this resampled frontier are composed of weight vectors obtained as the average of the efficient frontiers given a certain expected value. The performance of resampled efficient ensembles is better than the performance of those ensembles obtained using quadratic optimization techniques [82, 84].

Finally, the most appropriate ensemble is selected from the set of ensembles lying on the efficient frontier using the Sharpe ratio (*SR*) [85]. This ratio is a single statistical performance measure of variance-adjusted expected return defined as

$$SR = \frac{E[\xi_O]}{\sigma_O}. \tag{7.7}$$

The highest SR corresponds to the ensemble in the frontier obtaining the best performance. If a benchmark ensemble ξ_R exists, the performance of an ensemble in the frontier (P_e) is computed as follows:

$$P_e = \frac{1}{(|E[\xi_O] - \xi_R|)\sigma_O}, \tag{7.8}$$

where the highest performance corresponds to the most appropriated ensemble.

The above computation of weights and the ensemble selection method are summarized as follows:

1. Estimate the efficient frontier using the training data and quadratic programming techniques. This frontier is composed of ensembles varying from minimum variance to the maximum expected value ensembles. Divide the difference between the minimum and maximum return in m ranks.
2. Estimate the covariance matrix, Σ, and expected values, $E[\xi_i]$, of the training data,

$$E[\xi_i] = \frac{1}{K}\sum_{j=1}^{K} \tilde{\xi}_{ij}, \tag{7.9}$$

$$\Sigma = (\sigma_{i,j}), \tag{7.10}$$

   where K is the number of views.
3. Resample, using the training inputs in step 2, taking D draws for the input multivariate distribution. The number of draws D reflects the degree of uncertainty in the training data. Compute a new covariance matrix from the sampled series. The estimation error will result in different covariance matrices and mean vector from those in step 2.
4. Compute the efficient frontier for the inputs derived in step 3. Calculate the optimal ensemble weights for m equally distributed points along the frontier.
5. Repeat step 3 and step 4 P times. Calculate the averaged ensemble weights for each observation,

$$\overline{w}_i^{\text{resampled}} = \frac{1}{P}\sum_{im=1}^{P} w_{im}, \tag{7.11}$$

   where w_{im} denotes the weight vector for the mth ensemble along the frontier for the ith observation.

6. Evaluate the frontier of averaged ensembles by the variance–covariance matrix from the original training data to obtain the resampled frontier.
7. Select the ensemble from the frontier that exhibits the highest performance Equation (7.7 or Equation 7.8 as required).

7.2 Temporal Ensemble Learning

In this section, the model is extended to also take into account the evolution of observations over time (e.g., from still images to videos). The key idea is to include temporal information in the estimation of the parameters $E[\xi_1], \ldots, E[\xi_N]$ and Σ. These parameters are computed considering that each view of a given observation provides the same information to the final ensemble. However, because of the dynamic nature of data sequences, local observations should be taken into account more prominently than distant ones. In this way, the modification of the algorithm consists of using time series analysis to predict the expected values of observations rather than considering simple averages over views.

To express the dynamic structure of the data (observations and ensembles), a weighting process is used. Further, the dynamic structure of the variance within observations is also considered. There are two models to deal with these kinds of variations: exponentially weighted moving average (EWMA) and generalized autoregressive conditionally heteroscedastic (GARCH). Both models assume that serial correlation is present in the dynamics of the observations. As a result, both models assign higher weights to recent values than to the older ones. In this chapter, the EWMA model is used mainly due to its simplicity (less parameters to estimate) and the ability to cope with changes in the standard deviation of the incoming data [86, 87].

EWMA uses a decay factor that weighs the change of each past observation. More recent observations receive higher weights than older ones. Using EWMA, the input parameters of the optimization process are derived as follows:

$$E[\xi_i] = \frac{1}{\sum_{j=1}^{K} \lambda^{j-1}} \sum_{j=1}^{K} \lambda^{j-1} \tilde{\xi}_{ij}, \tag{7.12}$$

$$\Sigma = (\sigma_{nm}) = (1-\lambda) \sum_{j=1}^{K} \lambda^{j-1} (\tilde{\xi}_{nj} - E[\xi_n])(\tilde{\xi}_{mj} - E[\xi_m]), \tag{7.13}$$

where λ is the decay factor. This factor determines both the degree of weighting of recent observations and the speed with which the volatility measure will return to a lower level after a large return. A lower decay gives a higher weighting to recent values. K is the number of past observations unlike in the previous section where K was the number of different views available for each observation. Parameter K

can be set to infinity since the weighting procedure will rapidly reduce to zero for distant observations. Since $0 < \lambda < 1$, $\lambda^n \longrightarrow 0$ when $n \longrightarrow \infty$, the model will eventually place a zero weight on observations far in the past.

7.3 Learning Color Invariants for Region Detection

In this section, the method is applied to color-based region detection. In other words, the detection of object patches in images is recorded under varying imaging conditions using a set of color models composed of both color variant and invariant models. The goal is to derive color invariance by learning from color models to obtain diversified color invariant ensembles.

Every possible transformed color model is considered as an observation of the same object (color region) and each view corresponds to a different imaging condition such as lighting, viewing, and illumination variations. Further, each pixel within the region corresponds to different sampling values of the observation. Hence, the proper interpretation of the algorithm is as follows: O is the data distribution of the final combination of color (invariant) planes/models and $E[\xi_O]$ and σ_O its central value and variance respectively. $E[\xi_i]$ is the expected value of the ith color (invariant) plane estimated using the multiview procedure. That is, considering first the data distribution from pixels of each view and then the average value of the views. Finally, w_i denotes the contribution of the ith color model to the final ensemble.

During **training** (i.e., estimating $E[\xi_O]$, σ_O and w_i), the following steps are performed:

- Select a set of training images containing the object to be detected, imaged under different acquisition conditions (e.g., varying illumination).
- Select a region of interest for each training image (ith) and for each color model (jth) estimate $\tilde{\xi}_{ij}$ using the data distribution of pixels in the training region.
- Estimate the correlation matrix Σ of these values. This matrix contains information regarding the relative variations of each color model when the acquisition conditions vary.
- Estimate the weights $\mathbf{w}$ using the Monte Carlo method considering the central value of each color model for each view and the covariance matrix as input data.
- Compute $E[\xi_O]$ and σ_O using Equations 7.2 and 7.3, respectively.
- Finally, compute the SNR_O ratio of the model as follows:

$$SNR_O = \frac{E[\xi_O]}{\sigma_O}. \tag{7.14}$$

Then, during **classification**, the following steps are performed:

- Convert the image into the color models (the same as during training) and apply the weights **w** obtained in the training phase to combine them. This leads to a gray-level image.
- Estimate the signal-to-noise ratio, *SNR*, by dividing, at each pixel, the local mean value by the local standard deviation. The *SNR* is estimated using a rectangular region ($M \times N$ pixels) at each pixel.
- Compute the error between the SNR_O and the local *SNR* for each pixel. The lower the error, the more similar the colors are.
- Threshold the error image e to obtain the final binary mask C:

$$C(x,y) = \begin{cases} 1 & \text{if } e(x,y) < T \\ 0 & \text{otherwise.} \end{cases} \tag{7.15}$$

  The appropriate value of T is obtained using automatic thresholding techniques such as the isodata method [88].

Finally, if **temporal adaptation** is required, the following steps are performed:

- Use the classification procedure to classify pixels in the first image.
- Use the current result to estimate the central value of each color model for that frame. Add these central values to the historical data.
- Estimate input parameters to the optimization process (Σ and $E[\xi_1], \ldots, E[\xi_N]$) using the *EWMA* process outlined in Section 7.2.
- Select the optimal ensemble from the frontier considering the same *SNR* and reference as in the initial training stage.
- Use the new ensemble to process the incoming image.

To provide robustness against confounding imaging conditions (e.g., illumination, shading, highlights, and interreflections), different color models exhibiting different photometric invariance properties have been discussed in the previous chapters. For instance, for the dichromatic reflection model, normalized color *rgb* is to a large extent invariant to a change in camera viewpoint, object position, and the direction and intensity of the incident light. See Table 7.3 for an overview of color models and their invariance properties. In addition to the models described in previous chapters, the illumination invariant ($\Im$) proposed in Reference 89 is included. This color invariant requires a calibration parameter, the invariant direction which is an intrinsic parameter of the camera. Currently, this invariant direction can be found either by following the calibration procedure outlined in

Table 7.2 Derivation of opponent color space, normalized rgb, HSV and CIELab color spaces from RGB values.

Opponent Color Space

$$\begin{pmatrix} O_1 \\ O_2 \\ O_3 \end{pmatrix} = \begin{pmatrix} \frac{1}{\sqrt{2}} & \frac{-1}{\sqrt{2}} & 0 \\ \frac{1}{\sqrt{6}} & \frac{1}{\sqrt{6}} & \frac{-2}{\sqrt{6}} \\ \frac{1}{\sqrt{3}} & \frac{1}{\sqrt{3}} & \frac{1}{\sqrt{3}} \end{pmatrix} \begin{pmatrix} R \\ G \\ B \end{pmatrix}$$

Normalized *rgb*

$$r = \frac{R}{R+G+B}$$
$$g = \frac{G}{R+G+B}$$
$$b = \frac{B}{R+G+B}$$

HSV

$$\begin{pmatrix} V \\ V_1 \\ V_2 \end{pmatrix} = \begin{pmatrix} \frac{1}{3} & \frac{1}{3} & \frac{1}{3} \\ \frac{-1}{\sqrt{6}} & \frac{-1}{\sqrt{6}} & \frac{2}{\sqrt{6}} \\ \frac{1}{\sqrt{6}} & \frac{-2}{\sqrt{6}} & \frac{1}{\sqrt{6}} \end{pmatrix} \begin{pmatrix} R \\ G \\ B \end{pmatrix}$$

$$H = \arctan \frac{V_2}{V_1} \quad S = \sqrt{V_1^2 + V_2^2}$$

CIE *Lab*

$$\begin{pmatrix} X \\ Y \\ Z \end{pmatrix} = \begin{pmatrix} 0.490 & 0.310 & 0.200 \\ 0.177 & 0.812 & 0.011 \\ 0.000 & 0.010 & 0.990 \end{pmatrix} \begin{pmatrix} R \\ G \\ B \end{pmatrix}$$

$$L = 116\left(\frac{Y}{Y_0}\right) - 16$$

$$a = 500\left[\left(\frac{X}{X_0}\right) - \left(\frac{Y}{Y_0}\right)\right]$$

$$b = 200\left[\left(\frac{Y}{Y_0}\right) - \left(\frac{Z}{Z_0}\right)\right]$$

X_0, Y_0, and Z_0 are the coordinates of a reference white point.

Table 7.3 Invariance of color models (derived in Table 7.2) for different types of lighting variations, that is, light intensity (LI) or light color (LC) change and/or shift [92].

Taxonomy of color spaces	LI change	LI shift	LI change and shift	LC change	LC change and shift
RGB	–	–	–	–	–
O_1, O_2	–	+	–	–	–
O_3, Intensity, L	–	–	–	–	–
Saturation (S)	–	+	+	–	–
Hue (H)	+	+	+	–	–
r, g, a, b	+	–	–	–	–
$\Im$	+	+	+	+	+

Invariance is indicated with '+' and lack of invariance with '–'.

Reference 89 or by using a procedure that determines the invariant direction from a single image [90] or from a set of images [91]. The former consists in acquiring images of a Macbeth color checker under different daytime illuminations and then obtaining the invariant direction by analyzing the log chromaticity plot generated from these images. The latter considers the entropy of a single image to compute the invariant direction. Then, the method consists in generating invariant images using all the possible invariant directions within a range. The optimum direction is the one minimizing the entropy of its corresponding illumination-invariant image [90].

Considering all the color models in Table 7.3, a set is obtained of both color variants and invariants to achieve both distinctiveness and repeatability, respectively. The next step is to obtain a nonredundant subset. The covariance matrix Σ provides information about correlation between color models. This analysis can be done using principal component analysis (PCA) [93]. Then, correlation between color models is represented by the loadings of each color model (Fig. 7.2). The input data to PCA is the matrix containing the expected values for each view of each color model ($\overline{\xi_{ij}}$). The closer two points are in the loading space, the more correlated they are (and their corresponding color models). The number of principal components depends on the data and the amount of variation. The selection of color models that represent each cluster (e.g., *S* or *b* in Figure 7.2) is computed by the Hartigan's test for unimodality [94]. In this way, an orthogonal (variant/invariant) and nonredundant (decorrelated) color model subset is obtained.

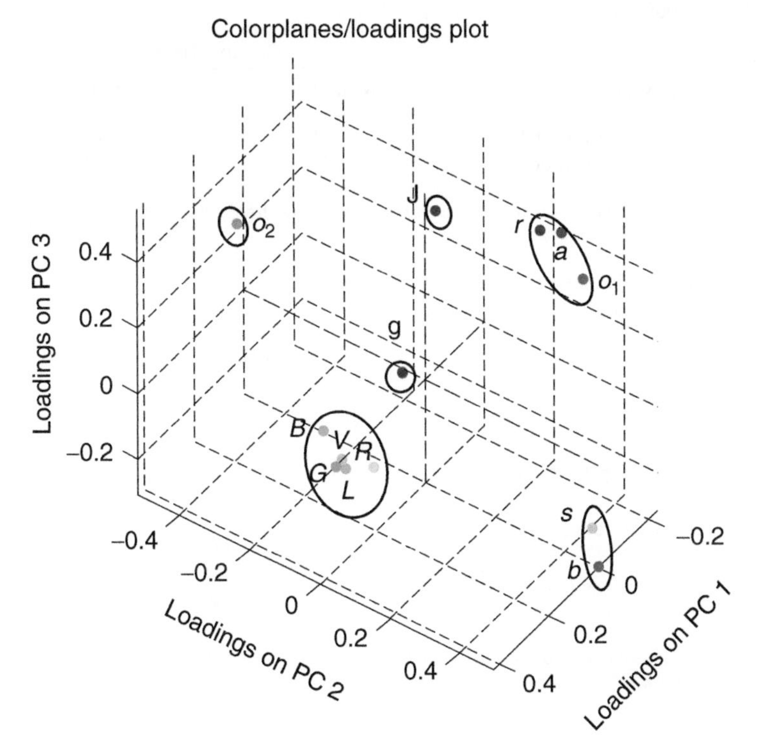

Figure 7.2 PCA is used to reduce redundancy within the training data. The analysis is done using the loadings plot of each color model. This example corresponds to the training set from the face database.

7.4 Experiments

In this section, the algorithm is applied to two different databases: (1) the Caltech Face database [95] and (2) a road sequence taken by an onboard camera. The first application is to detect facial skin in the Caltech image dataset. The other one is to detect roads under uncontrolled imaging conditions. We used 13 color models ($\Im$, R, G, B, r, g, O_1, O_2, L, a, b, S, V). The third opponent color O_3 is excluded since it provides intensity information that is already provided by V. Further, the hue component H from the HSV color space is excluded because of its instability being close to the achromatic axis [96]. Further, the calibration required to compute $\Im$ is done using the approach proposed in Reference 91. Finally, the reference white point for deriving CIELab color space is set to the D65 white point ($X_0 = 0.9505$, $Y_0 = 1.0000$, $Z_0 = 1.0888$) [25].

7.4.1 Error Measures

Quantitative evaluations are provided using pixel-based measures; see Table 7.4, from which the following error measures are computed: quality, detection accuracy, detection rate, and effectiveness (see Table 7.5. Each of these measures provides a different insight into the performance of a method. Quality takes

Table 7.4 The contingency table.

Contingency table		Ground Truth Nontarget	Ground Truth Target
Detection	Nontarget	TN	FN
Result	Target	FP	TP

Algorithms are evaluated based on the number of pixels correctly and incorrectly classified.

Table 7.5 Pixel-wise measures used to evaluate the performance of different algorithms.

Pixel-wise measure	Definition
Quality ($\hat{g}$)	$\hat{g} = \frac{TP}{TP + FP + FN}$
Detection accuracy (DA)	$DA = \frac{TP}{TP + FP}$
Detection rate (DR)	$DR = \frac{TP}{TP + FN}$
Effectiveness (F)	$F = \frac{2DADR}{DA + DR}$

These measures are defined using the entries of the contingency table (Table 7.4.)

into account the completeness of the extracted data as well as its correctness. Detection accuracy, also known as *precision*, is the probability that the result is valid. Detection rate, or recall, is the probability that the ground truth data is detected. Effectiveness is a single measure that trades off the detection accuracy versus detection rate. Further, the performance of our method is compared, on each dataset, to existing algorithms. Pair-wise comparisons between algorithms are computed by the Wilcoxon statistical significance test [97].

7.4.2 Skin Detection: Still Images

To detect skin pixels of faces, the Frontal Face Image Database of Caltech is used. This image dataset contains 450 face images taken from 27 different persons under different lighting, expressions and backgrounds. The appearance of the face in these images is clearly influenced by different illumination, shading, skin tone, and so on (Fig. 7.3). Ground truth is generated by manually segmenting all the images in the database. The training set is obtained by manually selecting 100 different patches from 100 different (randomly chosen) images. The unimodality test is used to discard inappropriate patches. Finally, 58 patches are used for training, representing 1% of the total of facial pixels in the database. Note that the covariance matrix (Σ) encapsulates variations not only in the illumination conditions but also in the appearance of the object (i.e., skin tone variations) since different instances of the same object class are considered at the same time. The color model set is computed using the procedure described in Section 7.3. The set of weights obtained are listed in Table 7.6.

Figure 7.3 Example images from The Frontal Face Image Database of Caltech [95].

Table 7.6 Set of weights obtained for the experiments.

	$\Im$	R	G	B	r	g	O_1	O_2	L	a	b	S	V
Skin	−0.017	—	—	—	—	0.022	—	0.013	0.176	0.652	0.154	—	—
Road	0.929	—	—	—	0.157	0.342	0.266	−0.024	—	−0.356	−0.082	−0.452	0.220

'—' corresponds to an unselected color model by the PCA procedure.

These weights reveal a dominance in a and b reflecting pale reddish color (i.e., skin). Example results are shown in Figure 7.4. For each original image (Fig. 7.4a) the weighted combination (Fig. 7.4b) and the skin data distribution (Fig. 7.4c) are provided. Further, all the skin pixels in the database are collected and the distribution of values of different color models is shown in Figure 7.5. For comparison, only one color model from each group in Table 7.3 is considered. As shown, the learning method leads to a unimodal distribution of pixels despite light color and skin tone variations. That is, lighting variations are compensated when color models are properly combined. Note that pixel values for other color models are not normally distributed, leading to erroneous mean and standard deviation values.

The performance of the method is compared to six other skin detection algorithms. Three of them use fixed boundaries in *RGB* [98], *CbCr* [99] and *HS* [100] color spaces. The fourth is a statistical approach using a mixture of Gaussians in *RGB* space. Note that these methods are particularly designed and fine-tuned to detect skin. The other two methods correspond to the (more generic) fusion schemes proposed by Jacobs [76] and Stokman and Gevers [77]. The same training set is used to train the different detection schemes. A summary of the results is listed in Table 7.7. Further, the results of the Wilcoxon test are shown in Table 7.8. The following conclusions can be derived from these results. First, the learning algorithm outperforms the others in terms of overall performance (quality and effectiveness) except for the *RGB*-based method. Nevertheless, the *RGB*-based, *HS* and *RGB* statistical method provides a better detection rate. This means that these methods provide higher invariance to skin-class variability at the expense of having low discriminative power. The learning method outperforms all the others, including the *RGB*-based method, in terms of detection accuracy. That is, the ratio, between skin pixels that are correctly classified and the number of skin pixels retrieved provided by our method, is higher. This is due to the resulting distribution of skin pixel values (Fig. 7.5). However, the overall performance of the method is lower than the *RGB* method because of the high variability in both skin appearance and lighting variations. This yields a data distribution in each view that is not unimodal except for very small patches of skin. Further, unobserved lighting conditions and user appearance (during training) shift the skin distribution (Fig. 7.4c) reducing the performance. Furthermore, although the *RGB*-based method fails in the presence of low intensity (due to illumination and shadows) there are only a few instances of this type, that is, only 3% of images in the image dataset showing severe intensity and shadow changes.

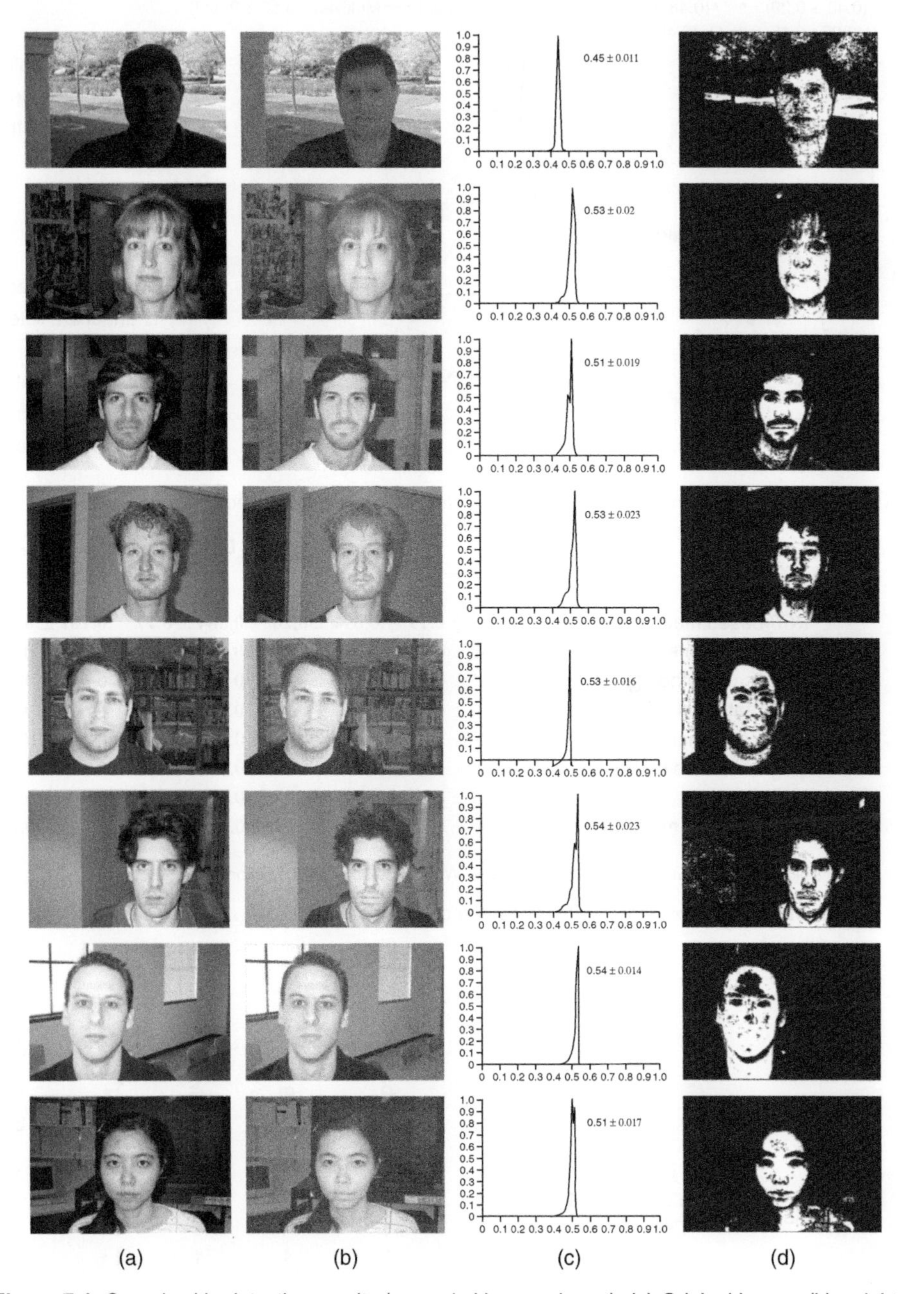

Figure 7.4 Generic skin detection results (second skin experiment). (a) Original image, (b) weighted combination of color models, (c) distribution of skin pixel values in the image, (d) skin detection results. *Source:* Reprinted with permission, © 2010 Springer.

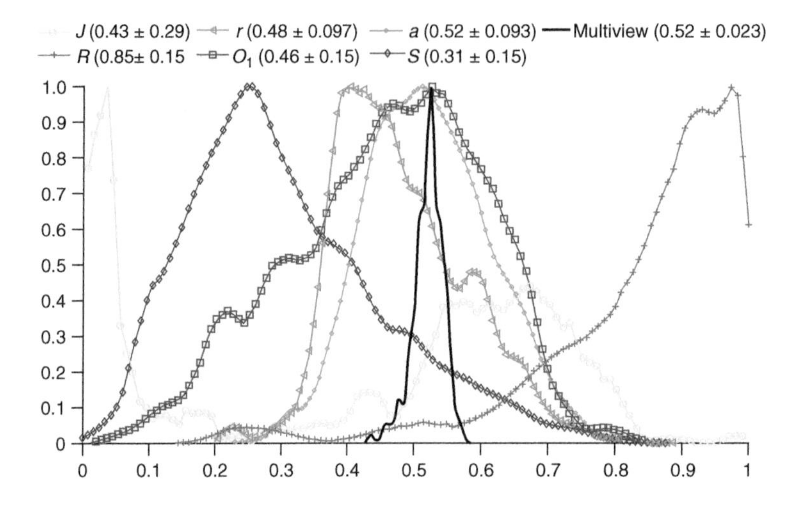

Figure 7.5 Distribution of all skin pixel values in the data set for different color models. Mean and standard deviation for each channel are listed in the legend. *Source:* Reprinted with permission, © 2010 Springer.

Table 7.7 Performance of different detection algorithms on Caltech face database.

	$\hat{g}$	Detection accuracy	Detection rate	F
RGB-based method [98, 101]	$\mathbf{0.640 \pm 0.19}$	0.694 ± 0.20	$\mathbf{0.884 \pm 0.17}$	$\mathbf{0.761 \pm 0.17}$
CbCr-based method [99]	0.259 ± 0.18	0.309 ± 0.21	0.548 ± 0.31	0.379 ± 0.23
HS-based method [100]	0.443 ± 0.21	0.514 ± 0.21	0.807 ± 0.28	0.585 ± 0.21
RGB Statistical [102]	0.510 ± 0.23	0.635 ± 0.23	0.723 ± 0.28	0.643 ± 0.22
Minimum variance [76]	0.189 ± 0.03	0.195 ± 0.03	0.190 ± 0.02	0.318 ± 0.05
Single-view fusion [77]	0.314 ± 0.24	0.365 ± 0.26	0.636 ± 0.34	0.430 ± 0.27
Multiview[a]	0.410 ± 0.23	0.703 ± 0.18	0.497 ± 0.20	0.550 ± 0.15
Multiview (our method)	0.589 ± 0.18	$\mathbf{0.756 \pm 0.22}$	0.718 ± 0.11	0.713 ± 0.17

[a]Bold values indicate maximum performance. Without color model selection.

Table 7.8 Wilcoxon test for the road detection experiment.

		RGB based	**CbCr based**	**HS based**	**RGB Statistical**	**Minimum Variance**	**Single View**	**Multi–View[a]**
Multi–view	$\hat{g}$	−1	1	1	1	1	1	1
	DA	1	1	1	1	1	1	−1
	DR	−1	1	−1	−1	1	1	1
	F	−1	1	1	1	1	1	1

A positive value indicates that the learning method outperforms the others.
Negative values indicate that our method does not perform significantly better. Bold values indicate when the learning method outperforms the others.
[a]Wilcoxon Test for the Skin Detection Experiment.

7.4.3 Road Detection in Video Sequences

The other application discussed in this chapter is RD. To detect roads in video, a sequence of more than 800 images is considered to analyze the dynamic nature of observations. This video sequence is recorded using an onboard camera. The aim is to detect the (not occluded) road in front of a moving vehicle using a color camera. The images used include different backgrounds, the occurrence of occluding and cluttered objects (vehicles), and different road appearances under varying illumination changes.

The training set consists of 15 different road patches that are manually selected from 15 different (randomly) selected images. The selection process avoids successive image indexes. These patches contain different illumination (i.e., shadows and highlights) and they represent less than 0.053% of the total amount of road pixels within the sequence. The selection of the most suitable color models is executed by the PCA procedure described in Section 7.3. The obtained weights for the ensemble are listed in Table 7.6 and shows a dominant weight for the invariant color model corresponding to an achromatic surface independent of illumination changes (e.g., sun casts and shadows), that is, roads.

Furthermore, the sequential nature of the data is also considered. Thus, once the optimal ensemble for the road is computed, it is adapted considering only images close in time. That is, the procedure described in Section 7.2 is used to estimate the input parameters ($E[\xi_1], \ldots, E[\xi_N]$ and Σ) to the optimization process. To estimate them, a temporal buffer is used considering the central value of the detected road in each frame for each selected color model (Fig. 7.6). Hence, the assumption is that the correlation between color models holds over time. To avoid possible outliers (false positives in the current result), robust statistics are used. The decay factor λ (Eq. 7.12) is empirically fixed to 0.5. Then, the optimal ensemble is recomputed at each frame considering these new values of $E[\xi_1], \ldots, E[\xi_N]$ and Σ.

To evaluate the improvement in performance when temporal information is taken into account, the error between the expected value of the road and the current value for two different updating techniques is considered (Fig. 7.7). The updating techniques are sample and hold, and *EWMA*. The former uses a fixed optimal ensemble estimated using training samples over all the image sequences. The latter uses a decay factor ($\lambda = 0.5$) to update the optimal ensemble accordingly to new data available. As shown in Figure 7.7, the error is significantly lower when the ensemble is updated over time. That is, if the ensemble is adapted considering new data available, then the road data distribution is modified accordingly to the new lighting conditions, leading to more accurate results. However, using a fixed ensemble (sample and hold), the variations due to unobserved (not in the training set) lighting conditions or road appearances lead to shifted road data distributions. Further, the analysis of the tracking error (ψ) and historical *SR* (S_h) for both methods (Table 7.9) suggests that the adaptive method has better performance in terms of following the road central value over all the images in the database.

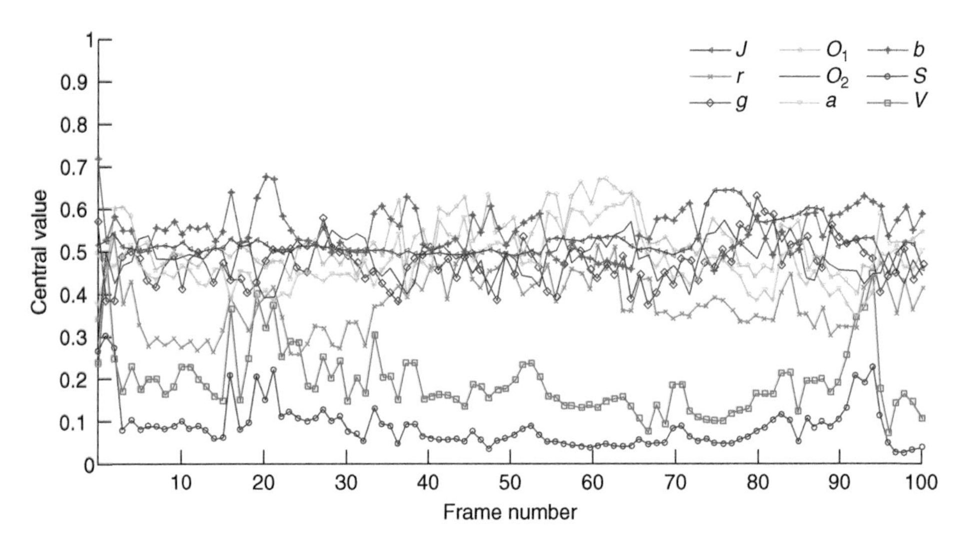

Figure 7.6 Central values of observations are estimated using robust statistics on results at each frame. These values are used to recompute the optimal ensemble.

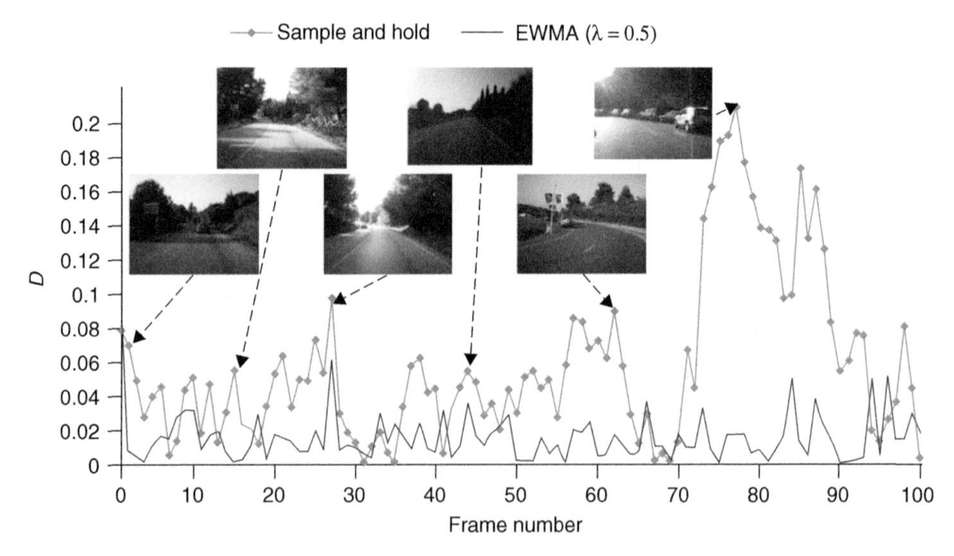

Figure 7.7 Comparison between errors in the expected road value at each frame. For clarity reasons 1 in every 100 frames are selected from the original video sequence). The error is higher when unobserved lighting conditions appear. *Source:* Reprinted with permission, © 2010 Springer.

Table 7.9 Tracking error (ψ) and historical *sr* (S_h) for the road experiment.

	ψ	S_h
Sample and hold	0.002212	0.001
EWMA ($\lambda = 0.5$)	**0.000257**	**8.331**

The bold values indicate the maximum performance.
The lower the ψ the better the performance, whereas the higher S_h the better the performance.

Furthermore, the video sequence is processed using three other methods. The first algorithm is the *HSI* RD algorithm proposed in Reference 103 and used in Reference 104. The *HSI* color space is used to process generic outdoor scenes under varying illumination [105, 106]. The second algorithm is the illuminant-invariant algorithm presented in Reference 91. The third algorithm is based on *2D* histograms in *rg* space [107]. Further, the two fusion methods proposed in References 76 and [77] are considered. Finally, three different instances of the learning method are considered: sample and hold without color model selection, sample and hold method using color model selection, and over time adaptive method using color model selection. Note that the *HSI* and illuminant-invariant algorithms are based on a frame-by-frame procedure. Further, these algorithms require various parameter settings. For fair comparison, a brute force approach is applied. In this way, a set of images is processed and evaluated using all possible values within the range of each parameter. The optimal set of parameter values is the one that maximizes the average performance. All algorithms (that need training) are trained using the same road pixels. Finally, all these state-of-the-art algorithms consider that the lowest part of the image corresponds to the road and that it is about 4 m away from the vehicle. Under this consideration, only detected results that are connected with a set of seeds placed at the bottom part of the image are retrieved as road pixels. The same set of seeds is used for all the methods.

The performance of all algorithms is outlined in Table 7.10. Various detection results of the learning method using temporal adaptation are shown in Figure 7.8. Further, the results of the Wilcoxon test are shown in Table 7.11. From the results, it can be concluded that when the learning method is adapted over time it performs significantly better than the others except for the detection accuracy for the *HSI* method and the nonadapted method. Results provided by the learning method are slightly overdetected compared to those provided by these two methods. However, regarding the overall performance (quality and effectiveness), the learning method performs best. This means that the learning algorithm achieves a higher trade-off between invariance (detection rate) and discriminative power (detection accuracy).

Table 7.10 Performance of different detection algorithms on road database.

	$\hat{g}$	Detection accuracy	Detection rate	F
***HSI*-based RD** [103]	0.673 ± 0.12	0.927 ± 0.12	0.729 ± 0.15	0.798 ± 0.09
Invariant RD [91]	0.798 ± 0.13	0.901 ± 0.15	0.866 ± 0.10	0.870 ± 0.10
rg model based [107]	0.272 ± 0.19	0.770 ± 0.23	0.410 ± 0.34	0.391 ± 0.29
Minimum variance [76]	0.137 ± 0.22	0.237 ± 0.30	0.193 ± 0.31	0.187 ± 0.28
Single-view fusion [77]	0.680 ± 0.14	0.936 ± 0.02	0.716 ± 0.15	0.801 ± 0.10
Multiview (our method)[a]	0.801 ± 0.36	0.714 ± 0.10	0.826 ± 0.05	0.746 ± 0.07
Multiview (our method)[b]	0.810 ± 0.09	$\mathbf{0.976 \pm 0.04}$	0.828 ± 0.09	0.893 ± 0.05
Multiview (our method)[c]	$\mathbf{0.915 \pm 0.06}$	0.963 ± 0.05	$\mathbf{0.949 \pm 0.05}$	$\mathbf{0.954 \pm 0.03}$

Bold values indicate the maximum performance.
[a]Without color model selection.
[b]Without temporal adaptation.
[c]With temporal adaptation.

Table 7.11 Wilcoxon test for the skin detection experiment.

		***HSI* RD**	**Invariant RD**	***rg* model based**	**Minimumvariance**	**Single view**	**Multi–view**[a]	**Multi–view**[b]
Multi–view	$\hat{g}$	−1	1	1	1	1	1	1
	DA	1	1	1	1	1	1	1
	DR	−1	1	−1	−1	1	1	1
	F	−1	1	1	1	1	1	1

Positive values indicate that the learning method performs significantly better. Negative values indicate that the method does not outperform the others. Bold values indicate when the learning method outperforms the others.
[a]Without color model selection.
[b]Without temporal adaptation.

The results reveal that the method produces false negatives (undetected road pixels) when highlights or lane markings are present. Further, the algorithm takes a few images to recover when an abundance of false positives are present. Hence, when the input data to estimate the ensemble is biased then the performance drops. This could be improved by adding more constraints (such as unimodality test) to the new data available. Furthermore, the performance may be improved by clustering detected road pixels to distinguish different lighting conditions in the same frame.

Figure 7.8 Results of the learning algorithm to detect roads. *Source:* Reprinted with permission, © 2010 Springer.

7.5 Summary

In this chapter, photometric invariance has been derived by learning from color models to obtain diversified color invariant ensembles using only positive examples. A method for combining color models is discussed to provide a multiview approach to minimize the estimation error. In this way, the method is robust to data uncertainty and produces properly diversified color invariant ensembles. Further, the learning method is extended to deal with temporal data by predicting the evolution of observations over time.

PART III

COLOR CONSTANCY

8 Illuminant Estimation and Chromatic Adaptation

Explicitly correcting an image for the color of the light source, producing a transformed version of the input image, is called *color constancy*. Human vision has the natural tendency to correct for the effects of the color of the light source, [108–111], but the mechanism involved with this ability is not yet fully understood. Early work by Land and McCann [13, 14, 112] resulted in the retinex theory. This theory posited that both the retina and the cortex are involved in the processing. Many computational models are derived on the basis of this perceptual theory, [113–115]. However, computational models can still not fully explain the observed color constancy of human observers. Kraft and Brainard [116] tested the ability of several computational theories to account for human color constancy, but found that each theory leaves considerable residual constancy. In other words, without the specific cues corresponding to the computational models, humans are still, to some extent, color constant [116]. Alternatively, observations on human color constancy cannot be readily applied to computational models either: Golz and Macleod [117, 118] showed that chromatic scene statistics influence the accuracy of human color constancy, but when mapped to computational models, the influence was found to be very weak at best [119]. Recent advances that are not pursued further in this book include the suggestion from computational color constancy that the optimal approach for specific images is based on the statistics of the scene [120, 121], and the suggestion from human color constancy that color memory, possibly in addition to contextual clues, could play an important role [122–125].

Although it would be interesting to bring the recent advances in human color constancy closer to the computational level, or to map the computational advances

Color in Computer Vision: Fundamentals and Applications, First Edition.
Theo Gevers, Arjan Gijsenij, Joost van de Weijer, and Jan-Mark Geusebroek.

onto human explanations, the focus in this part is on computational color constancy algorithms rather than perceptual plausible models. Consider, for example, the images in Figure 8.1. These images show the effects that different light sources can have on the perception of a scene. The goal of computational color constancy algorithms is to apply a correction to the target images so that they are identical to the canonical image (i.e., the image that is taken under a neutral or white light source).

Figure 8.1 An illustration of the effects of different light sources on the measured image. In these images, the only variable is the color of the light source. The purpose of computational color constancy algorithms is to correct these images so that they visually appear to be the same. *Source:* Images rendered using data taken from Reference 126.

The approach that is typically used for color constancy, and is also followed in this part, is outlined in Figure 8.2. For an input image that is recorded under unknown illumination, the chromaticity of the light source is estimated. Then, in the second step, this chromaticity is used to transform the input image so that it appears to be taken under a canonical light source. Finally, the output image, which is free of any deviations caused by the color of the light source, is returned. The methods discussed in this part are involved with the first step (illuminant estimation). The second step is called *chromatic adaptation*, and is described in brief in Section 8.2, but is not further explored here. Note that all illuminant estimation algorithms discussed in Chapters 9–11 are based on the assumption that the illuminant is spatially uniform, that is, it is assumed that the chromaticity of the light source is the same in every location of the image. Although it is easy to

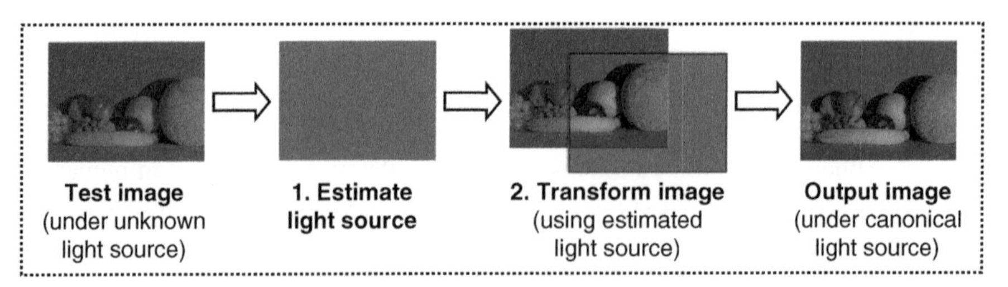

Figure 8.2 In the first step, the illuminant of an input image that is recorded under unknown illumination is estimated. Then, in the second step, this illuminant is used to correct the input image to generate an output image.

think of scenarios where this assumption is violated, for example, indoor images depicting multiple rooms with spectrally different light sources, or outdoor images showing parts of the scene in shadow and other parts in bright sunlight, the bulk of the images meets the single light source assumption. Relatively few methods that are able to deal with multiple light sources have been proposed. For instance, Finlayson et al. [127] and Barnard et al. [128] propose a retinex-based approach that explicitly assumes that surfaces that are illuminated by multiple light sources exist in the scene. Another retinex-based approach [129] uses stereo images to derive 3D information on the surfaces that are present in the images, to be able to distinguish material transitions from local light color changes, but the stereo information is often not available and is not trivial to obtain. Ebner [130] also proposed a method that is based on the assumption that the illuminant transition is smooth. This method uses the local space average color for local estimation of the illuminant by convolving the image with a kernel function (e.g., a Gaussian or exponential kernel). Finally, in Reference 131, human interaction is employed to specify locations in images that are illuminated by different light sources. All these methods are based on the assumption that the illuminant color smoothly varies from one color into the other. Although this line of research is interesting, it is not pursued further in this part.

8.1 Illuminant Estimation

Recall from Chapter 3 that under the assumption of Lambertian reflectance, the images values of an image $\mathbf{f}^{RGB} = (R, G, B)^T$ depend on the color of the light source $e(\lambda)$, the surface reflectance properties $s(\lambda, \mathbf{x})$ and the camera sensitivity function $\rho^c (c \in \{R, G, B\})$:

$$f^c(\mathbf{x}) = m^b(\mathbf{x}) \int_{\omega} e(\lambda)\rho^c(\lambda)s(\lambda, \mathbf{x})\mathrm{d}\lambda, \tag{8.1}$$

where ω is the visible spectrum, λ is the wavelength of the light, $\mathbf{x}$ is the spatial coordinate, and m^b is the Lambertian shading term that contributes to the overall light reflected at location $\mathbf{x}$. Illuminant spectra, camera sensitivity functions, and surface reflectance functions are usually given by m discrete samples within the visible spectrum ω. Hence, the continuous Equation 8.1 is often replaced by the digital form:

$$f^c(\mathbf{x}) = m^b(\mathbf{x}) \sum_{i=1}^{m} e(\lambda_i)\rho^c(\lambda_i)s(\lambda_i, \mathbf{x})\Delta\lambda, \tag{8.2}$$

where λ_i are the sample points and $\Delta\lambda$ is the sample width.

In order to create a more realistic model while still adhering to the simple assumptions of the Lambertian reflectance model, Shafer [26] proposes to add a "diffuse" light term. The diffuse light is considered to have low intensity and to be coming from all directions in an equal amount:

$$f^c(\mathbf{x}) = \int_\omega e(\lambda)\rho^c(\lambda)s(\lambda,\mathbf{x})d\lambda + \int_\omega a(\lambda)\rho^c(\lambda), \qquad (8.3)$$

where $a(\lambda)$ is the term that models the diffuse light. Using this equation, objects under daylight can be modeled more accurately, since daylight consists of both a point source (the sun) and diffuse light coming from the sky. However, the assumption that diffuse light is equal in all directions does not often hold in practice. A more realistic approximation is to consider the diffuse light to be dependent on the position in the image, according to

$$f^c(\mathbf{x}) = \int_\omega e(\lambda)\rho^c(\lambda)s(\lambda,\mathbf{x})d\lambda + \int_\omega \overline{a(\lambda,\mathbf{x})}\rho^c(\lambda), \qquad (8.4)$$

where we assume the dependence of the position to be low-frequent, which is indicated by the overline.

Typically, illuminant estimation is involved with estimating the illuminant using Equation 8.1. Only a few methods are based on Equation 8.4, and these are discussed in Chapter 10. However, even assuming the simplified version of image formation as given by Equation 8.1, the problem of illuminant estimation remains difficult to solve. As can be seen in this equation, the image values $\mathbf{f}$ are dependent on the intrinsic surface properties s as well as the spectrum of the scene illuminant. When either of the two changes, so do the image values, while in fact we only want to observe a change in image values when the surface properties change. For instance, in fundamental computer vision tasks such as image and scene segmentation, and object recognition and tracking, changing illumination could cause major difficulties if it is not properly taken into account. Further, color constancy is a fundamental process in the formation of images using a digital camera: if image colors are not suitably color corrected during capturing of the digital images, the image will not match the photographer's observation of the scene. Hence, in order to obtain a stable digital reproduction of a scene, it is important to dismiss the effects of the light source as much as possible. To still maintain a natural representation of the scene, the effects of the colored light source are usually replaced by a canonical light source, that is, a white or neutral light source.

Note that if an image consists of n different surfaces, then using Equation 8.2 results in $3n$ knowns: one known value for every color channel and for every surface (assuming an image is composed of three color channels, for example, red R, green G, and blue B). From these known image values, we want to recover the n true surface reflectance properties as well as the single illuminant. However, since the illuminant, surfaces, and camera sensitivity function are given by m discrete samples, the number of parameters to solve for totals $m(n+1)$. When assuming $m \geq 3$, it becomes clear that the number of knowns is outnumbered by the number of unknowns: $3n < m(n+1)$, regardless of the number of different surfaces in an image. Even when we assume that the digital spectrum of light source does not fully need to be recovered (but rather the representation of the image under a neutral light source using chromatic adaptation techniques described in the next section), the number of unknowns $(3n+3)$ still outnumbers the number of knowns. Hence, it becomes obvious that illuminant estimation is an underconstrained problem that cannot be solved without further assumptions.

8.2 Chromatic Adaptation

After the color of the light source is estimated, the image has to be transformed. This transformation will change the appearance of all colors, so that the image appears to be recorded under a white light source (e.g., D65). This can be achieved by *chromatic adaptation*, [132]. Most adaptation transforms are modeled using a linear scaling of the cone responses, and the simplest form independently scales the three color channel [33]:

$$\begin{pmatrix} R_c \\ G_c \\ B_c \end{pmatrix} = \begin{pmatrix} d_R & 0 & 0 \\ 0 & d_G & 0 \\ 0 & 0 & d_B \end{pmatrix} \begin{pmatrix} R_e \\ G_e \\ B_e \end{pmatrix}, \tag{8.5}$$

where $d_i = \frac{e_i}{\sqrt{3 \cdot (e_R^2 + e_G^2 + e_B^2)}}$, $i \in \{R, G, B\}$. Even though this model is merely an approximation of illuminant change and might not accurately be able to model photometric changes due to disturbing effects such as highlights and interreflections, it is widely accepted as the color correction model [133, 134, 50] and it underpins many color constancy algorithms described in the subsequent chapters.

A more accurate representation would be to first sharpen the cone responses before transformation, for example, Bradford transform [135] or CMCCAT2000 [136]. The latter is defined as

$$\begin{pmatrix} X_c \\ Y_c \\ Z_c \end{pmatrix} = M_{CMC}^{-1} \begin{pmatrix} d_X & 0 & 0 \\ 0 & d_Y & 0 \\ 0 & 0 & d_Z \end{pmatrix} M_{CMC} \begin{pmatrix} X_e \\ Y_e \\ Z_e \end{pmatrix}, \tag{8.6}$$

where d_X, d_Y and d_Z are computed from the tristimulus values of the true and the white illuminants, by multiplying the corresponding XYZ vectors with M_{CMC}. The matrix M_{CMC} is given by Li et al. [136]:

$$M_{CMC} = \begin{pmatrix} 0.7982 & 0.3389 & -0.1371 \\ -0.5918 & 1.5512 & 0.0406 \\ 0.0008 & 0.0239 & 0.9753 \end{pmatrix}. \tag{8.7}$$

Note that this transform is defined for tristimulus values *XYZ*, so an *RGB* image will have to be converted to *XYZ* before applying this transform, and back to *RGB* after the transform.

As stated earlier, under some conditions the diagonal model is too strict. Such situations can be troublesome for color constancy algorithms based on this model. To overcome this, Finlayson et al. [34] accounted for this shortcoming by adding an offset term to the diagonal model, resulting in the diagonal-offset model:

$$\begin{pmatrix} R^c \\ G^c \\ B^c \end{pmatrix} = \begin{pmatrix} \alpha & 0 & 0 \\ 0 & \beta & 0 \\ 0 & 0 & \gamma \end{pmatrix} \begin{pmatrix} R^u \\ G^u \\ B^u \end{pmatrix} + \begin{pmatrix} o_1 \\ o_2 \\ o_3 \end{pmatrix}. \tag{8.8}$$

Deviations from the diagonal model are reflected in the offset term $(o_1, o_2, o_3)^T$. Ideally, this term will be zero, which is the case when the diagonal model is valid.

Interestingly, by means of the offset, the diagonal model also takes diffuse lighting into account as approximated by Equation 8.3. To obtain position-dependent diffuse lighting of Equation 8.4, the following model called *local-diagonal-offset model* can be used:

$$\begin{pmatrix} R^c \\ G^c \\ B^c \end{pmatrix} = \begin{pmatrix} \alpha & 0 & 0 \\ 0 & \beta & 0 \\ 0 & 0 & \gamma \end{pmatrix} \begin{pmatrix} R^u \\ G^u \\ B^u \end{pmatrix} + \begin{pmatrix} \overline{o_1(\mathbf{x})} \\ \overline{o_2(\mathbf{x})} \\ \overline{o_3(\mathbf{x})} \end{pmatrix}. \tag{8.9}$$

This model is more robust against deviations from the diagonal model (e.g., saturated colors), diffuse light (assuming the dependence of the position is low-frequent) and veiling illumination. Methods using this modified version of the diagonal model are described in Chapter 10.

9 Color Constancy Using Low-level Features

The first type of illuminant estimation algorithms discussed in this book are static methods, or methods that are applied to input images with a fixed parameter setting. Two subtypes are distinguished: a) methods that are based on low-level statistics and b) methods that are based on the physics-based dichromatic reflection model.

9.1 General Gray-World

The best-known and most often used assumption of this type is the gray-world assumption [137]: *the average reflectance in a scene under a neutral light source is achromatic*. In the original work, the hypothesis is used to derive that the average reflectance for short-wave, middle-wave and long-wave regions is equal, but a stronger definition of achromatic reflectance of a scene is often employed ([139, 138]):

$$\frac{\int s(\lambda, \mathbf{x})d\mathbf{x}}{\int d\mathbf{x}} = g(\lambda) = k, \tag{9.1}$$

which avoids making further assumptions. The constant k is between 0 for no reflectance (black) and 1 for total reflectance (white) of the incident light, and the integral is over the domain of the scene. For such a scene with achromatic

Color in Computer Vision: Fundamentals and Applications, First Edition.
Theo Gevers, Arjan Gijsenij, Joost van de Weijer, and Jan-Mark Geusebroek.

reflectance, it holds that the reflected color is equal to the color of the light source, since

$$\frac{\int f^c(\mathbf{x})d\mathbf{x}}{\int d\mathbf{x}} = \frac{1}{\int d\mathbf{x}} \iint_{\omega} e(\lambda)s(\lambda,\mathbf{x})\rho^c(\lambda)d\lambda d\mathbf{x}, \tag{9.2}$$

$$= \int_{\omega} e(\lambda)\rho^c(\lambda)\left(\frac{\int s(\lambda,\mathbf{x})d\mathbf{x}}{\int d\mathbf{x}}\right)d\lambda, \tag{9.3}$$

$$= k\int_{\omega} e(\lambda)\rho^c(\lambda)d\lambda = ke^c, \tag{9.4}$$

where the theorem of Fubini is used to exchange the order of integration. The normalized light source color is computed with $\hat{\mathbf{e}} = (\hat{e}^R, \hat{e}^G, \hat{e}^B)^T = k\mathbf{e}/|k\mathbf{e}|$.

Alternatively, instead of computing the average color of all pixels, it has been shown that segmenting the image and computing the average color of all segments may improve the performance of the gray-world algorithm [140, 141]. This preprocessing step can lead to improved results because the gray-world is sensitive to large uniformly colored surfaces, as this often leads to scenes where the underlying assumption fails. Segmenting the image before computing the scene average color will reduce the effects of these large uniformly colored patches. Related methods attempt to identify the intrinsic gray surfaces in an image, that is, they attempt to find the surfaces under a colored light source that would appear gray if rendered under a white light source [142–144]. When *accurately* recovered, these surfaces contain a strong clue for the estimation of the light source. Finally, van de Weijer et al. [145] proposed a method using similar principles, based on a hypothesis they call the *green-grass hypothesis*: the average reflectance of a semantic class in an image is equal to the average reflectance of the semantic topic in the database. This hypothesis is captured by the following equation:

$$\sum_{\mathbf{x}\in T^s} \mathbf{f}(\mathbf{x}) = k \operatorname{diag}\left(\mathbf{d}^s\right)\mathbf{e}^s, \tag{9.5}$$

$$\mathbf{d}^s = \sum_{\mathbf{x}\in D^s} \mathbf{F}(\mathbf{x}), \tag{9.6}$$

where T^s is the set of indexes to pixels in image $\mathbf{f}$ assigned to semantic topic s, $\mathbf{F}$ is the collection of all pixels in the training data set, D^s are the indexes to all pixels in the training data set assigned to semantic topic s, and $\mathbf{e}^s$ is the estimate of the illuminant color based on topic s. Pixels in any input image are first classified to either of the semantic classes considered in the training set. Then, using Equation 9.5 an illuminant hypothesis is cast. Finally, using the semantic likelihood of the classified regions, the hypotheses are combined into one final estimate (e.g., by selecting the hypothesis with the highest probability according to the likelihood of the semantic content).

Another well-known assumption is the white-patch assumption [14]: *the maximum response in the RGB channels is caused by a perfect reflectance*. A surface with perfect reflectance properties will reflect the full range of light that it captures. Consequently, the color of this perfect reflectance is exactly the color of the light source. In practice, the assumption of perfect reflectance is alleviated by considering the color channels separately, resulting in the *max-RGB* algorithm. This method estimates the illuminant by computing the maximum response in the separate color channels:

$$\max_{\mathbf{x}} f^c(\mathbf{x}) = ke^c. \tag{9.7}$$

It should be noted that the max-*RGB* method does not require the maxima of the separate channels to be on the same location. An illustration of this can be seen in Figure 9.1. The maxima of the three color channels of the image with the ball happen to coincide and correspond to a white patch. However, the maxima of the three color channels of the image with the papers come from three different pixels. Since the names max-*RGB* and white-patch are both often used to denote the same algorithm, this observation can be confusing and should be taken into account when working with this algorithm.

Related algorithms apply some sort of smoothing to the image, before the illuminant estimation [130, 146]. This preprocessing step has similar effects on

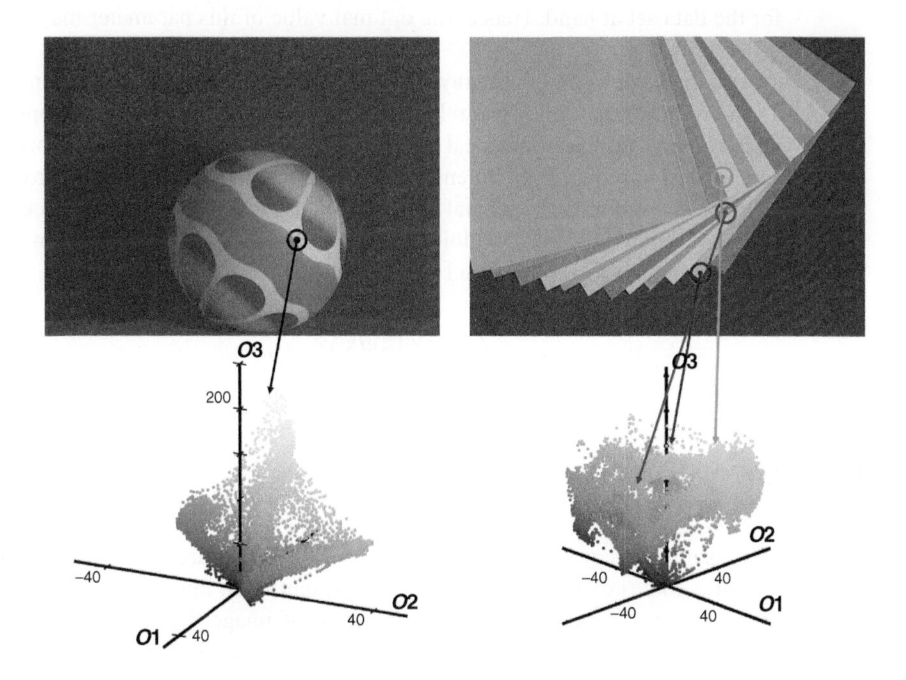

Figure 9.1 Two illustrations of the max-*RGB* method. The image with the papers shows that the maximum responses in the three color channels do not have to correspond to a white patch nor do they even have to correspond to the same pixel. *Source:* The top images are taken from Reference 44.

the performance of the white-patch algorithm as segmentation on the gray-world. In this case, the effect of noisy pixels (with an accidental high intensity) is reduced, improving the accuracy of the white-patch method. An additional advantage of the local space average color [130] (LSAC) method is that it can provide a pixel-wise illuminant estimate. Consequently, it does not require the image to be captured under a spectrally uniform light source. An analysis of the *max-RGB* algorithm is presented in References 147, 148, where it is shown that the dynamic range of an image, in addition to the preprocessing strategy, can have a significant influence on the performance of this method.

In Reference 138, the white-patch and the gray-world algorithms are shown to be special instantiations of the more general Minkowski framework:

$$\mathcal{L}^c(p) = \left(\frac{\int (f^c)^p(\mathbf{x})d\mathbf{x}}{\int d\mathbf{x}} \right)^{\frac{1}{p}} = ke^c. \tag{9.8}$$

Substituting $p = 1$ in Equation 9.8 is equivalent to computing the average of $\mathbf{f}(\mathbf{x})$, that is, $\mathcal{L}(1) = (\mathcal{L}^R(1), \mathcal{L}^G(1), \mathcal{L}^B(1))^T$ equals the gray-world algorithm. When $p = \infty$, Equation 9.8 results in computing the maximum of $\mathbf{f}(\mathbf{x})$, that is, $\mathcal{L}(\infty)$ equals the white-patch algorithm. In general, to arrive at a proper value, p is tuned for the data set at hand. Hence, the optimal value of this parameter may vary for different data sets.

As a final extension of the gray-world algorithm, local averaging is considered. The norm computation as given by Equation 9.8 is a global averaging operation, which ignores the important local correlation between pixels. This local correlation can be used to reduce the influence of noise. Local smoothing as a preprocessing step was proved to be beneficial for color constancy algorithms, as discussed in Barnard's study [141]. To exploit this local correlation, a local smoothing with a Gaussian filter, $\mathbf{G}^\sigma$ is introduced [146], with standard deviation σ:

$$\left(\frac{\int (f^c)^p(\mathbf{x})dx}{\int d\mathbf{x}} \right)^{\frac{1}{p}} = ke^c. \tag{9.9}$$

9.2 Gray-Edge

The assumptions of the above color constancy methods are based on the distribution of colors (i.e., pixel values) that are present in an image. The incorporation of higher order image statistics (in the form of image derivatives) is proposed in Reference 139, resulting in the *gray-edge* hypothesis: *The average of the reflectance differences in a scene is achromatic*:

$$\frac{\int |s^\sigma_{\mathbf{x}}(\lambda, \mathbf{x})d\mathbf{x}}{\int d\mathbf{x}} = g(\lambda) = k. \tag{9.10}$$

The subscript $\mathbf{x}$ indicates the spatial derivative at scale σ. With the gray-edge assumption, the light source color can be computed from the average color derivative in the image given by

$$\frac{\int (f^c)_{\mathbf{x}}(\mathbf{x})\mathrm{d}\mathbf{x}}{\int \mathrm{d}\mathbf{x}} = \frac{1}{\int \mathrm{d}\mathbf{x}} \iint_{\omega} e(\lambda)s_{\mathbf{x}}(\lambda,\mathbf{x})\rho^c(\lambda)\mathrm{d}\lambda\mathrm{d}\mathbf{x}, \tag{9.11}$$

$$= \int_{\omega} e(\lambda)\rho^c(\lambda)\left(\frac{\int s_{\mathbf{x}}(\lambda,\mathbf{x})\mathrm{d}\mathbf{x}}{\int \mathrm{d}\mathbf{x}}\right)\mathrm{d}\lambda, \tag{9.12}$$

$$= k\int_{\omega} e(\lambda)\rho^c(\lambda)\mathrm{d}\lambda = ke^c, \tag{9.13}$$

where $|(f^c)_{\mathbf{x}}(\mathbf{x})| = |C_{\mathbf{x}}(\mathbf{x})|$ and $C = \{R, G, B\}$. The gray-edge hypothesis originates from the observation that the color derivative distribution of images forms a relatively regular, ellipsoid-like shape, of which the long axis coincides with the light source [149]. In Figure 9.2, the color derivative distribution is depicted for three images. The color derivatives are rotated to the opponent color spaces as follows:

$$O1_{\mathbf{x}} = \frac{R_{\mathbf{x}} - G_{\mathbf{x}}}{\sqrt{2}}, \tag{9.14}$$

$$O2_{\mathbf{x}} = \frac{R_{\mathbf{x}} + G_{\mathbf{x}} - 2B_{\mathbf{x}}}{\sqrt{6}}, \tag{9.15}$$

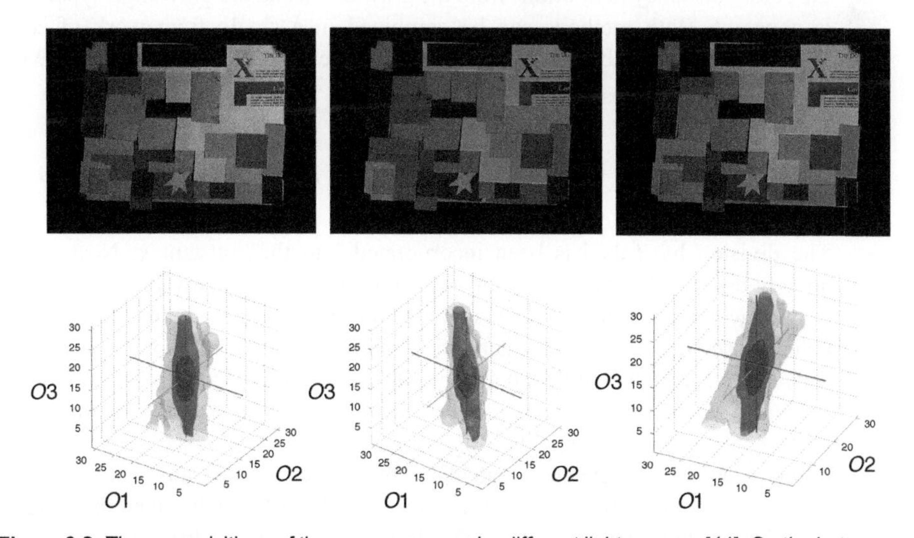

Figure 9.2 Three acquisitions of the same scene under different light sources [44]. On the bottom row, the color derivative distributions are shown, where the axes are the opponent color derivatives and the surfaces indicate derivative values with equal occurrence and darker surfaces indicating a more dense distribution. Note the shift in the orientation of the distribution of the derivatives with the changing of the light source. *Source:* Reprinted with permission, © 2007 IEEE.

$$O3_{\mathbf{x}} = \frac{R_{\mathbf{x}} + G_{\mathbf{x}} + B_{\mathbf{x}}}{\sqrt{3}}. \tag{9.16}$$

In the opponent color space, $O3$ coincides with the white light direction. For the scene rendered under white light (the leftmost image), the distribution of the derivatives is centered along the $O3$ axis, that is, the white-light axis. Once the color of the light source is changed, as in the images in the center and the right, the distribution of the color derivatives no longer coincides with the white-light axis. In other words, color constancy based on the gray-edge assumption can be interpreted as skewing the color derivative distribution such that the average derivative is in the $O3$ orientation.

Similar to the gray-world-based color constancy methods, the gray-edge hypothesis can be adapted to incorporate the Minkowski norm:

$$\left(\frac{\int |(f^c)^{\sigma}_{\mathbf{x}}(\mathbf{x})|^p d\mathbf{x}}{\int d\mathbf{x}} \right)^{\frac{1}{p}} = ke^c. \tag{9.17}$$

Color constancy based on this equation assumes that the pth *Minkowski norm of the derivative of the reflectance in a scene is achromatic*. Two special cases are distinguished. For $p = 1$, the illuminant is derived by a normal averaging operation over the derivatives of the channels. For $p = \infty$, the illuminant is computed from the maximum derivative in the scene. The resemblance between the color constancy derivations from the gray-world and the gray-edge hypothesis is apparent. Both methods can be combined in a single framework of color constancy methods based on low-level features derived from the following general hypothesis:

$$\left(\int \left| \frac{\partial^n (f^c)_{\sigma}(\mathbf{x})}{\partial \mathbf{x}^n} \right|^p d\mathbf{x} \right)^{\frac{1}{p}} = k(e^c)^{n,p,\sigma}. \tag{9.18}$$

The division by $\int d\mathbf{x}$ has been incorporated into the constant k. Next to the already discussed hypotheses (gray-world, max-RGB, Minkowski norm and the gray-edge), it is obvious that this framework also includes higher order based color constancy. Higher order derivatives have correspondences with the center-surround mechanisms of the human eyes for color constancy such as exploited in the well-known center-surround retinex algorithm, eg., [113, 150]. The influence of the color intensities could be weighted according to their distance to the center of the receptive field generally calculated by a difference of Gaussian function.

The illuminant estimation of Equation 9.18 describes a framework for low-level-based illuminant estimation. This framework produces different estimations for the illuminant color based on three variables:

1. The order, n, of the image structure is the parameter determining if the method is a gray-world or a gray-edge algorithm. The gray-world methods

are based on the *RGB* values, whereas the gray-edge methods are based on the spatial derivatives of order n. Usually, higher-order-based color constancy methods up to order $n = 2$ are investigated.

2. The Minkowski-norm p that determines the relative weights of the multiple measurements from which the final illuminant is estimated. A high Minkowski norm emphasizes larger measurements, whereas a low Minkowski norm equally distributes weights among the measurements.
3. The scale of the local measurements as denoted by σ. For first or higher order estimation, this local scale is combined with the differentiation operation computed with a Gaussian derivative. For zero-order gray-world methods, this local scale is imposed by a Gaussian smoothing operation.

An overview of the instantiations of the illuminant estimation given by the framework of Equation 9.18, which are usually considered, are given in Table 9.1.

Table 9.1 Overview of the different illuminant estimation methods together with their hypotheses. These illuminant estimations are all instantiations of equation 9.18.

Name	Symbol	Equation	Hypothesis
Gray-world	$\mathbf{e}^{0,1,0}$	$\left(\int f^c(\mathbf{x})d\mathbf{x}\right) = ke^c$	The average reflectance in a scene is achromatic
Max-RGB	$\mathbf{e}^{0,\infty,0}$	$\left(\int \lvert f^c(\mathbf{x})\rvert^{\infty}d\mathbf{x}\right)^{\frac{1}{\infty}} = ke^c$	The maximum reflectance in a scene is achromatic
Shades of gray	$\mathbf{e}^{0,p,0}$	$\left(\int \lvert f^c(\mathbf{x})\rvert^{p}d\mathbf{x}\right)^{\frac{1}{p}} = ke^c$	The pth Minkowski norm of a scene is achromatic
General gray-world	$\mathbf{e}^{0,p,\sigma}$	$\left(\int \lvert (f^c)^{\sigma}(\mathbf{x})\rvert^{p}d\mathbf{x}\right)^{\frac{1}{p}} = ke^c$	The pth Minkowski norm is achromatic after local smoothing
Gray-edge	$\mathbf{e}^{1,p,\sigma}$	$\left(\int \lvert (f^c)^{\sigma}_{\mathbf{x}}(\mathbf{x})\rvert^{p}d\mathbf{x}\right)^{\frac{1}{p}} = ke^c$	The pth Minkowski norm of the image derivative is achromatic
Max edge	$\mathbf{e}^{1,\infty,\sigma}$	$\left(\int \lvert (f^c)^{\sigma}_{\mathbf{x}}(\mathbf{x})\rvert^{\infty}d\mathbf{x}\right)^{\frac{1}{\infty}} = ke^c$	The maximum reflectance difference in a scene is achromatic
Second-order gray-edge	$\mathbf{e}^{2,p,\sigma}$	$\left(\int \lvert (f^c)^{\sigma}_{\mathbf{xx}}(\mathbf{x})\rvert^{p}d\mathbf{x}\right)^{\frac{1}{p}} = ke^c$	The pth Minkowski norm of the Second-order derivative is achromatic

An advantage of the color constancy methods based on Equation 9.1 is that they are all based on low computational demanding operations. In fact, the pth Minkowski norm of (smoothed) RGB values or derivatives can be computed extremely fast (even real time on dedicated hardware). Furthermore, the method does not require an image database taken under a known light source for calibration as is necessary for more complex color constancy such as discussed in the subsequent chapters.

After the gray-edge method was introduced, several extensions followed. First, the gray-edge was enhanced with an illuminant constraint by Chen et al. [151]. Further, Chakrabarti et al. [152] explicitly modeled the spatial dependencies between pixels. The advantage of this approach compared to the gray-edge is that it is able to learn the dependencies between pixels in an efficient way, but the training phase does rely on an extensive database of images. Finally, Gijsenij et al. [153] noted that different types of edges might contain various amounts of information. They extended the gray-edge method to incorporate a general weighting scheme (assigning higher weights to certain edges), resulting in the weighted gray-edge. Physics-based weighting schemes are proposed, concluding that specular edges are favored for the estimation of the illuminant. The introduction of these weighting schemes resulted in more accurate illuminant estimates, but at the cost of complexity (both in computation and implementation).

9.3 Physics-Based Methods

Most methods are based on the simpler Lambertian model following Equation 3.23, but some methods adopt the dichromatic reflection model of image formation, following Equation 3.21. These methods use information about the physical interaction between the light source and the objects in a scene, and are called physics-based methods. These approaches exploit the dichromatic model to constrain the illuminants. The underlying assumption is that all pixels of one surface fall on a plane in the *RGB* color space. If multiples of such planes are found, corresponding to various *different* surfaces, then the color of the light source is estimated using the intersection of those planes. Various approaches that use specularities or highlights [154–157] have been proposed. The principle behind such methods is that if pixels are found where the body reflectance factor m_b in Equation 3.21 is (close to) zero, then the color of these pixels are similar or identical to the color of the light source. However, all these methods suffer from some disadvantages: retrieving the specular reflections is challenging and color clipping can occur. The latter effectively eliminates the usability of specular pixels (which are more likely to be clipped than other pixels).

A different physics-based method is proposed by Finlayson and Schaefer [158]. This method uses the dichromatic reflection model to project the pixels of a single surface into chromaticity space. Then, the set of possible light sources is modeled by using the planckian locus of black-body radiators. This planckian locus is intersected with the dichromatic line of the surface to recover the color of the light source. This method, in theory, allows for the estimation of the illuminant even when there is only one surface present in the scene. However, it does require all pixels in the image to be segmented, so that all unique surfaces are identified.

Alternatively, the colors in an image can be described using a multilinear model consisting of several planes simultaneously oriented around an axis defined by the illuminant [159, 160]. This eliminates the problem of presegmentation, but does rely on the observation that a representative color of any given material can be identified. In Reference 161, these requirements are relaxed, resulting in a two Hough transform voting procedure.

9.4 Summary

Color constancy methods discussed in this chapter are methods based on low-level information. These methods are not dependent on training data and the parameters are not dependent on the input image, and are therefore called *static*. Advantages of such methods are a simple implementation (often, merely a few lines of code are required) and fast execution. Further, the accuracy of the estimations can be quite high, provided the parameters are selected appropriately. This last requirement is also one of the biggest weaknesses of such methods, since inaccurate parameter selection can severely reduce the performance. Moreover, the selection of the optimal parameters is quite opaque, especially without prior knowledge of the input data. The physics-based methods discussed suffer less from parameter selection than the framework presented in Equation 9.18, but are also less accurate (even for properly selected parameters).

10 Color Constancy Using Gamut-Based Methods

The gamut mapping algorithm was introduced by Forsyth [162]. It is based on the assumption that *in real-world images, for a given illuminant, one observes only a limited number of colors*. Consequently, any variations in the colors of an image (i.e., colors that are different from the colors that can be observed under a given illuminant) are caused by a deviation in the color of the light source. This limited set of colors that can occur under a given illuminant is called the *canonical gamut* $\mathcal{C}$, and it is found in a training phase by observing as many surfaces under one known light source (called the *canonical illuminant*) as possible.

The flow of the gamut mapping is illustrated in Figure 10.1. In general, a gamut mapping algorithm takes as input an image taken under an unknown light source (i.e., an image of which the illuminant is to be estimated), along with the precomputed canonical gamut (see steps 1 and 2 in Fig. 10.1). The precomputed canonical gamut is obtained by aggregating all colors of the training images into one gamut. The training images are acquired under the same illuminant or corrected so that they appear to be acquired under the same illuminant. The combined set of training colors is called *canonical gamut*. Next, the algorithm consists of three important steps:

1. Estimate the gamut of the unknown light source by assuming that the colors in the input image are representative of the gamut of the unknown light source. So, all colors of the input image are collected in the input gamut $\mathcal{I}$. The gamut of the input image is used as a feature in Figure 10.1.

Color in Computer Vision: Fundamentals and Applications, First Edition.
Theo Gevers, Arjan Gijsenij, Joost van de Weijer, and Jan-Mark Geusebroek.

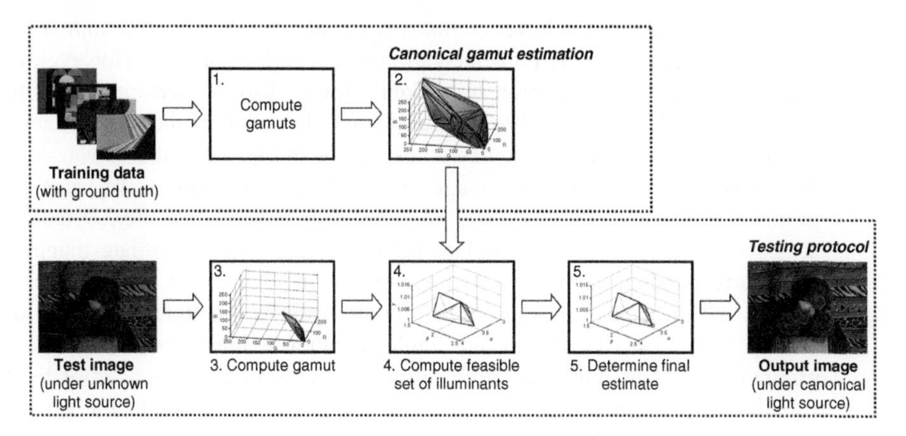

Figure 10.1 Overview of gamut-based algorithms. The training phase consists of learning a model given the features of a wide variety of input images (step 1), resulting in the canonical gamut (step 2). The testing protocol consists of applying the learned model to the computed features of the input image (steps 3 and 4). Finally, one illuminant estimate is selected from the feasible set of illuminants (step 5) and this estimate is used to correct the input image.

2. Determine the set of *feasible mappings* $\mathcal{M}$, that is, all mappings that can be applied to the gamut of the input image and that result in a gamut that lies completely within the canonical gamut. Under the assumption of the diagonal mapping, a unique mapping that converts the gamut of the unknown light source to the canonical gamut exists. However, since the gamut of the unknown light source is simply estimated by using the gamut of *one* input image, in practice several mappings are obtained. Every mapping i in the set $\mathcal{M}$ should take the input gamut completely inside the canonical gamut:

$$\mathcal{M}_i \mathcal{I} \in \mathcal{C}. \tag{10.1}$$

   This corresponds to step 4 in Figure 10.1, where the learned model (e.g., the canonical gamut) together with the input features (e.g., the input gamut) are used to derive an estimate of the color of the light source.

3. Apply an estimator to select one mapping from the set of feasible mappings (step 5 in Fig. 10.1). The selected mapping can be applied to the canonical illuminant to obtain an estimate of the unknown illuminant. The original method [162] used the heuristic that the mapping resulting in the most colorful scene, that is, the diagonal matrix with the largest trace, is the most suitable one. Simple alternatives are the average of the feasible set or a weighted average [163].

These are the basic steps of gamut mapping algorithms. Several extensions have been proposed. Difficulties in implementation are addressed in References 164, 165, where it is shown that the gamut mapping algorithm can also be computed in chromaticity space $(\frac{R}{B}, \frac{G}{B})$. The main advantage of working in chromaticity space is the lower complexity of the problem. The *2D* approach is easier to visualize and the implementation in *2D* is less complex. However, the performance of this *2D* approach is slightly lower than the performance of the *3D* approach. This is related to the perspective distortion of the possible set of illuminants (the set of feasible mappings, step 4 in Fig. 10.1) that is caused by the conversion of the original image to 2D-chromaticity values. To solve this problem, Finlayson and Hordley [165, 166] proposed to map the *2D* feasible set back to three dimensions before selecting the most appropriate mapping. This corresponds to a slightly modified step 4 in Figure 10.1. Alternatives to address the difficulties in implementation are proposed in References 167 and 168. In Reference 167, an efficient implementation is introduced using convex programming. This implementation reformulates the problem as a set of linear equations, which is shown to result in a performance similar to the original method. Finally, in Reference 168 a simpler version of the gamut mapping is proposed using a cube rather than the full convex hull of the pixel values. This implementation not only has the advantage of simple implementation but it can also be tuned to optimize the maximum error over a set of images rather than the mean or median error.

Another extension of the gamut mapping algorithm deals with dependency on the diagonal model. One of the disadvantages of the original method is that a null solution can occur if the diagonal model fails. In other words, if the diagonal model does not fit the input data accurately, then it is possible that no feasible mapping that maps the input data into the canonical gamut with one single transform can be found. This results in an empty solution set. One heuristic approach to avoid such situations is to incrementally augment the input gamut until a nonempty feasible set is found [141, 169]. Another heuristic approach is to extend the size of the canonical gamut. Finlayson [164] increases the canonical gamut by 5%, while Barnard [163] systematically enlarges the canonical gamut by learning this gamut not only with surfaces that are illuminated by the canonical light source but also with surfaces that are captured under different light sources that are mapped to the canonical illuminant using the diagonal model. Hence, a possible failure of the diagonal model is captured by augmenting the canonical gamut. Another strategy is to simulate specularities during computation of the canonical gamut, potentially increasing the performance of the gamut mapping method even in situations where there is no null solution [170, 171]. Alternatively, to avoid this null solution, an extension of the diagonal model called *diagonal-offset model* is proposed [34]. This model allows for translation of the input colors in addition to the regular linear transformation, effectively introducing some slack into the model. All these modifications are implemented in step 5 of Figure 10.1.

Finally, an interesting extension is proposed by Finlayson et al. [172] and it is called the *gamut-constrained illuminant estimation*. In essence, it is also designed to avoid the null solution that occurs when the diagonal model fails. This method effectively reduces the problem of illuminant estimation to *illuminant classification*, by considering only a limited number of possible light sources. One canonical gamut is learned for every possible light source. Then, the unknown illuminant of the input image is estimated by matching the input gamut to each of the canonical gamuts, selecting the best match as the final estimate. This approach makes it possible to intrinsically add prior knowledge to the system by limiting the possible light sources. When no prior knowledge is available, a generic solution can be supplied by modeling a variety of real-world and synthesized light sources.

10.1 Gamut Mapping Using Derivative Structures

As discussed above, gamut mapping is based on the assumption that only a limited set of colors is observed under a certain illuminant. Multiple phenomena in nature (e.g., blurring) and imaging conditions (e.g., scaling) can cause the mixture of colors. Therefore, if two colors are observed under a certain illuminant, then all colors in between can also be observed under this illuminant, since the set of all possible colors that can be seen under a certain illuminant form a convex hull (i.e., gamut). In Reference 173, the gamut theory is extended by proving that the above is not only true for *image values* but also for every *linear combination of image values*. Hence, the correct estimate of an illuminant will also map every gamut that is constructed by a linear combination of image values back into the canonical gamut constructed with the same linear operation.

10.1.1 Diagonal-Offset Model

The original gamut mapping [162] is designed for scenes that are composed of Lambertian reflectances. For such scenes, the diagonal model is often sufficient to correct for the color of the light source. However, under more realistic conditions, the diagonal model can be too strict, so the gamut mapping will find no solution (this situation is called the *null solution problem*). This could be caused by saturated colors, the presence of surfaces that were not represented in the canonical gamut, or scattering in the lens (veiling illumination), for instance. To overcome this, Finlayson et al. [34] proposed to use the diagonal-offset model (Eq. 8.8), and described an alternate implementation for the gamut mapping using this diagonal-offset model. The remainder of this chapter is based on Equations 8.8 and 8.9 for color correction of images.

10.1.2 Gamut Mapping of Linear Combinations of Pixel Values

In Reference 162, it is shown that the image values form a gamut, and that the transformations of the gamuts under illuminant changes follow the model given in Equation 8.5. Further, in Section 10.1.1 it is shown that transformations of the gamuts under illuminant changes can also be modeled by Equations 8.8 and 8.9. Here we will look at the image gamuts that are formed by a *linear combination* of image values.

Consider a set of image values:

$$\mathbf{F} = \{\mathbf{f}_1, \mathbf{f}_2, \ldots, \mathbf{f}_n\}, \tag{10.2}$$

where $f = \{R,G,B\}$, and an image feature g which is a linear combination of image values $\mathbf{g} = \mathbf{w}^T\mathbf{F}$.

If we consider the *von Kries Model*, the relation between the image values of an object taken under two different light sources is modeled by the diagonal model $f = Df'$. Then, for the feature g the following holds:

$$\begin{aligned} \mathbf{g} = \mathbf{w}^T\mathbf{F} &= w_1\mathbf{f}_1 + w_2\mathbf{f}_2 + \cdots + w_n\mathbf{f}_n, \\ &= w_1\mathbf{D}\mathbf{f}'_1 + w_2\mathbf{D}\mathbf{f}'_2 + \cdots + w_n\mathbf{D}\mathbf{f}'_n, \\ &= \mathbf{D}\left(w_1\mathbf{f}'_1 + w_2\mathbf{f}'_2 + \cdots + w_n\mathbf{f}'_n\right), \\ &= \mathbf{D}\left(\mathbf{w}^T\mathbf{F}'\right) = \mathbf{D}\mathbf{g}', \end{aligned} \tag{10.3}$$

proving that for measurements g also the diagonal models hold. The above is of importance because it shows that gamut mapping can also be performed on all measurements g that are a linear combination of the image values f.

Next, if we consider the diagonal-offset model given by $f = Df' + o$ then,

$$\begin{aligned} \mathbf{g} = \mathbf{w}^T\mathbf{F} &= w_1\mathbf{f}_1 + w_2\mathbf{f}_2 + \cdots + w_n\mathbf{f}_n, \\ &= w_1\left(\mathbf{D}\mathbf{f}'_1 + \mathbf{o}\right) + \cdots + w_n\left(\mathbf{D}\mathbf{f}'_n + \mathbf{o}\right), \\ &= \mathbf{D}\left(\mathbf{w}^T\mathbf{F}'\right) + \left(\sum_{i=1}^{n} w_i\right)\mathbf{o} = \mathbf{D}\mathbf{g}' + \left(\sum_{i=1}^{n} w_i\right)\mathbf{o}. \end{aligned} \tag{10.4}$$

Hence, to estimate the illuminant change between g' and g we have to estimate both the diagonal matrix $\mathbf{D}$ and the offset o. However, in the special case that $\sum_{i=1}^{n} w_i = 0$, the offset term o cancels out.

A similar reasoning can be applied to the local-diagonal-offset model of Equation 8.9. In this case, we have to ensure that all image values $\mathbf{f}_n$ that are linearly combined in g are taken from a local neighborhood where the offset $\overline{\mathbf{o}}$ can be considered constant. Hence, to perform gamut mapping under the local-diagonal-offset model the linear combination g has to satisfy two restrictions: the weights w should sum up to zero and the values $\mathbf{f}_n$ should come from a local neighborhood. Both these restrictions are satisfied by image derivative filters: the sum over the weights of the filter is equal to zero, and since it is a filter the values are taken from a local neighborhood. This makes image derivatives especially attractive for gamut mapping since, contrary to zero-order image value gamuts, they allow estimation of illuminant models under the more general local-diagonal-offset model.

In Reference 173, the gamut mapping based on the statistical nature of images in terms of their derivative structure is investigated. The derivative structure of an image is described (in a complete sense) by means of the n-jet (see References 174 and 175). In Reference 173, gamuts up to the second-order structure are considered, which is given by

$$\{\mathbf{f}, \mathbf{f}_x, \mathbf{f}_y, \mathbf{f}_{xx}, \mathbf{f}_{xy}, \mathbf{f}_{yy}\}, \tag{10.5}$$

where the derivatives are computed for image *f* by a convolution with a Gaussian at the scale of the derivative filter,

$$\mathbf{f} \otimes \frac{\partial}{\partial x} G^{\sigma} = \frac{\partial}{\partial x} \left(\mathbf{f} \otimes G^{\sigma} \right). \tag{10.6}$$

Since these derivative filters are all linear filters, it follows from Equation 10.3 that the gamuts of the n-jet behave similarly under illuminant variations as a normal zero-order gamut.

10.1.3 N-Jet Gamuts

The basic steps of the gamut mapping algorithm are identical when using derivative (n-jet) images. However, when using derivatives, during the construction of the gamuts (both the canonical gamut and the input gamut), the values that are captured in the gamut are symmetric (e.g., if a transition from surface a to surface b is present, then the transition from surface b to surface a should also be included in the gamut). Further, note that the diagonal model can consist of strictly positive elements only. For the pixel-based gamut mapping this restriction is imposed naturally, but the first and second-order gamuts can contain negative as well as positive values. Hence, during implementation one should make sure that the diagonal mappings that are found contain strictly positive elements only. Further note that the complexity of the algorithm based on pixel and derivative information remains the same (and hence the difference in runtime can be neglected).

In Figure 10.2, a few examples of gamuts of the different n-jet images are shown. From these images, it can be derived that the pixel-based gamut (i.e., the gamut of $\mathbf{f}$), the edge-based gamuts (i.e., the gamuts of $\mathbf{f}_x$ and $\mathbf{f}_y$), as well as the gamuts using higher-order statistics (i.e., the gamuts of $\mathbf{f}_{xx}$, $\mathbf{f}_{xy}$ and $\mathbf{f}_{yy}$) are considerably different although they were computed from the same scene where the only difference is a change in the color of the light source.

10.2 Combination of Gamut Mapping Algorithms

It is beneficial to incorporate additional information into illuminant estimation [176]. This can either be done by means of supplemental algorithms [177, 178] or by using higher order statistics in combination with pixel values [120]. In this section, the goal is to exploit these two different ways of combining derivative-based gamut mapping algorithms to provide additional information to estimate the illuminant.

The use of additional information introduces two mutually exclusive opportunities to increase the performance. First, the uncertainty of the estimates of the gamut mapping algorithm can be reduced. Second, the probability of finding the correct illuminant estimate can be increased. In general, the gamut mapping algorithm produces a set of illuminant estimates, called the *feasible set*. From this

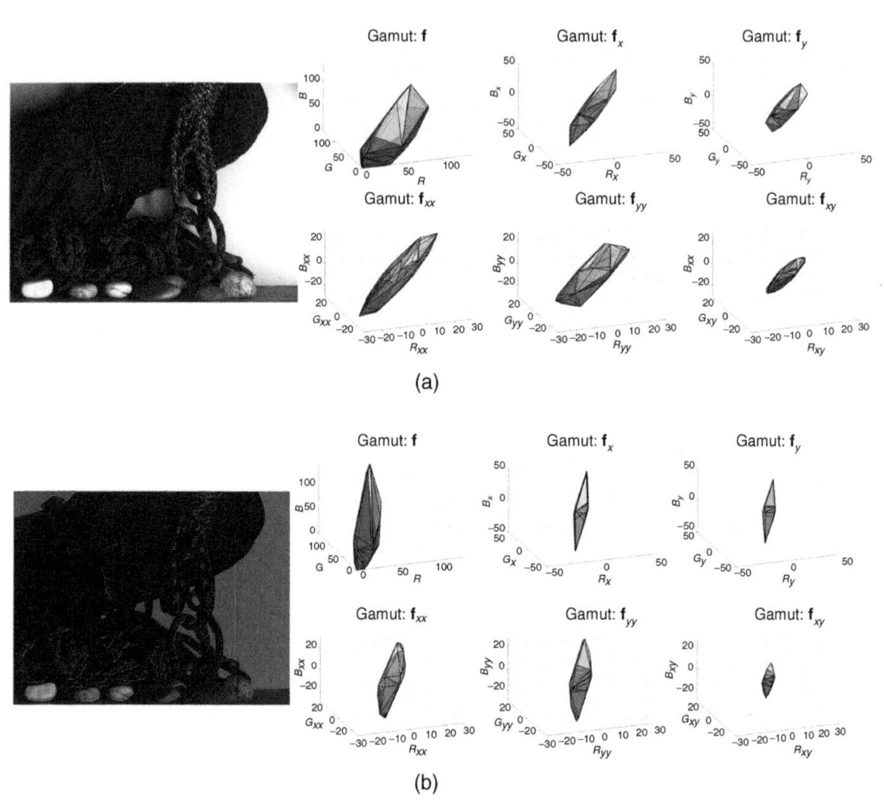

Figure 10.2 Examples of the gamuts of the different n-jet images for a scene taken under two different light sources (images from Reference 44). What is shown is the gamut of the corresponding image, using information that is present in either pixel values ($\mathbf{f}$), edges ($\mathbf{f}_x$ and $\mathbf{f}_y$), or higher order statistics ($\mathbf{f}_{xx}$, $\mathbf{f}_{xy}$ and $\mathbf{f}_{yy}$). Comparing the gamuts of the two images for one type of information (e.g., $\mathbf{f}_x$ of image (a) with $\mathbf{f}_x$ of image (b)) clearly shows the discriminative power of the different n-jets.

feasible set, one final illuminant estimate is selected using some method. If the size of the feasible set is large, then the possibility of selecting the wrong estimate is relatively large, that is, the uncertainty of the final estimate is relatively high. On the other hand, a smaller feasible set results in a lower probability that the correct illuminant is contained inside this set. If multiple feasible sets, which are all different from one another, can be found by using the different n-jet images, then we can choose to either increase or decrease the size of the final feasible set. Intuitively, a smaller feasible set results in a more accurate illuminant estimate than a larger feasible set.

The first approach, proposed in Section 10.2.1, is by combining the feasible sets obtained by the different algorithms. The second fusion method, described in Section 10.2.2, considers the different gamut mapping algorithms as separate algorithms and combines the final estimates of the algorithms.

10.2.1 Combining Feasible Sets

Each gamut mapping algorithm produces a *feasible set* which contains all diagonal mappings that map the gamut of the input image inside the canonical gamut. Hence, the feasible set is a set of possible light sources. Since all gamut mapping algorithms produce such a set, these sets can be used for the combination, instead of selecting only one mapping per algorithm. Since each feasible set represents all illuminant estimates that are considered possible, a natural approach of combining the feasible sets is to consider only those estimates that are present in all feasible sets, that is, an intersection of the feasible sets. Another approach of combining the feasible sets is to consider every estimate that is present in all feasible sets, that is, the union of the feasible sets:

$$\hat{\mathcal{M}}_{\text{intersect}} = \bigcap_i \mathcal{M}_i, \tag{10.7}$$

$$\hat{\mathcal{M}}_{\text{union}} = \bigcup_i \mathcal{M}_i, \tag{10.8}$$

where $\hat{\mathcal{M}}_{\text{intersect}}$ is the intersection of all feasible sets, $\hat{\mathcal{M}}_{\text{union}}$ is the union, and $\mathcal{M}_i$ is the feasible set produced by algorithm i. Then, on these combined feasible sets, an estimator is applied similar to step three of the gamut mapping algorithm.

10.2.2 Combining Algorithm Outputs

As a second possibility, the use of additional information is used in a later stage. Several methods can be considered. Bianco et al. [177] propose a number of alternatives, of which a regular average of the outputs is the simplest combination strategy and the *No-N-Max* method is the most effective. The latter method is a simple average of the outputs, excluding the N estimates that have the largest distance from the other estimates, where N is an adjustable parameter. Let D_j be the sum of the distances of the estimate of method j to all other considered algorithms:

$$D_j = \sum_{i=1}^{n} d(\mathbf{e}_j, \mathbf{e}_i), \tag{10.9}$$

where $d(\mathbf{e}_k, \mathbf{e}_k) = 0$. Then, all n estimates are ordered on the basis of their corresponding D-value, that is, $D_i < D_j < D_k \Rightarrow \mathbf{e}_i < \mathbf{e}_j < \mathbf{e}_k$. Finally, the No-$N$-Max committee can be computed as

$$\hat{\mathbf{e}}_{\text{No-}N\text{-Max}} = \frac{\sum_{i=1}^{n-N} \mathbf{e}_i}{n - N}, \tag{10.10}$$

where it should be noted that $\mathbf{e}_i$ is the ith estimate in the ranked list of illuminant estimates. Further, $\hat{\mathbf{e}}$ is the result of the combination of the n algorithms, and N

is the number of estimates that are excluded. Hence, $N = 0$ is equal to a simple average of all estimates.

10.3 Summary

This chapter describes gamut-based methods. In addition to the traditional gamut mapping, an extension that incorporates the differential nature of images is described. The main advantages of gamut-based methods are the elegant underlying theory and the potential high accuracy. However, proper implementation requires some effort and appropriate preprocessing can severely influence the accuracy.

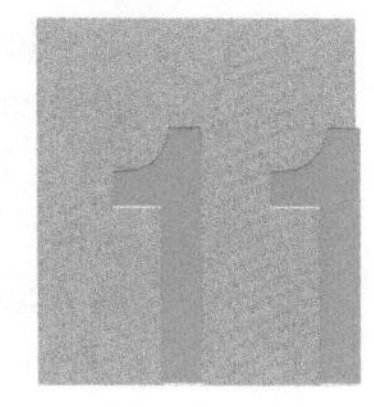

11 Color Constancy Using Machine Learning

The third type of algorithms estimates the illuminant using a model that is learned on training data. Indeed, gamut-based methods in Chapter 10 can be considered learning based too, but since this approach has been influential in color constancy research it has been discussed separately.

Initial approaches using machine learning techniques are based on neural networks [179]. The input to the neural network consists of a large binarized chromaticity histogram of the input image, the output is two chromaticity values of the estimated illuminant. Although this approach, when trained correctly, can deliver accurate color constancy even when only a few distinct surfaces are present, the training phase requires a large amount of training data. Similar approaches apply support vector regression [180–182] or linear regression techniques such as ridge regression and kernel regression [183–185] to the same type of input data. Alternatively, thin-plate spline interpolation is proposed in Reference 186 to interpolate the color of the light source over a nonuniformly sampled input space (i.e., training images).

11.1 Probabilistic Approaches

Color-by-correlation [187] is generally considered to be a discrete implementation of gamut mapping, but it is actually a more general framework that includes other

Color in Computer Vision: Fundamentals and Applications, First Edition.
Theo Gevers, Arjan Gijsenij, Joost van de Weijer, and Jan-Mark Geusebroek.

low-level statistics-based methods such as gray-world and white-patch as well. The canonical gamut is replaced with a correlation matrix. The correlation matrix for a known light source $\mathbf{e}_i$ is computed by first partitioning the chromaticity space into a finite number of cells, followed by computation of the probabilities of occurrence of the coordinates under illuminant $\mathbf{e}_i$. One correlation matrix is computed for every possible illuminant that is considered. Then, the information that is obtained from the input image is matched to the information in the correlation matrices to obtain a probability for every considered light source. The probability of illuminant $\mathbf{e}_i$ indicates the likelihood that the current input image was captured under this light source. Finally, using these probabilities, one light source is selected as scene illuminant, for example, using maximum likelihood [187] or Kullback-Leibler divergence [188].

Other methods using low-level statistics are based on the Bayesian formulation. Several approaches that model the variability of reflectance and light source as random variables are proposed. The illuminant is then estimated from the posterior distribution conditioned on the image intensity data [189–191]. However, the assumptions of independent reflectance that is Gaussian distributed proved to be too strong (unless learned for and applied to a specific application such as outdoor object recognition [192]). Rosenberg et al. [193] replace these assumptions with nonparametric models using the assumption that nearby pixels are correlated. Further, Gehler et al. [194] show that competitive results to state of the art can be obtained when precise priors for illumination and reflectance are used.

11.2 Combination Using Output Statistics

Despite the large variety of available methods, none of the color constancy methods can be considered as universal. All algorithms are based on error-prone assumptions or simplifications, and none of the methods can guarantee satisfactory results for all images. To still be able to obtain good results on a full set of images rather than on a subset of images, multiple algorithms can be combined to estimate the illuminant. The first attempts at combining color constancy algorithms are based on combining the output of multiple methods [177, 178, 195]. In Reference 195, three color constancy methods are combined using both linear (a weighted average of the illuminant estimates) and nonlinear (a neural network based on the estimates of the considered methods) fusion methods are considered. It is shown that a weighted average, optimizing the weights in a least mean square sense, results in the best performance, outperforming the individual methods that are considered *and* nonlinear combination methods such as a multi-layer perceptron neural network. If n algorithms are combined, then the weighted average is defined as

$$\overline{\mathbf{e}} = \sum_{i=1}^{n} w_i \mathbf{e}_i, \tag{11.1}$$

where $\sum_{i=1}^{n} w_i = 1$. The average is just a special instance of the weighted average: $w_1 = w_2 = \cdots = w_n$.

Other general statistics-based combination methods are evaluated in Reference 177. These strategies include the simple mean value of all estimates, the mean value of the two closest estimates, and the mean value of all methods excluding the N most remote estimates (i.e., excluding the estimates with the largest distance to the other estimates, denoted No-N-Max). This latter strategy resulted in the best performance, and is explained in Section 10.2.2.

In Reference 178, a statistics-based method is combined with a physics-based method using a similar approach as the weighted average defined in Equation 11.1. However, the outputs of the two algorithms used are somewhat different from the output of a general color constancy algorithm. Both methods return likelihoods for a predefined set of light sources, where each element represents the probability that the corresponding illuminant is the illuminant that was used to create the current image. After combining these probability vectors a posteriori, the illuminant with the highest probability is selected as the final estimate. These results are more accurate than using either of the two methods alone, but the combination method is not as general as the committee proposed by Cardei and Funt [195], since the outputs of the used color constancy methods have to adhere to a specific (irregular) form.

11.3 Combination Using Natural Image Statistics

Instead of combining the output of multiple algorithms into a more accurate estimate, a totally different strategy is proposed by Gijsenij and Gevers [120, 121]. They use the intrinsic properties of natural images to select the most appropriate color constancy method for every input image. Their approach is based on the observation that all color constancy methods, the methods discussed in Chapter 9, in particular, are based on assumptions on the distribution of colors (edges) that are present in an image. For instance, the gray-world algorithm assumes that the average color in a scene taken under a neutral light source is achromatic, while the gray-edge algorithm assumes that the average *edge* is achromatic. Further, in Chapter 10, it was shown that the incorporation of spatial dependencies between colors (e.g., edges) produce more constrained gamuts, improving the accuracy of color constancy in general. This means that the set of possible adjacent color values (i.e., color edges) in real-world images is more restricted than the set of possible pixel values. Hence, the use of local spatial information will provide more stable gamuts than pixel values to compute color constancy. Furthermore, a higher accuracy is obtained when there is a large variety of edges in a scene (see also [173]). The same observation is valid for the gray-world algorithm in terms of the number of different surfaces [169, 196]. Hence, color constancy methods are largely dependent on the distribution of colors and color edges in an image. Natural image statistics can be used to describe these distributions.

11.3.1 Spatial Image Structures

Image structures are valuable identification cues in determining which type of scene the image is taken from. In Reference 197, the authors show that the power spectrum (distribution of edge responses) of an image is characteristic of the type of scene. Further, in Reference 198, it is shown that this distribution of edge responses can be modeled by a Weibull distribution. In the context of scene classification, features derived from the power spectrum and Weibull distributions have successfully been applied [198–200]. In Reference 121, the focus is on modeling natural image statistics using the *two-parameter integrated Weibull distribution* [198]:

$$w(x) = C \exp\left(-\frac{1}{\gamma}\left|\frac{x}{\beta}\right|^{\gamma}\right), \tag{11.2}$$

where x is the edge responses in a single color channel to the Gaussian derivative filter, C is a normalization constant, $\beta > 0$ is the scale parameter of the distribution, and $\gamma > 0$ is the shape parameter. The parameters of this distribution are indicative of the edge statistics of an (natural) image. In fact, the *contrast* of the image is indicated by β (i.e., the width of the distribution), and the *grain size* by γ (i.e., the peakedness of the distribution). Hence, a higher value for β indicates more contrast, while a higher value for γ indicates a smaller grain size (more fine textures).

To fit the Weibull distribution, edge responses are computed by a Gaussian derivative filter. There exists a high correlation between the Weibull parameters that are fitted through the distribution of edges for the first derivative, second derivative, and third derivative. Hence, a single filter type, although measured in different orientations, is sufficient to assess the spatial statistics of images [198].

In Figure 11.1, examples are shown of images with their corresponding edge distributions that are approximated by a Weibull fit. The intensity channel is chosen for ease of illustration because a six-dimensional edge distribution (i.e., β and γ for each R, G and B channel) is hard to visualize. The edge distributions and corresponding Weibull fits computed for separate color channels show similar plots. The images are examples on which the different color constancy algorithms using the corresponding type of information (i.e., pixel values, edges, or second-order transitions) performs best (based on the angular error).

The relationship between the images in Figure 11.1 and their corresponding color constancy algorithm becomes clear from the edge distributions that are shown together with the images in Figure 11.1. Pixel-based algorithms (i.e., zeroth order) perform better than higher order methods (i.e., first and second order) on images with only little texture. This reflects in an edge distribution that is densely sampled around the origin, that is, many edges with little or zero energy. Higher order methods require more edge information for an accurate illuminant estimate, which is reflected in an edge distribution that is less sharply peaked.

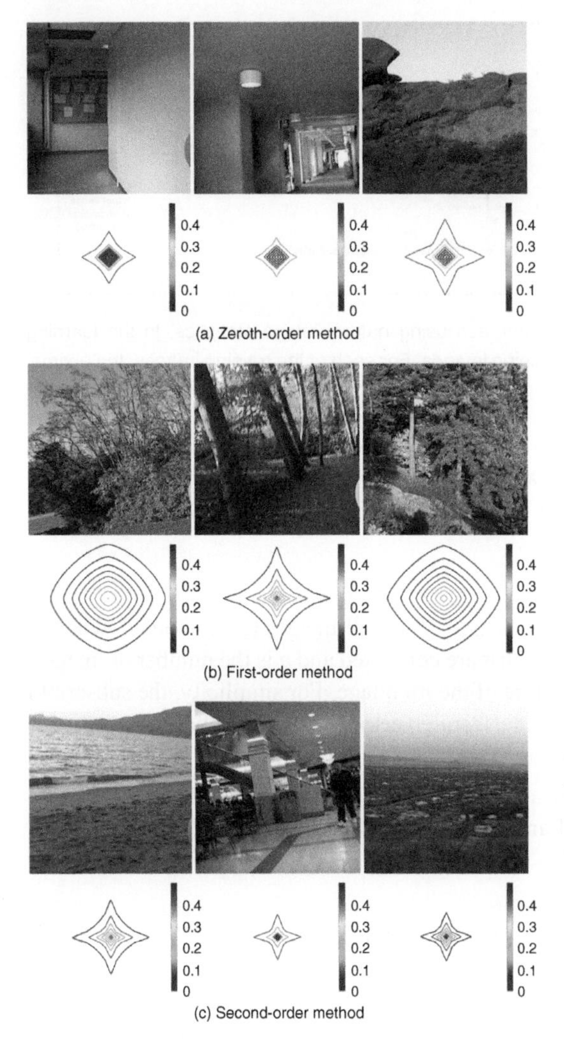

Figure 11.1 Examples of images that can be considered to be characteristic of the corresponding color constancy algorithms, that is, the corresponding color constancy algorithm will perform best on these type of images. Below each image, the distribution of edges in the intensity channel is plotted. *Source*: The images come from the data set published in Reference 201.

11.3.2 Algorithm Selection

Using the Weibull distribution as parameterization of the edge distribution, several characteristics are captured, such as the number of edges and the amount of texture and contrast. Gijsenij and Gevers use this parameterization (i.e., β and γ) to select the most appropriate color constancy algorithm for a given image. This algorithm aims at combining the estimates of several color constancy algorithms into a single more accurate estimate. To be precise, let M be the set of algorithms that are to be combined, where M_i denotes algorithm i. Further, the accuracy of the estimate of algorithm i on image j (i.e., the performance of algorithm i on image j) is denoted by $\epsilon_i(j)$. The algorithm consists of the following steps (see also Fig. 11.2):

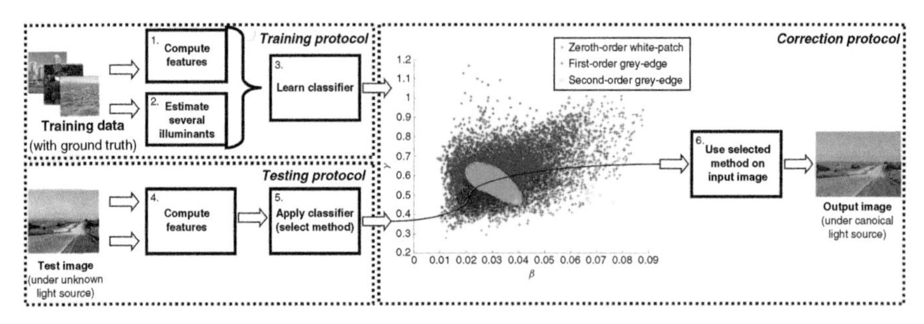

Figure 11.2 Overview of the combination approach using natural image statistics. In the learning phase, features are extracted from a set of training images. For each of the training images, the optimal color constancy method is determined by applying several methods and evaluating the performance using the ground truth (which should be available for the training images). Then, using the features and the optimal color constancy methods, a classifier is learned. In the testing phase, the features of the input image are used in combination with the learned classifier to determine the optimal color constancy method. Finally, the selected color constancy method is applied to the input image, and the input image is corrected accordingly.

- First, the image statistics $\omega \in \mathbb{R}^{p \times q}$ for all images are computed, where p is the number of features that are computed and q is the number of images, that is, ω_{ij} is the ith feature of the jth image. For simplicity, the subscript i is omitted, so ω_j denotes the feature vector representing the image statistics of the jth image. This step corresponds to block 1 in Figure 11.2.
- Then, all images that are in the training set are labeled (step 2 in Fig. 11.2). The label y_j of an image j is derived using the performance of the algorithms on image j:

$$y_j = \arg\min_i \{\epsilon_i(j)\}. \tag{11.3}$$

- Learn a classifier on the training data (see block 3 in Fig. 11.2). Although any classifier could be used, the authors in Reference 121 use an MoG classifier [202]. The likelihood of the observed image statistics ω_j for image j given color constancy algorithm y_j is computed as a weighted sum of k Gaussian distributions:

$$p(\omega_j|y_j) = \sum_{m=1}^{k} \alpha_m G(\omega_j, \mu_m, \Sigma_m). \tag{11.4}$$

Here, α_m are positive weights of the Gaussian components (with mean and variance defined as μ_m and Σ_m, respectively) such that $\sum_{m=1}^{k} \alpha_m = 1$. The parameters of the model are learned through training using the expectation-maximization (EM) algorithm.

- Apply the learned (MoG) classifier on the test data, and assign to the current image j the algorithm that maximized the posterior probability (steps 4–6 in Fig. 11.2). The selection of the most appropriate color constancy algorithm for the current image is done by computing the maximum posterior probability of the classifier. The corresponding color constancy algorithm is selected for the current image. The other algorithms are ignored.

Weibull parameters can be computed for each R, G and B channel separately. However, these color channels are highly correlated [203]. Therefore, the image is first transformed to a decorrelated color space before computing these parameters. To this end, in Reference 121 the opponent color space is used (see Section 4.2). Instead of using Weibull parameterization, various other features are explored in References 204–207 to predict the most appropriate algorithm for a given image. The most notable differences between these approaches is in the extraction of the features, that is, the first step of the algorithm.

11.4 Methods Using Semantic Information

Another type of approach using machine learning techniques attempts to estimate the illuminant using some sort of semantic information.

11.4.1 Using Scene Categories

Gijsenij and Gevers [120, 121, 208] propose to dynamically determine which color constancy algorithm should be used for a specific image, depending on the scene category. They propose that scene semantics can steer the process of color constancy. For instance, forest-like scenes show a similar edge distribution (see Figs. 11.2 and 11.3). Next, the method discussed in the previous section can be used to derive the best solving color constancy algorithm for such similar scenes, for example, the first-order grey-edge method generally is best suited for forest-like scenes.

Some categories have a larger variance in edge distribution than others. For instance, most of the images of the category *highway* have a low value for β and a low value for γ, indicating a low contrast and few edges. Images of the category *inside city*, on the other hand, generally have a large variance. However, even for this category, it can be observed that most images have lower values for γ, while β can take on a wider variation of values. Figure 11.3 shows the image statistics of four different scenes. Although there is some degree of overlap, the plot does show that images of the same semantic category generally have similar image statistics. Using this observation, a supervised selection of a color constancy algorithm for images of the same scene category can be achieved. By classifying an input image as one of these image categories (either supervised

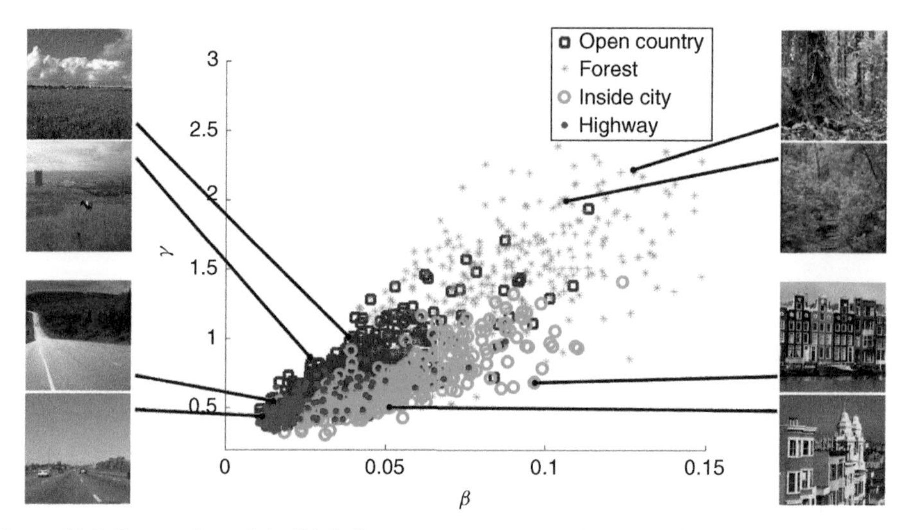

Figure 11.3 Scatter plots of the Weibull parameters based on O_3 derived from images coming from several categories (defined in Reference 209). From this plot, it can be seen that images from the same scene category have similar Weibull parameters (although there is some degree of overlap). Comparing these statistics to the statistics learned in Section 11.3 allows us to connect a specific color constancy algorithm to particular semantic categories.

by user intervention or unsupervised by a scene-recognition system as in [199–200, 210]), the corresponding color constancy algorithm can be applied to the image to obtain a performance that is similar to the proposed automatic selection algorithm.

From these observations, a supervised selection of a color constancy algorithm for images from all scene categories can be achieved. By classifying an input image as one of these image categories (either supervised by user intervention or unsupervised by a scene-recognition system as in [199–200, 210]), the corresponding color constancy algorithm can be applied to the image.

Another approach that uses scene categories is proposed by Bianco et al. [211], which makes explicit use of an indoor-outdoor classifier. When an input image is classified as an indoor image, they propose to use the shades-of-grey method. Classifying an image as an outdoor image should result in the second-order grey-edge method. In addition to these two classes, they propose to use an ‘‘unsure’’ class for images that have low probabilities for either of the two classes (indoor/outdoor). Such images are best solved by a general purpose method, for which they use a weighted average of multiple estimates. Instead of the arbitrary distinction between indoor and outdoor images, Lu et al. [212–213] propose to use a stage classifier that distinguishes medium-level semantic classes, called *stages* [214]. This results in a color constancy method that explicitly uses 3D scene information for the estimation of the color of the light source.

11.4.2 Using High-Level Visual Information

Rather than classifying images into a specific class and applying different color constancy methods depending on the corresponding class, van de Weijer et al. [145] propose to select the best illuminant estimate out of a set of *illuminant hypotheses*. Using prior knowledge about the world, possible illuminant estimates are evaluated on the basis of the likelihood of semantic content. In other words, the algorithm selects as final illuminant estimate the illuminant that will generate the most plausible output image, for example, an output image with blue rather than purple sky and green rather than reddish grass.

To be able to evaluate the likelihood of an illuminant hypothesis, a model has to be created that computes the probability of an image to occur under a white light source. For this purpose, images will be modeled as a mixture of semantic classes, such as sky, grass, road, and building. Each class is described by a distribution over visual words, which are described by three modalities—texture, color, and position. As an example, consider an image with sky and grass. This image will consist of visual words that are drawn from the distributions of sky and grass. Given these visual words, we will attempt to infer what classes are present in the image. Given the inferred classes and the visual words, the likelihood of the image can be computed, which is called the *semantic likelihood* of the image.

A generative model called probabilistic latent semantic analysis [215] is used in Reference 145. Images are modeled as a mixture of latent topics. The topics are semantic classes in the image such as sky, grass, road, building, etc. They are described by a distribution over visual words. As visual descriptors, 20×20 patches are extracted on a regular grid from the image. Each patch, or visual word, is described by three modalities:

- Texture, which is described with the SIFT descriptor [55].
- Color, which is described by the Gaussian averaged RGB value over the patch.
- Position, which is described by imposing an 8×8 grid of regular cells on the image.

Given a set of images $F = \{\mathbf{f}_1, \ldots, \mathbf{f}_N\}$ each described in a visual vocabulary $V = \{v_1, \ldots, v_M\}$, the words are taken to be generated by latent topics $Z = \{z_1, \ldots, z_K\}$. In the PLSA model, the conditional probability of a visual word v in an image $\mathbf{f}$ and an illuminant $\mathbf{e}$ is given by

$$P(v|\,\mathbf{f}, \mathbf{e}) = \sum_{z^{\mathbf{e}} \in Z^{\mathbf{e}}} P\left(v|\,z^{\mathbf{e}}\right) P\left(z^{\mathbf{e}}\right|\mathbf{f}\right), \tag{11.5}$$

where $z^{\mathbf{e}}$ indicates that the topic distribution has been computed from a data set that was taken under illuminant $\mathbf{e}$. Similar to the approach of Verbeek and Triggs [216], it is assumed that the three modalities are independent given the topics

$$P(v|z) = P\left(v^T|z\right) P\left(v^C|z\right) P\left(v^P|z\right), \tag{11.6}$$

where v^T, v^C, and v^P, are successively the texture, color, and position word. The distributions $P(z|\mathbf{f})$ and the various $P(v|z)$s are discrete, and can be estimated using an EM algorithm [215].

The goal is to compute the chance that an image was taken under a white light source, which, according to Bayes law, is proportional to

$$P(\mathbf{w}|\mathbf{f}) \propto P(\mathbf{f}|\mathbf{w})\, P(\mathbf{w})\,. \tag{11.7}$$

If a uniform distribution over the illuminants $p(\mathbf{w})$ is assumed, this can be rewritten using Equation 11.5:

$$P(\mathbf{w}|\mathbf{f}) \propto P(\mathbf{f}|\mathbf{w}) = \prod_{m=1}^{M} P\left(v_m|\mathbf{f}, \mathbf{w}\right), \tag{11.8}$$

$$= \prod_{m=1}^{M} \sum_{z^{\mathbf{w}} \in Z^{\mathbf{w}}} P\left(v_m|z^{\mathbf{w}}\right) P\left(z^{\mathbf{w}}|\mathbf{f}\right), \tag{11.9}$$

where $P\left(v_m|z^{\mathbf{w}}\right)$ means that the visual word topic distributions are learned from images taken under white light.

An overview of this approach is given in Figure 11.4. Given an input image that is recorded under an unknown illuminant, first a set of illuminant hypotheses is generated (step 1). Van de Weijer et al. [145] propose to use both bottom-up and top-down approaches (steps 2 and 3). Then, each of these hypotheses is used to correct the input image, and for each corrected image, the *semantic likelihood* is computed using Equation 11.9 (step 4). Consider for the sake of simplicity that texture descriptors do not change when the illuminant changes (in the final implementation, the texture descriptors are recomputed for each illuminant). When the illuminant color is varied (i.e., by applying one of the hypotheses to the input image), the color word v^C is changed and consequently $P(v^C|z)$, $P(v|z)$, and $P(z|\mathbf{f})$ are changed. The semantic likelihood that is used to assess the probability of the hypotheses is now given by the correspondence of $P(v^C|z)$ with the combined distribution of $P(v^T|z)P(v^P|z)$. This means that illuminants become more likely when the color words they generate are in accordance with the texture and position information (for instance, color words representing green are more likely together with texture words describing grass, and a skylike texture in the top of the image is more likely to be blue). In the depicted image in Figure 11.4, the method estimates the illuminant to be reddish, since after correction for this light source the image could be interpreted as green grass under a blue sky.

This approach is closely related to the work of Manduchi [217], who uses the color similarity between a test image and labeled classes in one training image taken under white light. These classes are not semantically meaningful, as they are labeled ''class I,'' ''class II,'' etc. Using a Gaussian color distribution to describe the classes, the color similarity is used to estimate the illuminant color of the test image. Each pixel is assigned to a class and an illuminant to optimize the likelihood

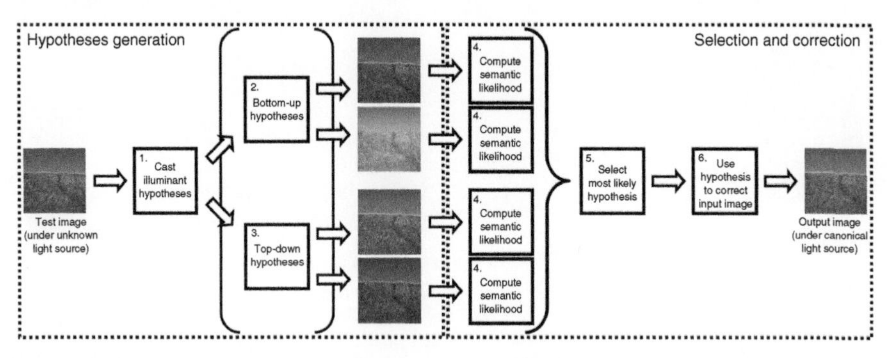

Figure 11.4 Overview of the combination approach using high-level visual features. First, several illuminant hypotheses are casted. This can be done using bottom-up *and* top-down approaches (as the authors suggest in Reference 145) or by using any other approach. Then, using Equation 11.9, the semantic likelihood of all hypotheses are computed. Finally, the most likely illuminant hypothesis is selected as the final estimate and used to correct the input image.

of the image. The method has the advantage that multiple illuminants are allowed within an image, but it is only demonstrated to succeed when a single training image, similar to the test image, is available. This might be due to the limited discriminative power of the class description, in which multimodality in color space as well as texture as position information are disregarded. Finally, another similar approach is proposed in Reference 218, where the term *memory color* is used to refer to colors that are specifically associated with object categories. These object-specific colors are used to refine the estimated illuminants.

11.5 Summary

Learning-based methods have the advantage over stand-alone methods that they can be tuned toward specific data (such as indoor or outdoor images). This chapter describes several methods that cannot operate without training phase. First, methods that learn low-level statistics are described, such as regression techniques and Bayesian approaches. Such approaches are often simple to implement, but the output is often rather nonintuitive since the model that is learned is quite opaque. After that, methods using higher level statistics and semantics are discussed. Since such methods select appropriate color constancy methods given an input image, they are more intuitive than other learning-based methods. Moreover, the accuracy of such approaches has been proved to be state of the art. However, the use of multiple single algorithms means that they are inherently slower than the single algorithms themselves.

12 Evaluation of Color Constancy Methods

Evaluation of illuminant estimation algorithms requires images with a scene illuminant that is known (ground truth). The general experimental setup is as follows. First, part of the data is used for training, if the algorithm requires this. Then, the color of the light source is estimated for every remaining image of the database and compared to the ground truth. Various publicly available data sets are discussed in Section 12.1. The comparison requires some similarity or distance measure, discussed in Section 12.2. Alternative setups exist, depending on the application. For instance, Funt et al. [219] describe an experiment to evaluate the usefulness of color constancy algorithms as a preprocessing step in object recognition. However, in this chapter, the intended application is correction of an input image for the color of the light source, that is, white balancing.

12.1 Data Sets

Two types of data that are used to evaluated color constancy methods can be distinguished: hyperspectral data and *RGB* images. Databases containing hyperspectral data sets are often smaller (less images) and contain less variation

Color in Computer Vision: Fundamentals and Applications, First Edition.
Theo Gevers, Arjan Gijsenij, Joost van de Weijer, and Jan-Mark Geusebroek.

than data sets with *RGB* images. The main advantage of hyperspectral data is that many different illuminants can be used to realistically render the same scene under various light sources, and consequently a systematic evaluation of the methods is possible. However, the simulation of illuminants generally does not include real-world effects such as interreflections and nonuniformity. Consequently, the evaluation of *RGB* images results in more realistic performance evaluations. Ideally, both types of data should be used for a thorough evaluation of color constancy methods [141, 169].

12.1.1 Hyperspectral Data

An often used hyperspectral database was composed by Barnard et al. [44]. This set consists of 1995 surface reflectance spectra and 287 illuminant spectra. These reflectance and illuminant spectra can be used to generate an extensive range of surfaces (i.e., *RGB* values), allowing for a systematic evaluation of color constancy performance. Another database that is specifically useful for the evaluation of the color constancy algorithm is created by Foster et al. [126, 220]. These two sets each contain eight natural scenes that can be converted into an arbitrary number of images using various illuminant spectra (not provided). Finally, a database by Parraga et al. [221] contains 29 hyperspectral images with low resolution (256×256 pixels).

12.1.2 RGB Data

Databases with *RGB* images are more informative on the performance of the algorithms under realistic circumstances. The first step toward realistic evaluation of color constancy methods involves isolated compositions of objects that are illuminated by 11 different light sources [44]. The 11 different lights include three different fluorescent lights, four different incandescent lights, and four incandescent lights combined with a blue filter, and are selected to span the range of natural and man-made illuminants as best as possible. The complete database contains 22 scenes with minimal specularities, 9 scenes with dielectric specularities, 14 scenes with metallic specularities, and 6 scenes with at least one fluorescent surface. Often, for illuminant estimation evaluation, a subset of 31 scenes that only consists of the scenes with minimal and with dielectric specularities is used. Even though these images encompass several different illuminants and scenes, the variation of the images is limited.

A more varied database was compiled by Ciurea and Funt [201]. This data set contains over 11,000 images, extracted from 2 hours of video recorded under a large variety of imaging conditions (including indoor, outdoor, desert, cityscape, and other settings). In total, the images are divided into 15 different clips taken at different locations. The ground truth is acquired by attaching a gray sphere to the camera, that is displayed in the bottom right corner of the image. Obviously, this gray sphere should be masked during experiments to avoid biasing the algorithms. Some examples of images that are in this data set are shown in Figure 12.1a.

(a) Example images of SFU data set

(b) Example images of color-checker-set

Figure 12.1 Some examples of the two data sets that are used for the experiments. (a) SFU data set (b) Color-checker set

The main disadvantage of this set is the correlation that exists between some of the images. Since the images are extracted from video sequences, some images are rather similar in content. This should especially be taken into account when dividing the images into training and test sets. Another issue of this data set is that an unknown postprocessing procedure is applied to the images by the camera, including gamma correction and compression. A similar data set is recently proposed in Reference 222. Although the number of images in this set (83 outdoor images) is not comparable to the previous set, the images are not correlated and are available in *XYZ*-format, and can be considered to be of better quality. Further, an extension of the data set is proposed in Reference 223, where an additional 126 images with varying environments (e.g., forest, seaside, mountain snow, and motorways) are introduced.

Gehler et al. [194] introduced a new database consisting of 568 images, both indoor and outdoor. The ground truth of these images is obtained using a MacBeth Color Checker that is placed in the scene. The main advantage of this database is the quality of the images (which are free of correction), but the variation of the images is not as large as the data set containing over 11,000 images. Some examples of images that are in this data set are shown in Figure 12.1b. Finally, Shi and Funt generated a set of 105 high dynamic range images [147, 148]. These images use four color checkers to capture the ground truth and are constructed from multiple exposures of the same scene.

12.1.3 Summary

A summary of available data sets is presented in Table 12.1. Generally, a distinction can be made between real-world *RGB* images and images with controlled illumination conditions. The latter type of data, including hyperspectral images, should mainly be used to aid the development of new algorithms and for the systematic analysis of methods. Conclusions about the performance with respect to existing methods based on such data sets should be avoided as much as possible, since it is relatively easy to tune any algorithm to obtain a high

Table 12.1 Summary of data sets with advantages and disadvantages.

Data Set	Pros	Cons
SFU hyperspectral set [44]	Large variety	Best-case assessment of performance
(1995 surface spectra)	Allows for systematic evaluation	
Foster et al. [126, 220]	High quality hyperspectral images	Limited amount of data
(8 + 8 images)	Real-world natural scenes	
Bristol set [221]	Hyperspectral images	Low quality images
(28 images)	Real-world natural scenes	
SFU set [44]	Scenes with varying characteristics	Laboratory setting
(223+98+149+59 images)	Captured with calibrated camera	
Gray-ball SFU set [201]	Largest set available	Correlation exists between images
(11,346 images)	Large variety of images	Images are postprocessed
Barcelona set [222]	Uncorrelated images	Few images
(83 + 126 images)	High quality *XYZ*-data available	Short time frame
Color checker set [194]	High quality images	Medium variety
(568 images)	Uncorrected data	
HDR images [147, 148]	High dynamic range images	Few images
(105 images)	Uncorrected data	

performance on such data sets. The real-world *RGB* images are more suited to compare algorithms, as such data are probably the target data of the intended application of most color constancy algorithms.

12.2 Performance Measures

Performance measures evaluate the performance of an illuminant estimation algorithm by comparing the estimated illuminant to a ground truth, which is known a priori. Since color constancy algorithms can only recover the color of the light source up to a multiplicative constant (i.e., the intensity of the light source is not estimated), distance measures compute the degree of resemblance in normalized *RGB*:

$$r = \frac{R}{R+G+B}, \tag{12.1}$$

$$g = \frac{G}{R+G+B}, \tag{12.2}$$

$$b = \frac{B}{R+G+B}. \tag{12.3}$$

Various distance measures can be derived. Gijsenij et al. [224] propose a taxonomy of different measures for color constancy algorithms. They distinguish between mathematics-based measures, perceptual measures, and color constancy-specific measures.

12.2.1 Mathematical Distances

In color constancy research, two frequently used performance measures are the Euclidean distance and the angular error, of which the latter is probably more widely used. The angular error measures the angular distance between the estimated illuminant $\mathbf{e}_e$ and the ground truth $\mathbf{e}_u$, and is defined as

$$d_{\text{angle}}(\mathbf{e}_e, \mathbf{e}_u) = \cos^{-1}\left(\frac{\mathbf{e}_e \cdot \mathbf{e}_u}{||\mathbf{e}_e|| \cdot ||\mathbf{e}_u||}\right), \tag{12.4}$$

where $\mathbf{e}_e \cdot \mathbf{e}_u$ is the dot product of the two illuminants and $|| \cdot ||$ is the Euclidean norm of a vector.

The Euclidean distance d_{euc} is actually a special instantiation of the more general Minkowski family of distances, denoted d_{Mink}:

$$d_{\text{Mink}}(\mathbf{e}_e, \mathbf{e}_u) = (|r_e - r_u|^p + |g_e - g_u|^p + |b_e - b_u|^p)^{\frac{1}{p}}, \tag{12.5}$$

where p is the corresponding Minkowski norm. Three special cases are well-known: the Manhattan distance (d_{man}) for $p = 1$, the Euclidean distance (d_{euc}) for $p = 2$, and the Chebychev distance (d_{sup}) for $p = \infty$.

12.2.2 Perceptual Distances

If the goal of color constancy algorithms is to obtain an output image that is identical to a reference image, that is, an image of the same scene taken under a canonical, often white, light source, then perceptual distance measures are an obvious choice for evaluation. For this purpose, the estimated color of the light source and the ground truth are first transformed into different (human vision) color spaces, after which they are compared. In Reference 224, the distance is measured in the (approximately) perceptually uniform color spaces CIELAB and CIELUV [225]. More information on these color spaces is presented in Chapter 4. Further, in addition to the Euclidean distance between CIELAB colors, the CIEDE2000 [226] is computed, since the metric is shown to be more uniform and is considered to be state of the art in industrial applications. Finally, the authors in Reference 224 propose a weighted Euclidean distance measure (denoted *PED* for *perceptual Euclidean distance*), inspired by the nonuniformity of the spectral sensitivity of the human eye:

$$\mathrm{PED}(\mathbf{e}_e, \mathbf{e}_u) = \sqrt{w_R(r_e - r_u)^2 + w_G(g_e - g_u)^2 + w_B(b_e - b_u)^2}, \tag{12.6}$$

where $w_R + w_G + w_B = 1$. This distance measure allows assigning higher weights to deviations in one color channel, since a deviation in this specific color channel might have a stronger effect on the perceived difference between two images than a deviation in another channel.

12.2.3 Color Constancy Distances

Finally, two color-constancy-specific distances are discussed. The first is the color constancy index (CCI) [108], also called Brunswik ratio [227], and is generally used to measure perceptual color constancy [110, 126]. It is defined as the ratio of the amount of adaptation that is obtained by a human observer versus no adaptation at all:

$$\mathrm{CCI} = \frac{b}{a}, \tag{12.7}$$

where b is defined as the distance from the estimated light source to the true light source and a is defined as the distance from the true light source to a white reference light. During evaluation, several different color spaces are used to compute the values a and b.

The second is called the *gamut intersection* [224], which makes use of the gamuts of the colors that can occur under a given light source. It is based on the assumption underlying the gamut mapping algorithm, that is *under a given light source, only a limited number of colors are observed*. The difference between the full gamuts of two light sources is an indication of the difference between these two light sources. For instance, if two light sources are identical, then the gamuts of colors that can occur under these two light sources will coincide, while the similarity of the gamuts will be smaller if the difference between the two light sources is larger. The gamut intersection is measured as the fraction of colors that occur under the estimated light source, with respect to the colors that occur under the true, ground truth, light source.

$$d_{\mathrm{gamut}}(\mathbf{e}_e, \mathbf{e}_u) = \frac{\mathrm{vol}(\mathcal{G}_e \cap \mathcal{G}_u)}{\mathrm{vol}(\mathcal{G}_u)}, \tag{12.8}$$

where $\mathcal{G}_i$ is the gamut of all possible colors under illuminant i and $\mathrm{vol}(\mathcal{G}_i)$ is the volume of this gamut. The gamut $\mathcal{G}_i$ is computed by applying the diagonal mapping, corresponding to light source i, to a canonical gamut.

12.2.4 Perceptual Analysis

In most situations, for instance, when the application is to obtain an accurate reproduction of the image under a white light source, the distance measure should be an accurate reflection of the quality of the output image. In Reference 224, the distance measures discussed above are analyzed with respect to this requirement. For this purpose, several psychophysical experiments are performed, where human subjects were shown four images on a calibrated LCD monitor (see Fig. 12.2). The upper two images are identical reference images, representing the test scene. The lower two correspond to the resulting output of two different color constancy algorithms, applied to the original test scene (i.e., the scene under a certain colored light source). Subjects are instructed to compare the color reproduction of each of the lower images with the upper references. Both the global color impression of the scene and the colors of local image details are to be addressed. Subjects then indicated which of the two lower images has the best color reproduction. Various different color constancy algorithms were tested, resulting in a ranking based on the human observations. This ranking was then correlated against the different distance measures (each distance measure can be used to generate a ranking).

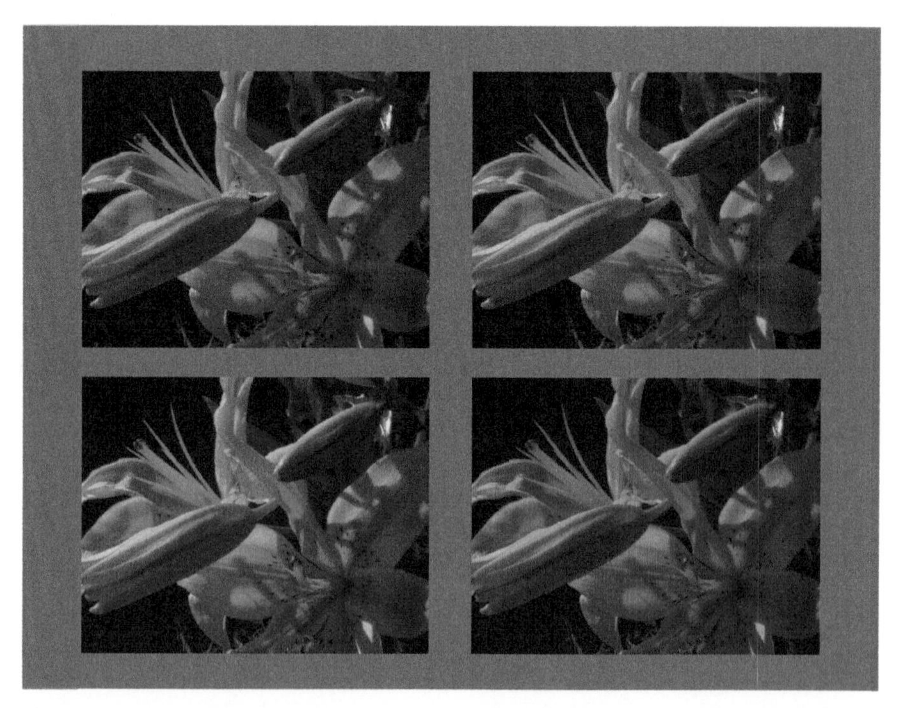

Figure 12.2 Screen capture of an experimental trial. Subjects indicate which of the two bottom images (resulting from two different color constancy algorithms) is the best match to the upper reference image. More details on the experimental setup can be found in Reference 224.

Table 12.2 Overview of the correlation coefficients ρ of several distance measures and using several color spaces with respect to the subjective measure derived from human observers.

Measure	ρ on Foster data [126]	ρ on Gray-ball SFU data [201]
d_{angle}	0.895	0.926
d_{man}	0.893	0.930
d_{euc}	0.890	0.928
d_{sup}	0.817	0.906
$d_{euc} - L^*a^*b^*$	0.894	0.921
$\Delta E^*_{00} - L^*a^*b^*$	0.896	0.916
$d_{euc} - L^*u^*v^*$	0.864	0.925
$d_{euc} - C + h$	0.646	0.593
$d_{euc} - C$	0.619	0.562
$d_{euc} - h$	0.541	0.348
$PED_{proposed}$	0.960	0.957
$CCI(d_{angle})$	0.895	0.931
$CCI(d_{euc,RGB})$	0.893	0.929
$CCI(d_{euc,L^*a^*b^*})$	0.905	0.921
$CCI(d_{euc,L^*u^*v^*})$	0.880	0.927
d_{gamut}	0.965	0.908

Subjects evaluate the quality of two different data sets. The first data set consists of eight hyperspectral images [126], each rendered under four different light sources. The second data set consists of 50 images from the Gray-ball SFU set [201]. The correlation coefficients for various distance measures on both data sets are shown in Table 12.2. It should be noted that the correlation coefficients are relatively stable across data sets. Further, it is shown that the often used angular error correlates *reasonably well* with the perceived quality of the output images (in fact, the angular error is on par with perceptual distance measures such as CIEDE2000). The main weakness of the angular error, however, is the fact that the angular error ignores the direction of the error completely. For instance, see the illustration in Figure 12.3, where compositions are shown of patches with

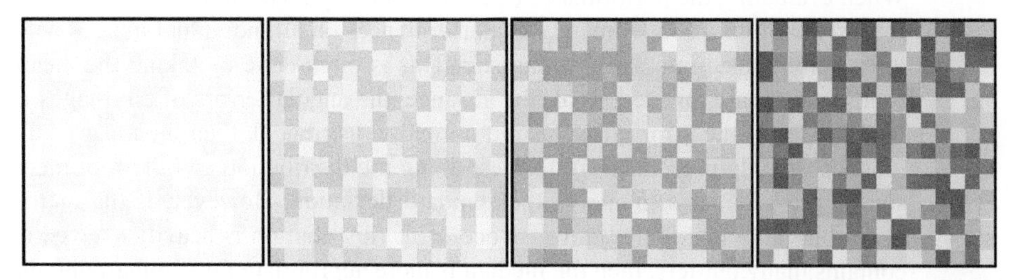

Figure 12.3 Illustration of the angular error. Shown are patches that have an angular deviation from white of 1°, 5°, 10°, and 20°, respectively. Notice the variation in colors for patches with similar deviation from a reference color (white).

the same deviation from white (from left to right, the deviation from white is 1°, 5°, 10°, and 20°). Notice the variation in colors for patches that have a similar deviation from a reference color (white). This indicates that two methods can have similar angular errors on the same image, but result in completely different output images. Obviously, this also is reflected in the perceived quality of the output images.

In order to take the direction of the error into account, the authors of Reference 224 propose the perceptual Euclidean distance (or PED), with which higher weights can be assigned to deviations in some color channels than others. An exhaustive search on the weight space was performed, resulting in a weight combination of $(w_R, w_G, w_B) = (0.26, 0.70, 0.04)$. The correlation of this distance measure with the perceptual quality of the output images is significantly higher than the angular error [224]. However, the optimal weight combination depends on the data set that is used. This means a psychophysical experiment that uses a small but representative subset of the complete data set to obtain the optimal weights is needed. Using these optimal weights, more reliable inferences can be made about the performance of color constancy algorithms. Without such psychophysical experiments, the angular error is a decent choice for a distance measure.

Finally, the experiments in Reference 224 are used to introduce the notion of *perceptual significance*. When comparing two algorithms, the errors of those two algorithms on a set of images are compared. However, the fact that one algorithm results in lower errors might not always justify the conclusion that one algorithm is better than the other. The degree of difference also has to be taken into account. One important question that should not be ignored is whether or not the obtained improvement is noticeable to a human observer. In Reference 224, it was shown that when using the angular error as distance measure, a relative improvement of at least 5–6% should be obtained to be noticeable to a human observer. For instance, if method A has an angular error of 10°, then an improvement of at least 0.6° is necessary; otherwise the improvement will not be visible to a human observer.

12.3 Experiments

When evaluating the performance of color constancy algorithms on a whole data set instead of on a single image, the performances on all individual images need to be summarized into a single statistic. This is often done by taking the mean, root mean square, or median of, for instance, the angular errors of all images in the data set. If the error measures are normally distributed, then the mean is the most commonly used measure for describing the distribution and the root mean square provides an estimate of the standard deviation. However, if the metric is not normally distributed, for instance, if the distribution is heavily skewed or contains many outliers, then the median is more appropriate for summarizing the underlying distribution [228].

12.3.1 Comparing Algorithm Performance

From previous work, it is known that the angular error is not normally distributed [196]. To test whether the perceptual Euclidean distance is normally distributed, a similar experiment as in Reference 196 is conducted. In Figure 12.4, the errors for the white-patch algorithm on the 11,000 images from the RGB images data set [201] are plotted, from which it is clear that both the angular error and the perceptual Euclidean distance are not normally distributed. The distributions of both metrics have a high peak at lower error rates, and a fairly long tail. For such distributions, it is known that the mean is a poor summary statistic, and hence, previously, it was proposed to use the median to describe the central tendency [196]. Alternatively, to provide more insight into the complete distribution of errors, one can calculate boxplots or compute the trimean instead of the median. Boxplots are used to visualize the underlying distributions of the error metric of a given color constancy method, as an addition to a summarizing statistic. This summarizing statistic can be the median, as proposed by Hordley and Finlayson [196], or it can be the trimean, a statistic that is robust to outliers (the main advantage of the median over a statistic such as the root mean square), but still has attention to the extreme values in the distribution [229, 230]. The trimean can be calculated as the weighted average of the first, second, and third quantile Q_1, Q_2 and Q_3, respectively:

$$TM = 0.25Q_1 + 0.5Q_2 + 0.25Q_3. \tag{12.9}$$

The second quantile Q_2 is the median of the distribution, and the first and third quantiles Q_1 and Q_3 are called *hinges*. In other words, the trimean can be described as the average of the median and the midhinge.

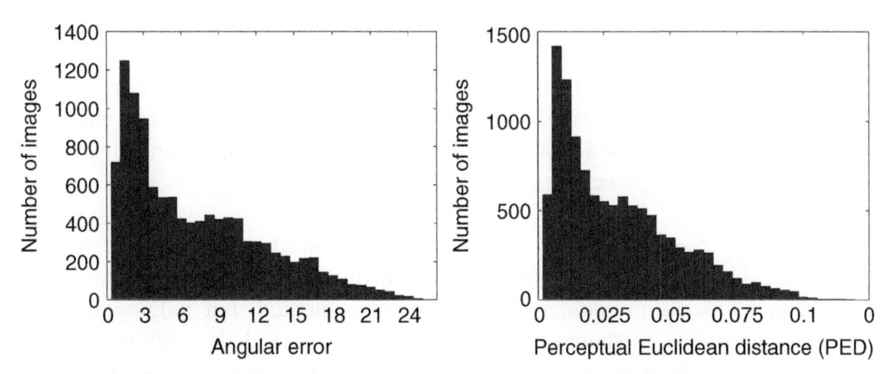

Figure 12.4 Distribution of estimated illuminant errors for the white-patch algorithm, obtained for a set of over 11,000 images.

12.3.2 Evaluation

Two data sets are used to evaluate various methods [231]. The two data sets evaluated are the gray-ball SFU set and the color checker set (note that the data used in this chapter is obtained from [232]). These sets are selected because of their size (they are the two largest sets available to date), their nature (the sets consist of real-world images in an unconstrained environment) and their benchmark status (the gray-ball SFU set is widely used, the recent color checker set has the potential to become widely used). For the exact details on the used data sets, refer to [231].

All algorithms are trained using the same setup, based on cross-validation. Training on the gray-ball SFU set is performed by dividing the data into 15 parts, where it is ensured that the correlated images (i.e., the images of the same scene) are grouped in the same part. Next, the method is trained on 14 parts of the data and tested on the remaining part. This procedure is repeated 15 times, so every image is in the test set exactly once and all images from the same scene will either be in the training set or in the test set at the same time. The color checker set adopts a simpler threefold cross-validation. The three folds are provided by the authors of the data set. This cross-validation-based procedure is also adapted to learn the optimal parameter setting for the static algorithms (optimizing p and σ) and the gamut-based algorithms (optimizing the filter size σ). Further, the regression-based method is implemented using LIBSVM [233], and is optimized for the number of bins of the binary histogram and for the Support Vector Regression (SVR) parameters. Finally, all combination-based methods are applied to a select set of static methods: using Equation 9.18 we systematically generated nine methods using pixel values, eight methods using first-order derivatives and seven methods using second-order derivatives. On the basis of the details of the corresponding methods, the following strategies are deployed. The No-N-Max combination method [177] is applied to a subset of six methods (finding the optimal combination of six methods using the same cross-validation-based procedure), the method using high level visual information [145] is applied to the full set of methods (setting the number of semantic topics to 30) and the method using natural image statistics [120, 121] is applied to a subset of three methods (one pixel-based, one edge-based, and one second-order derivative-based method, finding the optimal combination using the same cross-validation procedure).

12.3.2.1 Gray-Ball SFU Set The results on the SFU set are shown in Table 12.3. Some example results are shown in Figure 12.5. Pixel-based gamut mapping performs similar to the gray-edge method, but judging from these results, simple methods such as the white-patch and the gray-world are not suited for this data set with the current preprocessing strategy. As expected, combination-based methods outperform single algorithms, where the difference

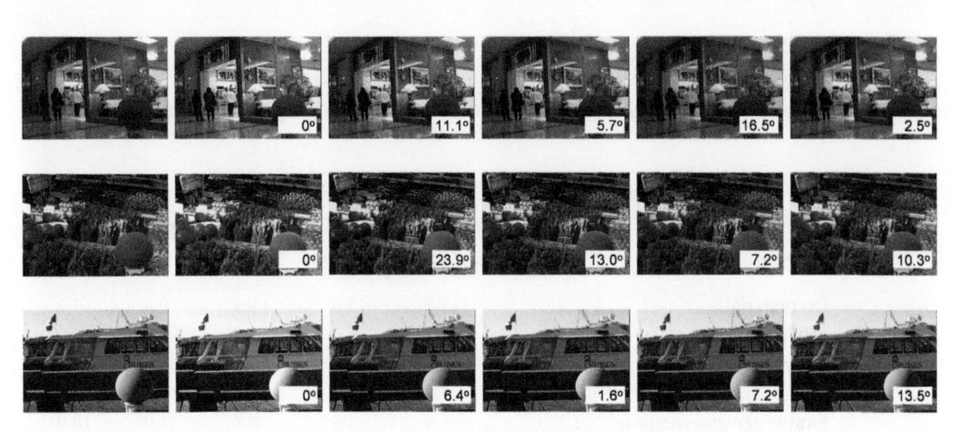

Figure 12.5 Some example results of various methods applied to several test images. The angular error is shown in the bottom right corner of the images. The methods used are, from left to right, perfect color constancy using ground truth, gray-world, second-order gray-edge, inverse intensity chromaticity space and using high level visual information.

Table 12.3 Performance of several methods on the *linear* gray-ball SFU set (11,346 images).

Method	Mean μ	Median	Trimean	Best, 25%	Worst, 25%
Do nothing	15.6°	14.0°	14.6°	2.1°	33.0°
White-patch ($\mathbf{e}^{0,\infty,0}$)	12.7°	10.5°	11.3°	2.5°	26.2°
Gray-world ($\mathbf{e}^{0,1,0}$)	13.0°	11.0°	11.5°	3.1°	26.0°
General gray-world ($\mathbf{e}^{0,p,\sigma}$)	12.6°	11.1°	11.6°	3.8°	23.9°
First-order gray-edge ($\mathbf{e}^{1,p,\sigma}$)	11.1°	9.5°	9.8°	3.2°	21.7°
Second-order gray-edge ($\mathbf{e}^{2,p,\sigma}$)	11.2°	9.6°	10.0°	3.4°	21.7°
Spatial Correlations (without regression)	12.7°	10.8°	11.5°	2.4°	26.0°
Spatial Correlations (with regression)	12.7°	5.3°	5.7°	1.2°	16.1°
Using inverse intensity chromaticity space	14.7°	11.0°	11.6°	3.2°	32.7°
Pixel-based gamut mapping	11.8°	8.9°	10.0°	2.8°	24.9°
Edge-based gamut mapping	13.7°	11.9°	12.3°	3.7°	26.9°
Intersection: complete 1-jet	11.8°	8.9°	10.0°	2.8°	24.9°
Regression (SVR)	13.1°	11.2°	11.8°	4.4°	25.0°
Statistical combination (No-*N*-Max)	10.3°	8.2°	8.8°	2.7°	21.2°
Using high level visual information	9.7°	7.7°	8.2°	2.3°	20.6°
Using natural image statistics	9.9°	7.7°	8.3°	2.4°	20.8°

between illuminant estimation using high level visual information and using natural image statistics is negligible (i.e., not statistically significant).

12.3.2.2 Color Checker Set The results on this data set are shown in Table 12.4 and some example results are shown in Figure 12.6. On this data set, the edge-based methods, that is, gray-edge, spatial correlations and edge-based gamut mapping,

Table 12.4 Performance of several methods on *linear* color checker.[a]

Method	Mean μ	Median	Trimean	Best, 25%	Worst, 25%
Do nothing	13.7°	13.6°	13.5°	10.4°	17.2°
White-patch ($\mathbf{e}^{0,\infty,0}$)	7.5°	5.7°	6.4°	1.5°	16.2°
Gray-world ($\mathbf{e}^{0,1,0}$)	6.4°	6.3°	6.3°	2.3°	10.6°
General gray-world ($\mathbf{e}^{0,p,\sigma}$)	4.7°	3.5°	3.8°	1.0°	10.1°
First-order gray-edge ($\mathbf{e}^{1,p,\sigma}$)	5.4°	4.5°	4.8°	1.9°	10.0°
Second-order gray-edge ($\mathbf{e}^{2,p,\sigma}$)	5.1°	4.4°	4.6°	1.9°	10.0°
Spatial correlations (without regression)	5.9°	5.1°	5.4°	2.4°	10.8°
Spatial correlations (with regression)	4.0°	3.1°	3.3°	1.1°	8.5°
Using inverse intensity chromaticity space	13.6°	13.6°	13.5°	9.5°	18.0°
Pixel-based gamut mapping	4.1°	2.5°	3.0°	0.6°	10.3°
Edge-based gamut mapping	6.7°	5.5°	5.8°	2.1°	13.7°
Intersection: complete 1-jet	4.1°	2.5°	3.0°	0.6°	10.3°
Bayesian	4.8°	3.5°	3.9°	1.3°	10.5°
Regression (SVR)	8.1°	6.7°	7.2°	3.3°	14.9°
Statistical combination (No-*N*-Max)	4.3°	3.4°	3.7°	1.4°	8.5°
Using high level visual information	3.5°	2.5°	2.6°	0.8°	8.0°
Using natural image statistics	4.2°	3.1°	3.5°	1.0°	9.2°

[a]Five hundred and Sixty-eight images, taken from Reference 232.

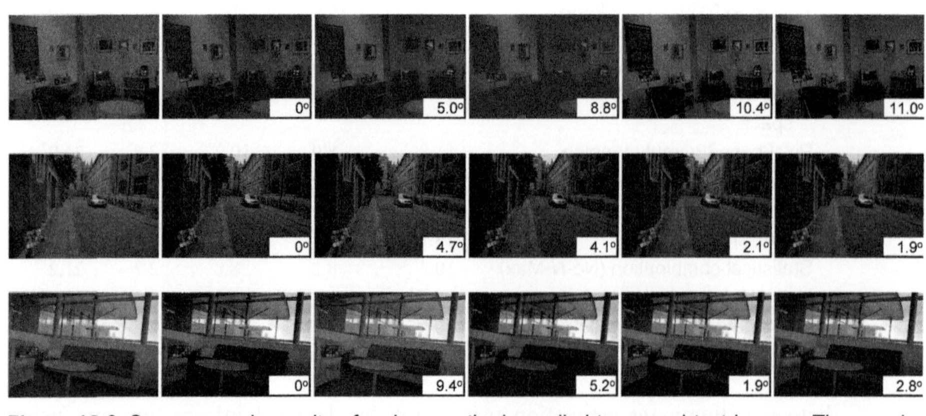

Figure 12.6 Some example results of various methods applied to several test images. The angular error is shown in the bottom right corner of the images. The methods used are, from left to right, perfect color constancy using ground truth, white-patch, first-order gray-edge, pixel-based gamut mapping and using natural image statistics.

perform significantly worse than pixel-based methods such as gamut mapping and general gray-world. However, it can be observed that the error on ''difficult'' images (i.e., images on which the method estimates an inaccurate illuminant, the *worst*, 25% column) for both types of algorithms is similar. This indicates that the performance of methods using low level information (either static algorithms or learning-based methods) is bounded by the information that is present. Using multiple algorithms is required to decrease the error of these ''difficult'' images, as can be seen by the performance of combination-based methods. Even though the increase in overall performance is not very high, methods using high level visual information and natural image statistics are statistically similar to the pixel-based gamut mapping, the largest improvement in accuracy is obtained on these difficult images (the mean angular error on the worst 25% of the images drops from 10.3° to 8.0° and 9.2°, respectively). Hence, to arrive at a robust color constancy algorithm that is able to accurately estimate the illuminant on any type of image, it is necessary to combine several approaches.

12.4 Summary

In this chapter, several often used approaches to illuminant estimates are evaluated. Further, in Chapters 9, 10, and 11, a wide range of methods is discussed.

Table 12.5 Summary of methods with advantages and disadvantages.

Method	Section	Pros	Cons
Static (using low level statistics)	Ch. 9.1, 9.2	Simple to implement Accurate for adequate parameters Fast execution	Opaque parameter selection Inaccurate for inferior parameters
Static (physics-based)	Ch. 9.3	No training phase Fast execution Few parameters	Difficult to implement Mediocre performance
Gamut-based	Ch. 10	Elegant underlying theory Potentially high accuracy	Requires training data Difficult to implement Requires proper preprocessing
Learning-based (using low level statistics)	Ch. 11.1, 11.2	Tunable for specific data set Simple to implement	Requires training data Slow execution
Learning-based (using higher level statistics)	Ch. 11.3	Potentially high accuracy Intuitive	Requires training data Inherently slower than single methods Difficult to implement
Learning-based (using semantics)	Ch. 11.4	Potentially high accuracy Incorporates semantics	Requires training data Difficult to implement Slow execution

Criteria that are important for computational color constancy algorithms are the requirement of training data, the accuracy of the estimation, the computational runtime of the method, the transparency of the approach, the complexity of the implementation, and the number of tunable parameters. A summary of the discussed methods is presented in Table 12.5.

Methods using low level statistics as discussed in Sections 9.1 and 9.2 are not dependent on training data and the parameters are not dependent on the input data. Such methods are called *static*. Existing methods include the gray-world, the white-patch and extensions to incorporate higher order statistics. Advantages of such methods are a simple implementation (often, merely a few lines of code are required) and fast execution. Further, the accuracy of the estimations can be quite high, provided the parameters are selected appropriately. On the other hand, inaccurate parameter selection can severely reduce the performance. Moreover, the selection of the optimal parameters is quite opaque, especially without prior knowledge on the input data. Physics-based methods discussed in Section 9.3 suffer less from the parameter selection, but are also less accurate (even for properly selected parameters).

Chapter 10 describes gamut-based methods, including an extension to incorporate the differential nature of images. The main advantage of gamut-based methods are the elegant underlying theory and the potential high accuracy. However, proper implementation requires some effort and appropriate preprocessing can severely influence the accuracy.

Finally, Chapter 11 describes methods than cannot operate without training phase. Sections 11.1 and 11.2 discuss methods that learn low level statistics, such as regression techniques and Bayesian approaches. Advantages of such methods are that they are (relatively) simple to implement and that they can be tuned toward specific data (such as indoor or outdoor images). Disadvantages are that the output is often rather nonintuitive since the model that is learned is quite opaque. On the other hand, methods using higher level statistics and semantics, as discussed in Sections 11.3 and 11.4, are often quite intuitive since it can be predicted beforehand which method will be selected for a specific input image. Moreover, the accuracy of such approaches has been proven to be state of the art. However, the use of multiple single algorithms means they are inherently slower than the single algorithms themselves.

PART IV

COLOR FEATURE EXTRACTION

13 Color Feature Detection

With contributions by Arnold W. M. Smeulders and Andrew D. Bagdanov

Differential-based features such as edges, corners, and salient points are widely used in a variety of applications such as matching, object recognition, and object tracking. Many applications are based on luminance-based features. In this chapter we discuss algorithms for the detection of color features in images. As we will see, this has several advantages over luminance-based features. First of all, we can apply the photometric invariance theory discussed in Chapter 6, which allows us to detect photometric invariant features. Secondly, color plays an important role in attributing saliency to images.

From a mathematical viewpoint the extension from luminance to color signals is an extension from scalar signals to vectorial signals. This change is accompanied by several mathematical obstacles. Straightforward application of existing luminance-based operators on the separate color channels, and subsequent combination of the results, often fails. For example, combining derivatives with a simple addition of separate channels results in cancellation in the case of opposing vectors [234, 235]. This is illustrated in Figure 13.1. For the blue-red and cyan-yellow edge on the right of Figure 13.1a, the vectors in the red and blue channel point in opposite directions and a summation could result in a zero-edge response, while a prominent edge is clearly present. Also, for more complex local features,

Color in Computer Vision: Fundamentals and Applications, First Edition.
Theo Gevers, Arjan Gijsenij, Joost van de Weijer, and Jan-Mark Geusebroek.

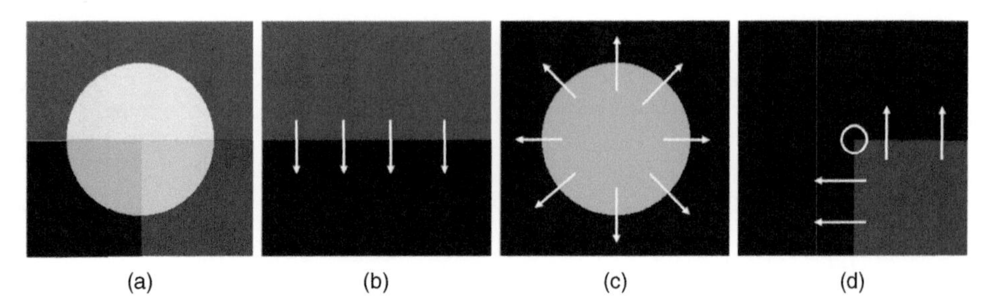

Figure 13.1 (a) Example image and (b) red channel, (c) green channel, and (d) blue channel of example image. Superimposed are arrows indicating the directions of the gradients in the separate channels and a circle indicating corner detections. Note that although only a single corner is detected in the separate channels, the original color image has four corner points.

such as corners and T-junctions, the combination of the channels poses problems. Applying a corner detector to the separate channels results in a single detected corner in the blue channel. However, there is no evidence for the cross-points on the border of the circle in any of the separate channels. As a result, a combination of corner information from the separate channels will fail. In the first part of the chapter we investigate how to combine the differential structure of the color channels in a principled way.

The second part of the chapter is on the detection of salient color features in images [149]. Throughout this chapter we use the term *saliency* to refer to events in images that have higher information content. Here we use information content in an information theoretical sense, from which we know that, for example, rare events are more informative than frequent events. In Figure 13.2b, the color gradient of an image containing a colorful bird on a dull background is depicted. Surprisingly, for the image of the standard color gradient, the edges between the bird and the background are dominated by the edges, which are caused by intensity variations in the background. For human observers, the transition between the bird and the background clearly pops out. In the second part of this chapter

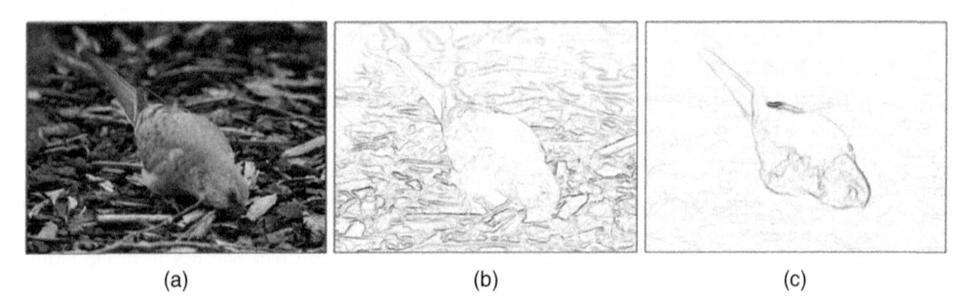

Figure 13.2 (a) Input color image, (b) color gradient image, and (c) color-boosted image derivative. The color gradient edges do not reflect the clear edge that is observed by humans between the background and the colorful bird. The salient edges detected by the color-boosted algorithm coincide with edges that are found important by humans.

we look into color image statistics with the aim of finding the edges in images that have the highest information content. The method, which will be explained, is called *color boosting*. The result of this method is given in Figure 13.2c: the important transitions between the bird and the background yield the most prominent responses.

13.1 The Color Tensor

Here we discuss the usage of the color tensor for color feature computation. We also look into how to combine the photometric invariance theory developed in Chapter 6 with the differential-based-features presented in this chapter.

Let us first look into several desired properties of color features. First, the features should target the photometric variation needed for their application. This ensures that accidental physical events, such as shadows and specularities, do not influence results. Second, features must be robust against noise and should not contain instabilities. Especially for the full photometric invariant features instabilities require caution. Third, the theory should be generally applicable to ensure that it can be applied to the vast literature on features for luminance images. We start from the observation that tensors are well suited to combine first-order derivatives for color images. Then we will show how to combine tensor-based features with photometric derivatives for photometric invariant feature detection and extraction. Finally, we show that for feature extraction applications, for which quasi-invariants are unsuited (see Chapter 6), an uncertainty measure that robustifies feature extraction can be introduced.

As we saw in the start of this chapter, simply summing the differential structure of various color channels may result in cancellation even when evident structure exists in the image [234]. This is further illustrated in Figure 13.3. Rather than adding the direction information (defined on $[0, 2\pi\rangle$) of the channels, it is more appropriate to sum the orientation information (defined on $[0, \pi\rangle$). Such a method is provided by tensor mathematics for which vectors in opposite directions reinforce one another. Tensors describe the local orientation rather than the direction. More precisely, the tensor of a vector and its 180° rotated counterpart vector are equal. It is for that reason that we use the tensor as a basis for color feature detection.

Given an image f, the structure tensor is given by Bigun et al. [236]:

$$\mathbf{G} = \begin{pmatrix} \overline{f_x^2} & \overline{f_x f_y} \\ \overline{f_x f_y} & \overline{f_y^2} \end{pmatrix}, \qquad (13.1)$$

where the subscripts indicate spatial derivatives and the bar $(\bar{\cdot})$ indicates convolution with a Gaussian filter. Note that there are two scales involved in the computation of the structure tensor: the scale at which the derivatives are computed and the tensor scale that is the scale at which the spatial derivatives are averaged. The structure tensor describes the local differential structure of images,

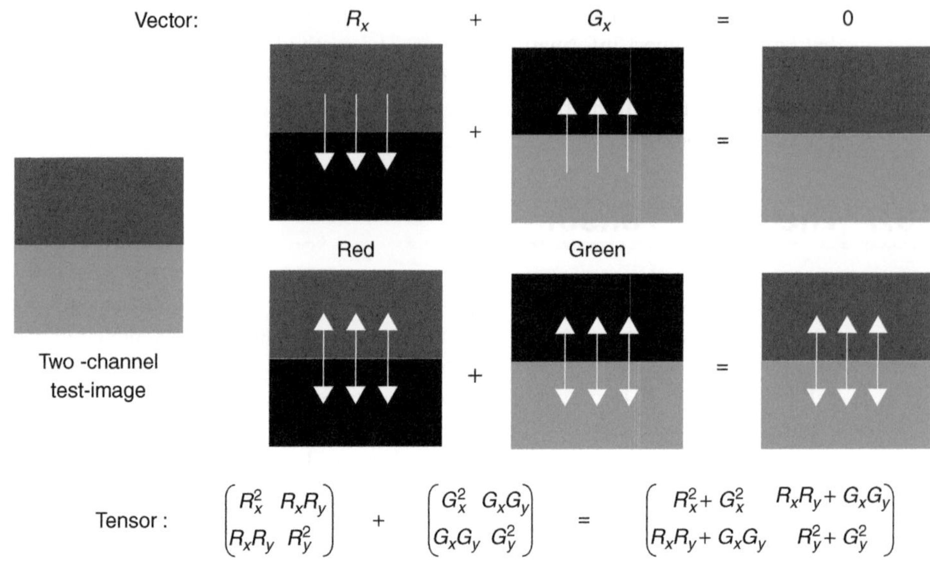

Figure 13.3 Example of edge detection in color images with two channels, R and G. In the case of a simple vector summation, no edge is detected because of the cancellation of the derivatives, R_x and G_x, in the red and green channels. Edge detection based on tensors will be successful because the structure tensors code orientation rather than direction. The structure tensors of the red and green channels reinforce each other and a clear edge is detected.

and is suited to finding features such as edges and corners [234, 237, 238]. For a multichannel image $\mathbf{f} = (f^1, f^2, \ldots, f^n)^T$, the structure tensor is given by

$$\mathbf{G} = \begin{pmatrix} \overline{\mathbf{f}_x \cdot \mathbf{f}_x} & \overline{\mathbf{f}_x \cdot \mathbf{f}_y} \\ \overline{\mathbf{f}_y \cdot \mathbf{f}_x} & \overline{\mathbf{f}_y \cdot \mathbf{f}_y} \end{pmatrix}. \tag{13.2}$$

In the case where $\mathbf{f} = (R, G, B)$, this yields the color tensor:

$$\mathbf{G} = \begin{pmatrix} \overline{R_x^2 + G_x^2 + B_x^2} & \overline{R_xR_y + G_xG_y + B_xB_y} \\ \overline{R_xR_y + G_xG_y + B_xB_y} & \overline{R_y^2 + G_y^2 + B_y^2} \end{pmatrix}. \tag{13.3}$$

Later in this chapter we derive a certainty measure for photometric derivatives. This measure can be used as a weight in the color tensor. For derivatives that are accompanied by a weighting function, w_x and w_y, which appoint a weight to every measurement $\mathbf{f_x}$ and $\mathbf{f_y}$, the structure tensor is defined by

$$\mathbf{G} = \begin{pmatrix} \dfrac{\overline{w_x^2 \mathbf{f}_x \cdot \mathbf{f}_x}}{\overline{w_x^2}} & \dfrac{\overline{w_x w_y \mathbf{f}_x \cdot \mathbf{f}_y}}{\overline{w_x w_y}} \\ \dfrac{\overline{w_y w_x \mathbf{f}_y \cdot \mathbf{f}_x}}{\overline{w_y w_x}} & \dfrac{\overline{w_y^2 \mathbf{f}_y \cdot \mathbf{f}_y}}{\overline{w_y^2}} \end{pmatrix}. \tag{13.4}$$

The elements of the tensor are known to be invariant under rotation and translation of the spatial axes. To prove the invariant, we use the fact that $\frac{\partial}{\partial x}\mathbf{Rf} = \mathbf{Rf}_x$, where $\mathbf{R}$ is a rotation operator,

$$\overline{(\mathbf{Rf}_x)^T \mathbf{Rf}_y} = \overline{\mathbf{f}_x^T \mathbf{R}^T \mathbf{Rf}_y} = \overline{\mathbf{f}_x^T \mathbf{f}_y}, \tag{13.5}$$

where we have rewritten the inner product according to $\mathbf{f} \cdot \mathbf{f} = \mathbf{f}^T \mathbf{f}$.

13.1.1 Photometric Invariant Derivatives

A good motivation for using color images is that photometric information can be exploited to understand the physical causes of features. For example, pixels can be classified as being from the same color but having different intensities, which is possibly caused by a shadow or a shading change in the image. Further, pixel differences can also indicate specular reflection. For many applications it is important to distinguish the scene incidental information from material edges. When color images are converted to luminance this photometric information is lost.

Photometric invariance in Equation 13.2 can be obtained by using invariant derivatives to compute the structure tensor. In Chapter 6, we derived photometric full- and quasi-invariants. Quasi-invariants differ from full invariants in that they vary with respect to a physical parameter. We also saw that full invariants can be computed from quasi-invariants by the normalization with a signal-dependent scalar. The quasi-invariants have the advantage that they do not exhibit the instabilities common to full photometric invariants. However, the applicability of the quasi-invariants is restricted to photometric invariant feature detection. For feature extraction full photometric invariance is desired.

We will briefly summarize the relevant results from Chapter 6 here. The dichromatic model (see Chapter 3 for more details) divides the reflection in the interface (specular) and body (diffuse) reflection component for optically inhomogeneous materials according to

$$\mathbf{f} = e(m^b\, \mathbf{c}^b + m^i\, \mathbf{c}^i), \tag{13.6}$$

in which $\mathbf{c}^b$ is the color of the body reflectance, $\mathbf{c}^i$ the color of the interface reflectance (i.e., specularities or highlights), m^b and m^i are scalars representing the corresponding magnitudes of reflection, and e is the intensity of the light source. For matte surfaces there is no interface reflection and the model further simplifies to

$$\mathbf{f} = em^b\, \mathbf{c}^b. \tag{13.7}$$

The photometric derivative structure of the image can be computed by computing the spatial derivative of Equation 13.6:

$$\mathbf{f}_x = em^b\, \mathbf{c}_x^b + (e_x m^b + em_x^b)\, \mathbf{c}^b + \left(em_x^i + e_x m^i\right)\, \mathbf{c}^i. \tag{13.8}$$

The spatial derivative is a summation of three weighted vectors, representing in sequence body reflectance, shading-shadow and specular changes. From Equation 13.7, it follows that for matte surfaces the shadow-shading direction is parallel to the *RGB* vector, $\mathbf{f}||\mathbf{c}_b$. The specular direction follows from the assumption that the color of the light source is known.

For matte surfaces (i.e., $m^i = 0$), the projection of the spatial derivative on the shadow-shading axis yields the shadow-shading variant containing all energy, which could be explained by changes due to shadow and shading. Subtraction of the shadow-shading variant $\mathbf{S}_x$ from the total derivative $\mathbf{f}_x$ results in the shadow-shading quasi-invariant:

$$\begin{aligned}\mathbf{S}_x &= \left(\mathbf{f}_x \cdot \hat{\mathbf{f}}\right)\hat{\mathbf{f}} = \left(em^b\left(\mathbf{c}_x^b \cdot \hat{\mathbf{f}}\right) + \left(e_x m^b + em_x^b\right)\left|\mathbf{c}^b\right|\right)\hat{\mathbf{f}},\\ \mathbf{S}_\mathbf{x}^\mathbf{c} &= \mathbf{f}_x - \mathbf{S}_x = em^b\left(\mathbf{c}_x^b - \left(\mathbf{c}_x^b \cdot \hat{\mathbf{f}}\right)\hat{\mathbf{f}}\right),\end{aligned} \tag{13.9}$$

which does not contain derivative energy caused by shadows and shading. The hat, $(\hat{\cdot})$, denotes unit vectors. The full shadow-shading invariant results from normalizing the quasi-invariant $\mathbf{S}_x^c$ by the intensity magnitude $|\mathbf{f}|$:

$$\mathbf{s}_x = \frac{\mathbf{S}_x^c}{|\mathbf{f}|} = \frac{em^b}{em^b\left|\mathbf{c}^b\right|}\left(\mathbf{c}_x^b - \left(\mathbf{c}_x^b\right)\cdot\hat{\mathbf{f}}\right), \tag{13.10}$$

which is invariant for m^b.

For the construction of the shadow-shading-specular quasi-invariant, we introduce the hue direction that is perpendicular to the light source direction $\hat{\mathbf{c}}^i$ and the shadow-shading direction $\hat{\mathbf{f}}$:

$$\hat{\mathbf{b}} = \frac{\hat{\mathbf{f}} \times \hat{\mathbf{c}}^i}{\left|\mathbf{f} \times \mathbf{c}^i\right|}. \tag{13.11}$$

Projection of the derivative, $\mathbf{f}_x$, on the hue direction results in the shadow-shading-specular quasi-invariant:

$$\mathbf{H}_x^c = \left(\mathbf{f}_x \cdot \hat{\mathbf{b}}\right)\hat{\mathbf{b}} = em^b\left(\mathbf{c}_x^b \cdot \hat{\mathbf{b}}\right) + \left(e_x m^b + em_x^b\right)\left(\mathbf{c}^b \cdot \mathbf{b}\right). \tag{13.12}$$

The second part of this equation is zero if we assume that shadow-shading changes do not occur within a specularity, since then either $(e_x m^b + em_x^b) = 0$ or $(\mathbf{c^b} \cdot \mathbf{b}) = (\mathbf{f} \cdot \mathbf{b}) = 0$. Subtraction of the quasi-invariant $\mathbf{H}_x^c$ from the spatial derivative $\mathbf{f}_x$ results in the shadow-shading-specular variant $\mathbf{H}_x$:

$$\mathbf{H}_x = \mathbf{f}_x - \mathbf{H}_x^c. \tag{13.13}$$

The full shadow-shading invariant is computed by dividing the quasi-invariant by the saturation. The saturation is equal to the norm of the color vector, $\mathbf{f}$, after the projection on the plane perpendicular to the light source direction (which is equal to subtraction of the part in the light source direction). The final invariant is thus

$$\mathbf{h}_x = \frac{\mathbf{H}_x^c}{\left|\mathbf{f} - \left(\mathbf{f} \cdot \hat{\mathbf{c}}^i\right)\hat{\mathbf{c}}^i\right|} = \frac{em^b}{em^b\left|\mathbf{c}^b - \left(\mathbf{c}^b \cdot \hat{\mathbf{c}}^i\right)\hat{\mathbf{c}}^i\right|}\left(\mathbf{c}_x^b \cdot \hat{\mathbf{b}}\right). \tag{13.14}$$

The expression $\mathbf{h}_x$ is invariant for both m^i and m^b.

13.1.2 Invariance to Color Coordinate Transformations

From a physical point of view, features that are invariant to rotation of the coordinate axes make sense. This starting point has been applied in the design of image geometry features, resulting in, for example, gradient and Laplace operators [62]. For the design of physically meaningful color features not only the invariance with respect to spatial coordinate changes is desired but also the invariance with respect to rotations of the color coordinate systems. Features based on different measurement devices that measure the same spectral space should yield the same results.

For color images, values are represented in the *RGB* coordinate system. In fact, the infinite-dimensional Hilbert space is sampled with three probes, which results in the red, green, and blue channels (Fig. 13.4). For operations on the color coordinate system to be physically meaningful they should be independent of orthonormal transformation of the three axes in Hilbert space. An example of an orthonormal color coordinate system is the opponent color space (Fig. 13.4b). The opponent color space spans the same subspace as the subspace defined by the *RGB* axes and hence both subspaces should yield the same features.

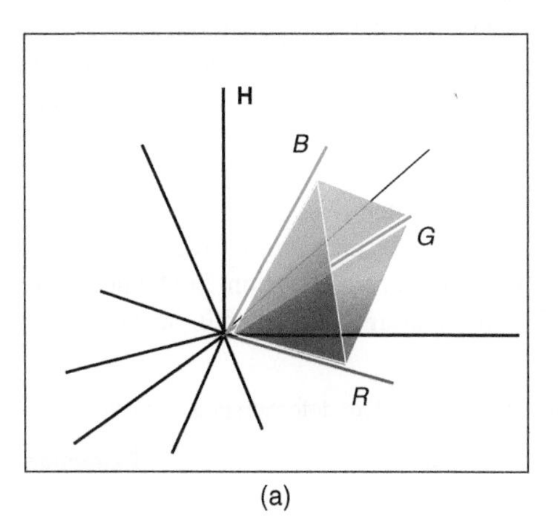

(a)

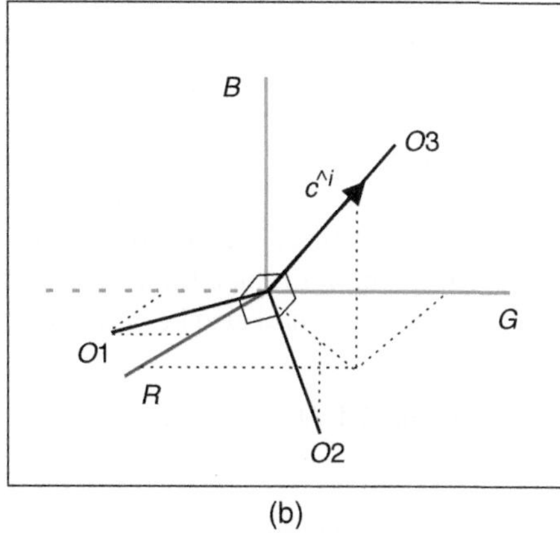

(b)

Figure 13.4 (a) The subspace of measured light in the Hilbert space of possible spectra. (b) The *RGB* coordinate system and an alternative orthonormal color coordinate system which spans the same subspace. *Source*: Reprinted with permission, © 2006 IEEE.

By projecting the local spatial derivative on three photometric axes in the *RGB* cube we have derived the photometric quasi-invariants. These can be combined with the structure tensor of Equation 13.2 for photometric quasi-invariant feature detection. We would like these features to be independent of the accidental choice of the color coordinate frame. As a consequence, a rotation of the color coordinates should result in the same rotation of the quasi-invariant derivatives. For example, for the shadow-shading quasi-variant $\mathbf{S}_x$ this can be proved by

$$\left((\mathbf{Rf}_x)^T \mathbf{R\hat{f}}\right)\left(\mathbf{R\hat{f}}\right) = \left(\mathbf{f}_x^T\mathbf{R}^T\mathbf{R\hat{f}}\right)\left(\mathbf{R\hat{f}}\right) = \mathbf{R}\left(\mathbf{f}_x^T\mathbf{\hat{f}}\right)\mathbf{\hat{f}} = \mathbf{RS}_x, \tag{13.15}$$

where $\mathbf{R}$ is the rotation operator. Similar proofs hold for the other photometric variants and quasi-invariants. The invariance with respect to color coordinate transformation of the shadow-shading full invariants follow from the fact that $|\mathbf{Rf}| = |\mathbf{f}|$. For the shadow-shading-specular full invariant, the rotational invariance is proved by the fact that the inner product between two vectors remains the same under rotations, and therefore $|\mathbf{Rf} - (\mathbf{Rf}\cdot\mathbf{R\hat{c}}^i)\mathbf{R\hat{c}}^i| = |\mathbf{R}(\mathbf{f} - (\mathbf{f}\cdot\hat{\mathbf{c}}^i)\hat{\mathbf{c}}^i)|$. Since the elements of the structure tensor are also invariant for color coordinate transformations (Eq. 13.5) the combination of the quasi-invariants and the structure tensor into a quasi-invariant structure tensor is also invariant for color coordinate transformations.

13.1.3 Robust Full Photometric Invariance

In Section 13.1.1, the quasi- and full invariant derivatives are described. The quasi-invariants outperform the full invariants in terms of discriminative power and are more robust to noise (Section 6.2.4). However, the quasi-invariants are not suited for applications that require feature extraction. These applications compare the photometric invariant values between various images and need full photometric invariance (Table 13.1). A disadvantage of full photometric invariants is that they are unstable in certain areas of the RGB cube. For example, the invariants for shadow shading and specularities are unstable near the gray axis. These instabilities greatly reduce the applicability of the invariant derivatives for which a small deviation of the original pixel color value may result in a large deviation of the invariant derivative. To counter this, we discuss a measure that describes the uncertainty of the photometric invariant derivatives, thereby allowing for robust full photometric invariant feature detection. Such uncertainty measures could also be derived for the full invariants described in Section 6.1.

Table 13.1 Applicability of the different invariants for feature detection and extraction.

	Detection	Extraction
Quasi-invariant	+++	-
Full invariant	+	+
Robust full invariant	++	++

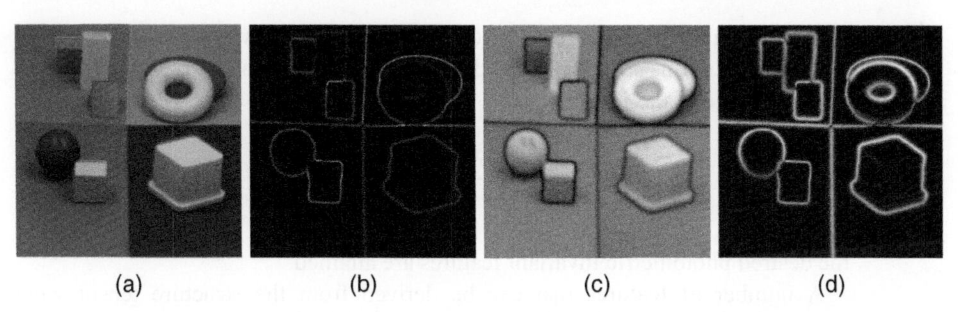

(a) (b) (c) (d)

Figure 13.5 (a) Test image, (b) hue derivative, (c) saturation, and (d) quasi-invariant.

We first derive the uncertainty for the shadow-shading full invariant from its relation to the quasi-invariant. We assume additive uncorrelated uniform Gaussian noise. Owing to the high-pass nature of differentiation we can assume the noise of the zero-order signal ($|\mathbf{f}|$) to be negligible compared to the noise on the first-order signal ($\mathbf{S}_x^c$). In Section 13.1.1, the quasi-invariant has been derived by a linear projection of the derivative $\mathbf{f}_x$ on the plane perpendicular to the shadow-shading direction. Therefore, uniform noise in $\mathbf{f}_x$ will result in uniform noise in $\mathbf{S}_x^c$. The noise in the full invariant can be written as

$$\tilde{\mathbf{s}}_x = \frac{\mathbf{S}_x^c + \sigma}{|\mathbf{f}|} = \frac{\mathbf{S}_x^c}{|\mathbf{f}|} + \frac{\sigma}{|\mathbf{f}|}. \tag{13.16}$$

The uncertainty of the measurement of $\tilde{\mathbf{s}}_x$ depends on the magnitude of $|\mathbf{f}|$. For small $|\mathbf{f}|$ the error increases proportionally. Therefore, it is better to weight the full shadow-shading invariant with the function $w = |\mathbf{f}|$ to robustify the color tensor based on the chromatic invariant.

For the shadow-shading-specular invariant, the weighting function should be proportional to the saturation since

$$\tilde{\mathbf{h}}_x = \frac{\mathbf{H}_x^c + \sigma}{|\mathbf{s}|} = \frac{\mathbf{H}_x^c}{|\mathbf{s}|} + \frac{\sigma}{|\mathbf{s}|}. \tag{13.17}$$

Hence, $w = |\mathbf{s}|$ should be used as the weighting function of the hue derivative $\tilde{\mathbf{h}}_x$, (Fig. 13.5). On locations where there is an edge, the saturation drops and with the saturation the certainty of the hue measurement. The quasi-invariant (Fig. 13.5d), which is equal to the weighted hue, is more stable than the full invariant derivative because of the incorporation of the certainty in the measurements. With the derived weighting function we can compute the robust photometric invariant tensor (Eq. 13.4).

13.1.4 Color-Tensor-Based Features

In this section we show the generality of the method by summing features that can be derived from the color tensor. In Sections 13.1.1 and 13.1.3 we described how to compute invariant derivatives. Depending on the task at hand you should use either quasi-invariants for detection or robust full invariants for extraction. The features in this section will be derived for $\mathbf{g}_x$. By replacing the inner product of $\mathbf{g}_x$ with one of the following:

$$\left\{ \overline{\mathbf{f}_x \cdot \mathbf{f}_x}, \overline{\mathbf{S}_x^c \cdot \mathbf{S}_x^c}, \frac{\overline{\mathbf{S}_x^c \cdot \mathbf{S}_x^c}}{|\mathbf{f}|^2}, \overline{\mathbf{H}_x^c \cdot \mathbf{H}_x^c}, \frac{\overline{\mathbf{H}_x^c \cdot \mathbf{H}_x^c}}{|\mathbf{s}|^2} \right\}, \tag{13.18}$$

the desired photometric invariant features are attained.

A number of features that can be derived from the structure tensor were proposed by scientists who where designing features for oriented patterns [175]. Oriented patterns (e.g., fingerprint images) are defined as patterns with a dominant orientation everywhere. For oriented patterns, other mathematics are needed than for regular object images. The local structure of object images is described by a step edge, whereas for oriented patterns the local structure is described as a set of lines (roof edges). Lines generate opposing vectors on a small scale. Hence, for geometric operations on oriented patterns, methods are needed for which opposing vectors reinforce one another. This is the same problem as encountered for all color images, where the opposing vector problem does not only occur for oriented patterns but also for step edges for which the opposing vectors occur in the different channels. Hence, similar equations were found in both fields. Apart from orientation estimation, a number of other estimators were proposed by oriented pattern research [236–240]. These operations are based on adaptations of the structure tensor and can also be applied to the color structure tensor. We will now look into several of these tensor-based features.

13.1.4.1 Eigenvalue-Based Features We start by describing features derived from the eigenvalues of the tensor. Eigenvalue analysis of the tensor leads to two eigenvalues that are defined by

$$\begin{aligned} \lambda_1 &= \tfrac{1}{2}\left(\overline{\mathbf{g}_x \cdot \mathbf{g}_x} + \overline{\mathbf{g}_y \cdot \mathbf{g}_y} + \sqrt{\left(\overline{\mathbf{g}_x \cdot \mathbf{g}_x} - \overline{\mathbf{g}_y \cdot \mathbf{g}_y}\right)^2 + \left(2\overline{\mathbf{g}_x \cdot \mathbf{g}_y}\right)^2}\right), \\ \lambda_2 &= \tfrac{1}{2}\left(\overline{\mathbf{g}_x \cdot \mathbf{g}_x} + \overline{\mathbf{g}_y \cdot \mathbf{g}_y} - \sqrt{\left(\overline{\mathbf{g}_x \cdot \mathbf{g}_x} - \overline{\mathbf{g}_y \cdot \mathbf{g}_y}\right)^2 + \left(2\overline{\mathbf{g}_x \cdot \mathbf{g}_y}\right)^2}\right). \end{aligned} \tag{13.19}$$

The direction of λ_1 indicates the prominent local orientation

$$\theta = \tfrac{1}{2}\arctan\left(\frac{2\overline{\mathbf{g}_x \cdot \mathbf{g}_y}}{\overline{\mathbf{g}_x \cdot \mathbf{g}_x} - \overline{\mathbf{g}_y \cdot \mathbf{g}_y}}\right). \tag{13.20}$$

The λs can be combined to give the following local descriptors:

- $\lambda_1 + \lambda_2$ describes the total local derivative energy.
- λ_1 is the derivative energy in the most prominent direction.

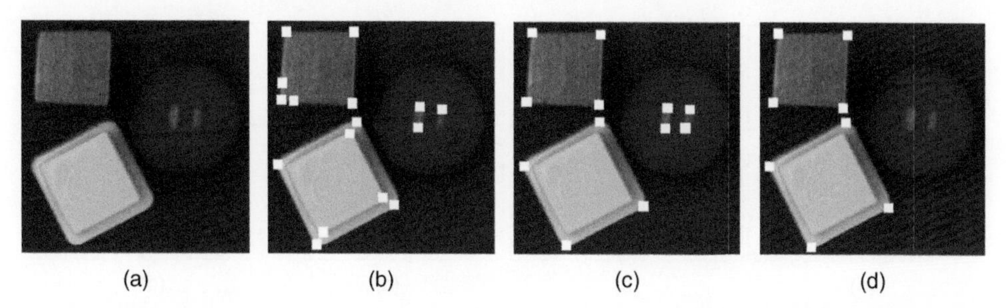

Figure 13.6 (a) Input image and Harris corner detector results based on (b) RGB gradient ($\mathbf{f}_x$), (c) shadow-shading quasi-invariant ($\mathbf{S}_x^c$), and (d) shadow-shading-specular quasi-invariant ($\mathbf{H}_x^c$).

- $\lambda_1 - \lambda_2$ describes the line energy [241]. The derivative energy in the prominent orientation is corrected for the energy contributed by the noise λ_2.
- λ_2 describes the amount of derivative energy perpendicular to the prominent local orientation that is used to select features for tracking [242].

13.1.4.2 Color Harris Detector An often applied feature detector is the Harris corner detector [53]. The color Harris operator H can be written as a function of the eigenvalues of the structure tensor:

$$\begin{aligned} H\mathbf{f} &= \overline{\mathbf{g}_x \cdot \mathbf{g}_x}\,\overline{\mathbf{g}_y \cdot \mathbf{g}_y} - \overline{\mathbf{g}_x \cdot \mathbf{g}_y}^2 - k\left(\overline{\mathbf{g}_x \cdot \mathbf{g}_x} + \overline{\mathbf{g}_y \cdot \mathbf{g}_y}\right)^2 \\ &= \lambda_1 \lambda_2 - k\left(\lambda_1 + \lambda_2\right)^2 . \end{aligned} \tag{13.21}$$

Corner detection results are given in Figure 13.6. As can be seen, the shadow-shading quasi-invariant detector does not detect shadow-shading corners, whereas the shadow-shading-specular quasi-invariant also ignores the specular corners.

13.1.4.3 Color Canny Edge Detection We illustrate the use of eigenvalue-based features by adapting the Canny edge detection algorithm to allow for vectorial input data. The algorithm consists of the following steps:

1. Compute the spatial derivatives, $\mathbf{f}_x$ and combine them if desired into a quasi-invariant (Eq. 13.9 or 13.12).
2. Compute the maximum eigenvalue (Eq. 13.19) and its orientation (Eq. 13.20).
3. Apply nonmaximum suppression on λ_1 in the prominent direction.

In Figure 13.7, the results of color Canny edge detection for several photometric quasi-invariants are shown. The results show that the luminance-based Canny (Fig. 13.7b), misses several edges that are correctly found by the *RGB*-based method in Figure 13.7c. Also, the removal of spurious edges by photometric

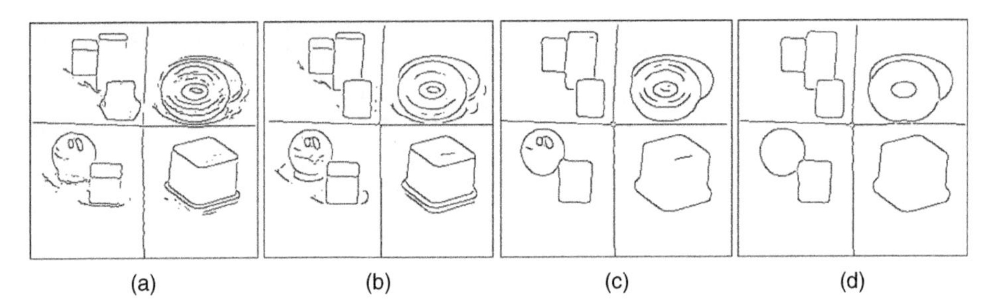

(a) (b) (c) (d)

Figure 13.7 Canny edge detection based on successively (a) luminance derivative (b) RGB derivatives, (c) the shadow-shading quasi-invariant, and (d) the shadow-shading-specular quasi-invariant.

invariance is demonstrated. In Figure 13.7d, the edge detection is robust to shadow and shading changes and only detects material and specular edges. In Figure 13.7e, only the material edges are depicted.

13.1.4.4 Color Symmetry Detectors The structure tensor of Equation 13.2 can also be seen as a local projection of the derivative energy on two perpendicular axes, namely, $\mathbf{u}_1 = (1 \quad 0)^T$ and $\mathbf{u}_2 = (0 \quad 1)^T$,

$$\mathbf{G}^{\mathbf{u}_1,\mathbf{u}_2} = \begin{pmatrix} \overline{(\mathbf{G}_{x,y}\mathbf{u}_1)\cdot(\mathbf{G}_{x,y}\mathbf{u}_1)} & \overline{(\mathbf{G}_{x,y}\mathbf{u}_1)\cdot(\mathbf{G}_{x,y}\mathbf{u}_2)} \\ \overline{(\mathbf{G}_{x,y}\mathbf{u}_1)\cdot(\mathbf{G}_{x,y}\mathbf{u}_2)} & \overline{(\mathbf{G}_{x,y}\mathbf{u}_2)\cdot(\mathbf{G}_{x,y}\mathbf{u}_2)} \end{pmatrix}, \tag{13.22}$$

in which $\mathbf{G}_{x,y} = (\mathbf{g}_x \quad \mathbf{g}_y)$. From the Lie group of transformation several other choices of perpendicular projections can be derived [237, 238]. They include feature extraction for circle, spiral, and starlike structures.

The star and circle detector is given as an example. It is based on $\mathbf{u}_1 = \frac{1}{\sqrt{x^2+y^2}}(x \quad y)^T$, which coincides with the derivative of circular patterns and $\mathbf{u}_2 = \frac{1}{\sqrt{x^2+y^2}}(-y \quad x)^T$, which denotes the perpendicular vector field that coincides with the derivative pattern of starlike patterns. These vectors can be used to compute the adapted structure tensor with Equation 13.22. Only the elements on the diagonal have nonzero entries and are equal to

$$\mathbf{H} = \begin{pmatrix} H_{11} & H_{12} \\ H_{21} & H_{22} \end{pmatrix}, \tag{13.23}$$

where $H_{12} = H_{21} = 0$ and

$$H_{11} = \overline{\frac{x^2}{x^2+y^2}\mathbf{g}_x\cdot\mathbf{g}_x} + \overline{\frac{2xy}{x^2+y^2}\mathbf{g}_x\cdot\mathbf{g}_y} + \overline{\frac{y^2}{x^2+y^2}\mathbf{g}_y\cdot\mathbf{g}_y} \tag{13.24}$$

$$H_{22} = \overline{\frac{x^2}{x^2+y^2}\mathbf{g}_y\cdot\mathbf{g}_y} - \overline{\frac{2xy}{x^2+y^2}\mathbf{g}_x\cdot\mathbf{g}_y} + \overline{\frac{y^2}{x^2+y^2}\mathbf{g}_x\cdot\mathbf{g}_x}. \tag{13.25}$$

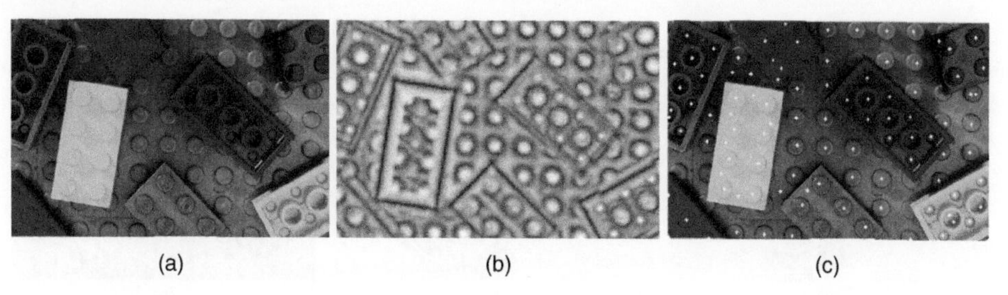

Figure 13.8 (a) Input image, (b) the circularity coefficient C, and (c) the detected circles.

Here, λ_1 describes the amount of derivative energy contributing to circular structures and λ_2 the derivative energy that describes a starlike structure. Similar to the proof given in Equation 13.5, the elements of Equation 13.23 can be proved to be invariant under transformations of the *RGB* space.

We apply the circle symmetry detector to an image containing Lego blocks (Fig. 13.8). Because we know that the color within the blocks remains the same, circle detection is done on the shadow-shading-specular variant, $\mathbf{H}_x$ (Eq. 13.13). The shadow-shading-specular variant contains all the derivative energy except for the energy that can only be caused by a material edge. With the shadow-shading-specular variant the circular energy λ_1 and the starlike energy λ_2 are computed according to Equation 13.23. Dividing the circular energy by the total energy yields a descriptor of local circularity (Fig. 13.8b):

$$C = \frac{\lambda_1}{\lambda_1 + \lambda_2}. \tag{13.26}$$

The superimposed maxima of C (Fig. 13.8c) gives a good estimation of the circle centers.

13.1.4.5 Color Curvature Curvature is another feature that can be derived from an adaptation of the structure tensor as proposed in Reference 240. The fit between the local differential structure and a parabolic model function can be written as a function of the curvature. Finding the optimum of this function yields an estimation of the local curvature. For vector data, the equation for the curvature is given by

$$\kappa = \frac{\overline{w^2 \mathbf{g}_v \cdot \mathbf{g}_v} - \overline{w^2} \cdot \overline{\mathbf{g}_w \cdot \mathbf{g}_w} - \sqrt{\left(\overline{w^2} \cdot \overline{\mathbf{g}_w \cdot \mathbf{g}_w} - \overline{w^2 \mathbf{g}_v \cdot \mathbf{g}_v}\right)^2 + 4\overline{w^2} \cdot \overline{w \mathbf{g}_v \cdot \mathbf{g}_w}^2}}{2\overline{w^2} \cdot \overline{w \mathbf{g}_v \cdot \mathbf{g}_w}}, \tag{13.27}$$

in which $\mathbf{g}_v$ and $\mathbf{g}_w$ are the derivatives in gauge coordinates.

The use of photometric invariant orientation and curvature estimation is demonstrated on a circle detection example. Circular object recognition is difficult

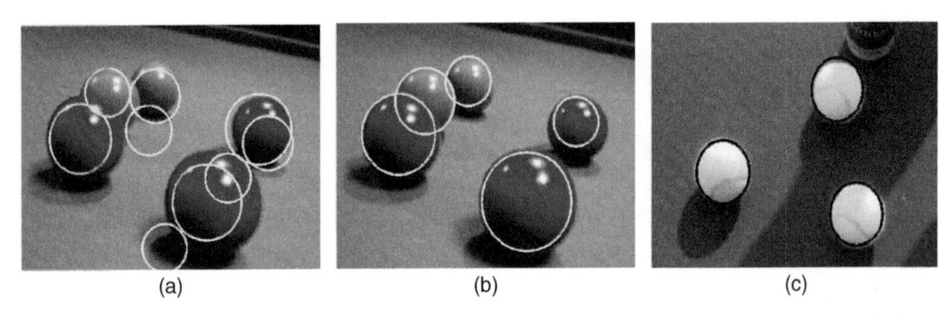

Figure 13.9 (a) Detected circles based on luminance, (b) detected circles based on shadow-shading-specular quasi-invariant, and (c) detected circles based on shadow-shading-specular quasi-invariant.

because of shadow, shading, and specular events that influence the feature extraction. We apply the following algorithm for circle detection:

1. Compute the spatial derivatives, $\mathbf{f}_x$, and combine them if desired into a quasi-invariant (Eq. 13.9 or 13.12).
2. Compute the local orientation with Equation 13.20 and curvature with Equation 13.27.
3. Compute the Hough space [243], $H(R, x^0, y^0)$, where R is the radius of the circle and x^0 and y^0 indicate the center of the circle. The computation of the orientation and curvature reduces the number of votes per pixel to 1. For example, for a pixel at position $\mathbf{x} = (x^1, y^1)$,

$$\begin{aligned} R &= \tfrac{1}{\kappa}, \\ x^0 &= x^1 + \tfrac{1}{\kappa}\cos\theta, \\ y^0 &= y^1 + \tfrac{1}{\kappa}\sin\theta. \end{aligned} \tag{13.28}$$

   Every pixel votes with the derivative energy $\sqrt{\mathbf{f}_x \cdot \mathbf{f}_x}$.
4. Compute the maxima in the Hough space. These maxima indicate the circle centers and their radii.

In Figure 13.9, the results of circle detection are given. The luminance-based circle detection is corrupted by photometric variation in the image. Nine circles had to be detected before the five balls were detected. For the shadow-shading-specular quasi-invariant based method, the five most prominent peaks in the Hough space coincide with reasonable estimates of the radii and center points of the circles. In Figure 13.9c, an outdoor example with a shadow partially covering the objects is given.

13.1.4.6 Color Optical Flow Optical flow can also be computed from the structure tensor. This was originally proposed by Simoncelli [244] and has been extended to color in References [245, 246]. The vector of a multichannel point over time stays constant [247, 248]:

$$\frac{d\mathbf{g}}{dt} = \mathbf{0}. \tag{13.29}$$

Differentiating yields the following set of equations

$$\mathbf{G_{x,y}}\ \mathbf{v} + \mathbf{g_t} = \mathbf{0}, \tag{13.30}$$

with $\mathbf{v}$ the optical flow. To solve the singularity problem and to robustify the optical flow computation we follow Simoncelli et al. [244] and assume a constant flow within a Gaussian window. Solving Equation 13.30 leads to the following optical flow equation:

$$\mathbf{v} = \overline{(\mathbf{G}_{x,y} \cdot \mathbf{G}_{x,y})^{-1}}\ \overline{\mathbf{G}_{x,y} \cdot \mathbf{g}_t} = \mathbf{M}^{-1}\mathbf{b}, \tag{13.31}$$

with

$$\mathbf{M} = \begin{pmatrix} \overline{\mathbf{g}_x \cdot \mathbf{g}_x} & \overline{\mathbf{g}_x \cdot \mathbf{g}_y} \\ \overline{\mathbf{g}_y \cdot \mathbf{g}_x} & \overline{\mathbf{g}_y \cdot \mathbf{g}_y} \end{pmatrix}, \tag{13.32}$$

and

$$\mathbf{b} = \begin{pmatrix} \overline{\mathbf{g}_x \cdot \mathbf{g}_t} \\ \overline{\mathbf{g}_y \cdot \mathbf{g}_t} \end{pmatrix}. \tag{13.33}$$

The assumption of color optical flow based on *RGB* is that *RGB* pixel values remain constant over time (Eq. 13.29). A change in brightness introduced because of a shadow or a light source with fluctuating brightness such as the sun results in a nonexistent optical flow. This problem can be overcome by assuming constant chromaticity over time. For photometric invariant optical flow, full invariance is necessary since optical flow estimation is based on comparing the (extracted) edge response of multiple frames. Consequently, photometric invariant optical flow can be attained by replacing the inner product of $\mathbf{g}_x$ by one of the following:

$$\left\{ \frac{\overline{\mathbf{S}_x^c \cdot \mathbf{S}_x^c}}{\overline{|\mathbf{f}|^2}}, \frac{\overline{\mathbf{H}_x^c \cdot \mathbf{H}_x^c}}{\overline{|\mathbf{s}|^2}} \right\}. \tag{13.34}$$

An example of color optical flow in a real-world scene is given in Figure 13.10. Multiple frames are taken from static objects while the light source position is changed. This results in a violation of the brightness constraint by changing the shading and moving the shadows. Since neither the camera nor the objects moved, the ground truth optical flow is zero. The violation of the brightness constraint disturbs the optical flow estimation based on *RGB* (Fig. 13.10b). The shadow-shading invariant optical flow estimation is much less disturbed by the violation of

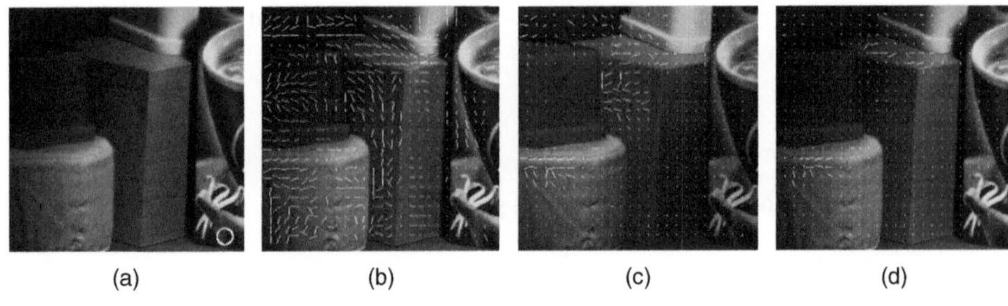

(a) (b) (c) (d)

Figure 13.10 (a) Frame 1 of object scene with filter size superimposed on it. (b) *RGB* gradient optical flow, (c) shadow-shading invariant optical flow, and (d) robust shadow-shading invariant optical flow.

the brightness constraint (Fig. 13.10c). However, flow estimation is still unstable around some of the edges. The robust shadow-shading invariant optical flow has the best results and is only unstable in low-gradient areas (Fig. 13.10d).

13.1.5 Experiment: Robust Feature Point Detection and Extraction

Here we compare full, quasi- and robust full invariants with respect to photometric changes, stability of the invariants, and robustness to noise. Further, the ability of invariants to detect and extract features is examined (see also Table 13.2). The experiment is performed with the photometric invariant Harris corner detector (Eq. 13.21) and is executed on the Soil-47 multiobject set [249], which consists of 23 images (Fig. 13.11a).

First, the feature detection accuracy of the invariants is tested. For each image and invariant, the 20 most prominent Harris points are extracted. Next, uncorrelated Gaussian noise is added to the data, and Harris point detection is computed 10 times per image. The percentage of points that do not correspond to the Harris points in the noiseless case are given in Table 13.2. The Harris point detector based on the quasi-invariant outperforms the alternatives. The instability of the full invariant can be partially repaired by the robust full invariant; however, for detection purposes the quasi-invariants remain the best choice.

(a) (b) (c)

Figure 13.11 (a) An example from the Soil-47 image. (b) Shadow-shading distortion with the shadow-shading quasi-invariant Harris points superimposed. (c) Specular distortion and the shadow-shading-specular Harris points superimposed.

Table 13.2 Percentage of falsely detected points and percentage of wrongly classified points.[a]

	Detection error %		Extraction error %	
Standard deviation noise	**5**	**20**	**5**	**20**
Shadow shading				
Quasi-invariant	<u>5.1</u>	<u>20.2</u>	100	100
Full invariant	11.7	50.1	8.7	56.6
Robust full invariant	6.4	37.7	<u>3.0</u>	<u>35.3</u>
Shadow-shading-specular				
Quasi-invariant	<u>9.7</u>	<u>46.6</u>	100	98.2
Full invariant	38.8	75.5	62.3	84.0
Robust full invariant	15.7	60.2	<u>9.8</u>	<u>66.6</u>

Underlined values indicate the lowest error.
[a]Classification is based on the extraction of invariant information. Uncorrelated Gaussian noise is added with standard deviation 5 and 20.

Next, feature extraction for the invariants is tested. Again, the 20 most prominent Harris points are detected in the noise-free image. For these points, the photometric invariant derivative energy is extracted by $\sqrt{\lambda_1 + \lambda_2 - 2\lambda_n}$, where λ_n is an estimation of the noise that contributes to the energy in both λ_1 and λ_2. To imitate photometric variations of images we apply the following photometric distortion to the images (compare with Eq. 13.6):

$$\mathbf{g}(\mathbf{x}) = \alpha(\mathbf{x})\mathbf{f}(\mathbf{x}) + \beta(\mathbf{x})\mathbf{c}^i + \eta(\mathbf{x}), \tag{13.35}$$

where $\alpha(\mathbf{x})$ is a smooth function resembling variation similar to shading and shadow effects, $\beta(\mathbf{x})$ is a smooth function which imitates specular reflections, and $\eta(\mathbf{x})$ is Gaussian noise. To test the shadow-shading extraction, $\alpha(\mathbf{x})$ is chosen to vary between 0 and 1, and $\beta(\mathbf{x})$ is 0. To test the shadow-shading-specular invariants, $\alpha(\mathbf{x})$ was chosen constant at 0.7 and $\beta(\mathbf{x})$ varied between 0 and 50. After photometric distortion, the derivative energy is extracted at the same 20 points. The extraction is considered correct if the deviation of the derivative energy between the distorted and the noise-free case is less then 10%. The results are given in Table 13.2. Quasi-invariants, which are not suited for extraction, have a 100% error. The full invariants have better results, but with decreasing signal-to-noise ratio its performance drops drastically. In accordance with the theory in Section 13.1.3 the robust full invariants successfully improve performance.

13.2 Color Saliency

Visual saliency is the quality that makes certain parts of a scene stand out from their surroundings. Designing computational models for saliency is an active field of research in computer vision [250]. It is known that color plays an important role in attributing saliency [251]. In this section we investigate how to compute salient color features.

Saliency is especially relevant for applications that are based on local feature-based image representation. For these applications, it is important to select the most salient local features. The more salient the local features, the better and more compact the final image description. Indexing objects and object categories as a collection of salient points has been successfully applied to several applications, such as image matching, content-based retrieval, learning, and recognition [55, 252, 253].

Examples of salient points are local features in the image that exhibit geometrical structure, such as T-junctions, corners, and symmetry points. Applications based on salient points are generally composed of three phases: (i) a feature detection phase locating the features, (ii) an extraction phase in which local descriptions are extracted at the detected locations, and (iii) a matching phase in which the extracted descriptors are matched against a database of descriptors. Here we investigate how we can exploit color saliency in the detection phase.

An example of a salient feature detector is the Harris corner detector [53]. In Figure 13.12, an example of the Harris detector based on several input signals is given. When looking at the results of the luminance and RGB-based detector,

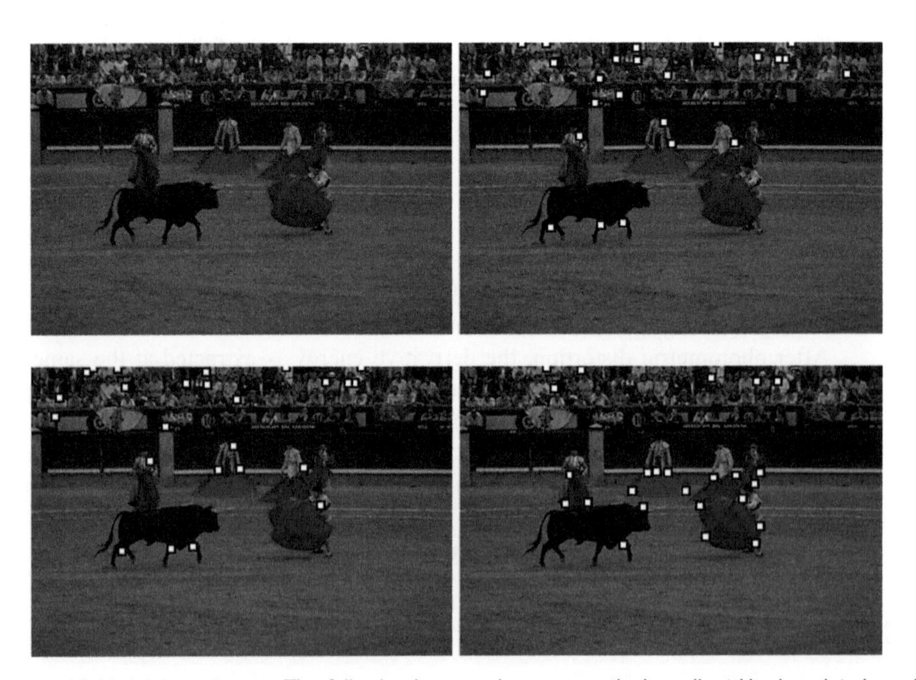

Figure 13.12 (a) Input image. The following images show, respectively, salient Harris points based on luminance (b), RGB (c), and color boosting (d). For each detector, the 25 most salient points are depicted. Only color-boosted detector finds points on the salient capes.

we see that they are almost identical. Apparently, the black and white changes that dominate the luminance-based detector also dominate the RGB detector. Both detectors fail to detect the color salient capes. In this section we develop a technique called *color boosting*, which focuses on the most informative color events in an image.

The color boosting technique is general in that it can be applied to existing salient point detectors, which are based on image derivatives. In general, salient point detectors are derived from a saliency map, which describes the saliency for every location in the image. For a color image, salient points are the maxima of the saliency map, which compares the derivative vectors in a neighborhood fixed by scale σ,

$$s = H^{\sigma}\left(\mathbf{f}_x, \mathbf{f}_y\right), \tag{13.36}$$

where H is the saliency function and the subscript indicates differentiation with respect to the parameter. This type of saliency maps includes References [53, 235, 237, 239, 254]. The impact of a derivative vector on the outcome of the local salience depends on its vector norm, $|\mathbf{f}_x|$. Hence, vectors with equal norm have an equal impact on local saliency. Rather than deriving saliency from the vector norm, the challenge is to adapt the saliency function in order that vectors with equal color distinctiveness have an equal impact on the saliency function.

13.2.1 Color Distinctiveness

The effectiveness of salient point detection depends on the distinctiveness of the extracted salient points. At the salient points' positions, local neighborhoods are extracted and described by local image descriptors. The distinctiveness of the descriptor defines the conciseness of the representation and the discriminative power of the salient points. The distinctiveness of interest points is measured by their information content.

The information content of the local patch can be measured by looking at the distinctiveness of the local color 1-jet descriptor

$$\mathbf{v} = \begin{pmatrix} R & G & B & R_x & G_x & B_x & R_y & G_y & B_y \end{pmatrix}^T. \tag{13.37}$$

The information content of this color descriptor includes the information content of more complex local color descriptors such as color differential invariant descriptors, since these complex descriptors are computed from the elements of Equation 13.37.

From information theory, it is known that the information content of an event is dependent on its frequency or probability. Events that occur rarely are more informative. The dependency of information content on its probability is given by

$$I(\mathbf{v}) = -\log\left(p\left(\mathbf{v}\right)\right), \tag{13.38}$$

where $p(\mathbf{v})$ is the probability of the descriptor $\mathbf{v}$. The information content of the descriptor, given by Equation 13.37, is approximated by assuming independent probabilities of the zeroth-order signal and the first- order derivatives:

$$p(\mathbf{v}) = p(\mathbf{f})\, p(\mathbf{f}_x)\, p(\mathbf{f}_y)\,. \tag{13.39}$$

To improve the information content of the salient point detector, defined by Equation 13.36, the probability of the derivatives, $p(\mathbf{f}_x)$, should be small.

We can now restate our aim in a more precise manner. The aim is to find a transformation $g : \Re^3 \rightarrow \Re^3$, for which

$$p(\mathbf{f}_x) = p(\mathbf{f}'_x) \leftrightarrow |g(\mathbf{f}_x)| = |g(\mathbf{f}'_x)|\,. \tag{13.40}$$

This implies that two vectors, $\mathbf{f}_x$ and $\mathbf{f}'_x$, with equal information content have equal impact on the saliency function. The transformation, attained by the function g, is called *color saliency boosting*. Similar equations hold for $p(\mathbf{f}_y)$. Once a color boosting function g has been found, the color-boosted saliency can be computed with

$$s = H^{\sigma}\left(g(\mathbf{f}_x), g(\mathbf{f}_y)\right). \tag{13.41}$$

The classic saliency of Equation 13.36 derives saliency from the orientations and gradient strength of the derivatives in a local neighborhood. After color boosting, the saliency is based on the orientations and the information content of these derivatives. Gradient strength has been replaced by information content, thereby better representing the aim of saliency detectors.

From Equation 13.40, the color boosting function g is found by analyzing the probabilities of the derivatives. The channels of $\mathbf{f}_x$, $\{R_x, G_x, B_x\}$, are correlated because of the physics of the world. Photometric events in the world, such as shading and reflection of the light source in specularities, influence *RGB* values in a well-defined manner. Before investigating the statistics of color derivatives, the derivatives need to be transformed to a color space that is uncorrelated with respect to these photometric events.

13.2.2 Physics-Based Decorrelation

Here we describe three color coordinate transformations which partition *RGB*-space differently. The transformation are the same as the ones used to obtain photometric invariance in 6.2. Here we use the same color transformations to decorrelate the spatial derivative, $\mathbf{f}_x$, into axes that are photometrically variant and photometrically invariant.

13.2.2.1 Spherical Color Spaces The spherical color transformation (Fig. 13.13a), is given by

$$\begin{pmatrix} \theta \\ \varphi \\ r \end{pmatrix} = \begin{pmatrix} \arctan\left(\frac{G}{R}\right) \\ \arcsin\left(\frac{\sqrt{R^2+G^2}}{\sqrt{R^2+G^2+B^2}}\right) \\ r = \sqrt{R^2+G^2+B^2} \end{pmatrix}. \tag{13.42}$$

The spatial derivatives are transformed to the spherical coordinate system by

$$S(\mathbf{f}_x) = \mathbf{f}_x^s = \begin{pmatrix} r\sin\varphi\,\theta_x \\ r\varphi_x \\ r_x \end{pmatrix} = \begin{pmatrix} \frac{G_x R - R_x G}{\sqrt{R^2+G^2}} \\ \frac{R_x RB + G_x GB - B_x(R^2+G^2)}{\sqrt{R^2+G^2}\sqrt{R^2+G^2+B^2}} \\ \frac{R_x R + G_x G + B_x B}{\sqrt{R^2+G^2+B^2}} \end{pmatrix}. \tag{13.43}$$

The scale factors follow from the Jacobian of the transformation. They ensure that the norm of the derivative remains constant under transformation, hence $|\mathbf{f}_x| = |\mathbf{f}_x^s|$. In the spherical coordinate system, the derivative vector is a summation of a shadow-shading variant part, $\mathbf{S}_x = (0, 0, r_x)^T$ and a shadow-shading quasi-invariant part, given by $\mathbf{S}_x^c = (r\sin\varphi\theta_x, r\varphi_x, 0)^T$.

13.2.2.2 Opponent Color Spaces The opponent color space (Fig. 13.13b) is given by

$$\begin{pmatrix} o1 \\ o2 \\ o3 \end{pmatrix} = \begin{pmatrix} \frac{R-G}{\sqrt{2}} \\ \frac{R+G-2B}{\sqrt{6}} \\ \frac{R+G+B}{\sqrt{3}} \end{pmatrix}. \tag{13.44}$$

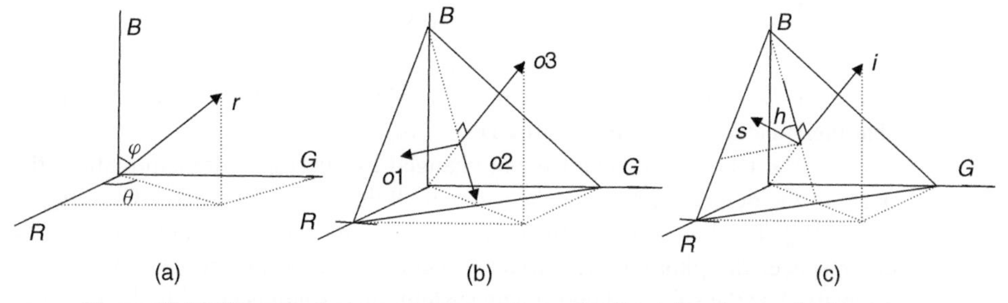

Figure 13.13 The spherical, opponent, and hue-saturation-intensity coordinate system.

For this, the transformation of the derivatives is as follows:

$$O(\mathbf{f}_x) = \mathbf{f}_x^o = \begin{pmatrix} o1_x \\ o2_x \\ o3_x \end{pmatrix} = \begin{pmatrix} \frac{1}{\sqrt{2}}(R_x - G_x) \\ \frac{1}{\sqrt{6}}(R_x + G_x - 2B_x) \\ \frac{1}{\sqrt{3}}(R_x + G_x + B_x) \end{pmatrix}. \tag{13.45}$$

The opponent color space decorrelates the derivative with respect to specular changes. The derivative is divided into a specular variant part, $\mathbf{O}_x = (0, 0, o3_x)^T$, and a specular quasi-invariant part $\mathbf{O}_x^c = (o1_x, o2_x, 0)^T$.

13.2.2.3 Hue-Saturation-Intensity Color Spaces The hue-saturation-intensity (Fig. 13.13c) is given by

$$\begin{pmatrix} h \\ s \\ i \end{pmatrix} = \begin{pmatrix} \arctan\left(\frac{o1}{o2}\right) \\ \sqrt{o1^2 + o2^2} \\ o3 \end{pmatrix}. \tag{13.46}$$

The transformation of the spatial derivatives into the *hsi* space decorrelates the derivative with respect to specular, shadow, and shading variations,

$$H(\mathbf{f}_x) = \mathbf{f}_x^h = \begin{pmatrix} s\,h_x \\ s_x \\ i_x \end{pmatrix} \tag{13.47}$$

$$= \begin{pmatrix} \dfrac{(R(B_x - G_x) + G(R_x - B_x) + B(G_x - R_x))}{\sqrt{2(R^2 + G^2 + B^2 - RG - RB - GB)}} \\ \dfrac{R(2R_x - G_x - B_x) + G(2G_x - R_x - B_x) + B(2B_x - R_x - G_x)}{\sqrt{6(R^2 + G^2 + B^2 - RG - RB - GB)}} \\ \dfrac{(R_x + G_x + B_x)}{\sqrt{3}} \end{pmatrix}.$$

The shadow-shading-specular variant is given by $\mathbf{H}_x = (0, 0, i_x)^T$ and the shadow-shading-specular quasi-invariant by $\mathbf{H}_x^c = (sh_x, s_x, 0)^T$.

Since the length of a vector is not changed by coordinate transformation, the norm of the derivative remains the same in all three representations $|\mathbf{f}_x| = |\mathbf{f}_x^c| = |\mathbf{f}_x^o| = |\mathbf{f}_x^h|$. For both the opponent color space and the hue-saturation-intensity color space, the photometric variant direction is given by the $L1$ norm of the intensity. For the spherical coordinate system, the variant is equal to the $L2$ norm of the intensity.

The three color spaces that we discussed decorrelate the color spaces with respect to various physical events. In the decorrelated color spaces, frequent physical variations, such as intensity changes, will only influence the photometric variant axes. We will examine the color derivative statistics in these decorrelated color spaces.

13.2.3 Statistics of Color Images

As discussed in Section 13.2.1, the information content of a descriptor depends on the probability of the derivatives. Here, we investigate the statistics of color derivatives in the decorrelated color spaces. From the statistics we aim to find a mathematical description of surfaces of equal probability, the so-called isosalient surfaces since a description of these surfaces leads to the solution of Equation 13.40.

The statistics of color images are shown for the Corel database [255], which consists of 40,000 images after the exclusion of black and white ones. In Figure 13.14, the distributions of the first-order derivatives, $\mathbf{f}_x$, are given for the various color coordinate systems. The isosalient surfaces show a remarkably simple structure, approximately similar to an ellipsoid. For all three color spaces, the third coordinate axis coincides with the axis of maximum variation (i.e., the intensity). For the opponent and the spherical coordinate system, the first and second coordinates are rotated, with rotation matrix R^{ϕ}, so that the first coordinate coincides with the axis of minimum variation

$$
\begin{aligned}
\left(r \sin \tilde{\varphi}\, \tilde{\theta}_x, r\tilde{\varphi}_x\right)^T &= R^{\phi}\left(r \sin \varphi \theta_x, r\varphi_x\right)^T, \\
\left(\tilde{o}1_x, \tilde{o}2_x\right)^T &= R^{\phi}\left(o1_x, o2_x\right)^T.
\end{aligned}
\tag{13.48}
$$

The tilde indicates the color space transformation with the aligned axes. Similarly, the aligned transformations are given by $\tilde{S}(\mathbf{f}_x) = \mathbf{f}_x^{\tilde{s}}$ and $\tilde{O}(\mathbf{f}_x) = \mathbf{f}_x^{\tilde{o}}$.

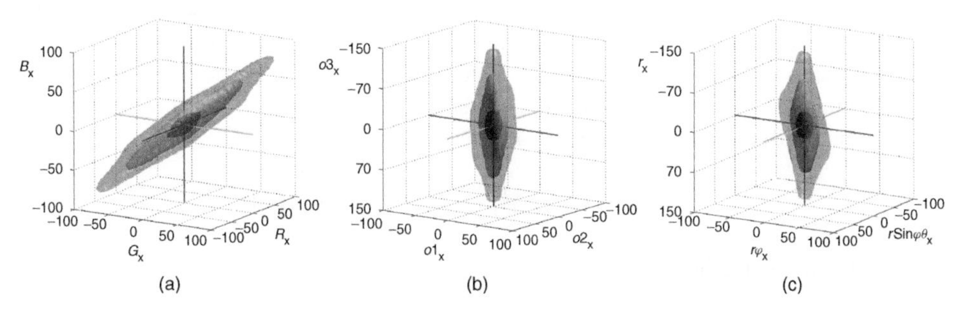

Figure 13.14 The histograms of the distribution of the transformed derivatives of the Corel image database in the (a) RGB coordinates, (b) the opponent coordinates, and (c) the spherical coordinates, respectively. The three planes correspond to the isosalient surfaces, which contain (from dark to light), respectively, 90%, 99%, 99.9% of the total number of pixels. *Source: Reprinted with permission,* © 2006 IEEE.

After alignment of the axes, isosalient surfaces of the derivative histograms can be approximated by ellipsoids:

$$\left(\alpha h_x^1\right)^2 + \left(\beta h_x^2\right)^2 + \left(\gamma h_x^3\right)^2 = R^2, \tag{13.49}$$

where $\mathbf{h}_x = h\left(\mathbf{f}_x\right) = \left(h_x^1, h_x^2, h_x^3\right)^T$ and h is one of the transformations $\tilde{S}$, $\tilde{O}$, or H.

We can compute the parameters α, β, and γ, which describe the derivative distribution from large data sets. Here, we show results for a subset of 1000 randomly chosen images from the Corel data set. We fit ellipses to the histogram of the data set as follows. First, the axes of the opponent and the spherical transformation are aligned by Equation 13.48. Next, the axes of the ellipsoid are derived by fitting the isosaliency surface, which contains 99% of the pixels of the histogram of the Corel data set. The results for the various transformations are summarized in Table 13.3. The relation between the axes in the various color spaces clearly confirms the dominance of the luminance axis in the *RGB* cube, since γ, the multiplication-factor of the luminance axis, is much smaller than the color-axes multiplication factors, α and β.

Table 13.3 The ellipsoid parameters for the corel data set computed for gaussian derivatives with $\sigma = 1$.

	$\mathbf{f}_x$	$\mathbf{f}_x^{\tilde{s}}$	$\mathbf{f}_x^{\tilde{o}}$	$\mathbf{f}_x^{h}$
α	0.577	0.851	0.850	0.858
β	0.577	0.515	0.524	0.509
γ	0.577	0.099	0.065	0.066

13.2.4 Boosting Color Saliency

We now return to our goal, that is, of incorporating color distinctiveness into salient point detection. Or, mathematically, to find the transformation for which vectors with equal information content have an equal impact on the saliency function. In the previous section, we saw that derivatives of equal saliency form an ellipsoid. Since Equation 13.49 is equal to

$$\left(\alpha h_x^1\right)^2 + \left(\beta h_x^2\right)^2 + \left(\gamma h_x^3\right)^2 = \left|\Lambda h\left(\mathbf{f}_x\right)\right|^2, \tag{13.50}$$

the following holds:

$$p\left(\mathbf{f}_x\right) = p\left(\mathbf{f}'_x\right) \leftrightarrow \left|\Lambda h\left(\mathbf{f}_x\right)\right| = \left|\Lambda h\left(\mathbf{f}'_x\right)\right|, \tag{13.51}$$

Figure 13.15 (a, c) Corel input images. (b, d) Results of Harris detector (red dots) and the Harris detector with color boosting (yellow dots). The red dots mainly coincide with black and white events, while the yellow dots are focused on colorful points.

where Λ is a 3×3 diagonal matrix with $\Lambda_{11} = \alpha$, $\Lambda_{22} = \beta$, and $\Lambda_{33} = \gamma$. Λ is restricted to $\Lambda_{11}^2 + \Lambda_{22}^2 + \Lambda_{33}^2 = 1$. The desired saliency boosting function (Eq. 13.40) is obtained:

$$g\left(\mathbf{f}_x\right) = \Lambda h\left(\mathbf{f}_x\right). \tag{13.52}$$

The oriented isosalient ellipsoids are transformed into spheres by a rotation of the color axes followed by a rescaling of the axis. The vectors of equal saliency are thereby transformed into vectors of equal length.

In Figure 13.15, results of the *RGB*-gradient-based and color-boosted Harris detector are depicted. From a color information point of view, the *RGB*-gradient-based method does a poor job. Most of the salient points have a black and white local neighborhood, with a low color saliency. The salient points, after color boosting, focus on more distinctive points. Color boosting can be applied to all derivative-based detectors. In the introduction of this chapter we saw an example of a color-boosted image gradient in Figure 13.2.

It should be noted that color boosting negatively influences the signal-to-noise of the detector. Depending on the task at hand, distinctiveness may be less desired than signal-to-noise. To balance both criteria, we introduce a parameter α, which allows for choosing between best signal-to-noise characteristics, $\alpha = 0$, and best information content, $\alpha = 1$:

$$g^{\alpha}\left(\mathbf{f}_x\right) = \alpha \Lambda h\left(\mathbf{f}_x\right) + (1 - \alpha)\, h\left(\mathbf{f}_x\right). \tag{13.53}$$

For $\alpha = 0$, this is equal to color-gradient-based salient point detection.

13.2.5 Evaluation of Color Distinctiveness

For the evaluation of salient point detectors two criteria are considered to be important: (i) *distinctiveness*, salient points should focus on events with a low probability of occurrence; and (ii) *repeatability*, salient point detection should be stable under the varying viewing conditions, such as geometrical changes and photometric changes. Most salient point detectors are designed according to these criteria [256]. In this and the following section, we look at how color boosting influences distinctiveness and repeatability. We start by analyzing color distinctiveness in this section.

We have chosen the Harris point detector (Section 13.1.4) to test color boosting. In Reference 256, the Harris detector has already been shown to outperform other detectors both on 'shape' distinctiveness and repeatability. It is computed with

$$H^{\sigma}\left(\mathbf{f}_x, \mathbf{f}_y\right) = \overline{\mathbf{f}_x \cdot \mathbf{f}_x}\,\overline{\mathbf{f}_y \cdot \mathbf{f}_y} - \overline{\mathbf{f}_x \cdot \mathbf{f}_y}^{2} - k\left(\overline{\mathbf{f}_x \cdot \mathbf{f}_x} + \overline{\mathbf{f}_y \cdot \mathbf{f}_y}\right)^2, \tag{13.54}$$

by substituting $\mathbf{f}_x$ and $\mathbf{f}_y$ by $g\left(\mathbf{f}_x\right)$ and $g\left(\mathbf{f}_y\right)$.

The color distinctiveness of salient point detectors is described by the information content of the descriptors extracted at the locations of the salient points. From the combination of Equations 13.38 and 13.39, it follows that the total information is computed by summing the information of the zeroth- and first-order part, $I(\mathbf{v}) = I(\mathbf{f}) + I(\mathbf{f}_x) + I(\mathbf{f}_y)$. The information content of the parts is computed from histograms with

$$I\left(\mathbf{f}\right) = -\sum_i p_i \log\left(p_i\right), \tag{13.55}$$

where p_i are the probabilities of the bins of the histogram of $\mathbf{f}$.

The results for 20 and 100 salient points per image are shown in Table 13.4. Next to the absolute information content we have also computed the relative information gain with respect to the information content of the color-gradient-based Harris detector. For this purpose, the information content of a single image is defined as

$$I = -\sum_{j=1}^{n} \log\left(p\left(v_j\right)\right), \tag{13.56}$$

where $j = 1, 2, \ldots n$ and n is the number of salient points in the image. Here, $p(v_j)$ is computed from the global histograms, which allows comparison of the results per image. The information content change is considered substantial for a 5% increase or decrease.

Table 13.4 The information content of salient point detectors.

	20 points			100 points		
Method	**Information content**	**Increase (%)**	**Decrease (%)**	**Information content**	**Increase**	**Decrease**
$\mathbf{f}_x$	20.4	—	—	20.0	—	—
$\lVert\mathbf{f}_x\rVert_1$	19.9	0	1.4	19.8	0	0.8
$\tilde{\mathbf{S}}_x^c$	22.2	45.5	10.1	20.4	9.1	17.7
$\mathbf{f}_x^{\tilde{s}}$	22.3	49.4	.6	20.8	13.1	1.3
$\tilde{\mathbf{O}}_x^c$	22.6	51.4	12.9	20.5	12.0	34.2
$\mathbf{f}_x^{\tilde{o}}$	23.2	62.6	0.0	21.4	21.5	0.9
$\mathbf{H}_x^c$	21.0	21.7	43.4	19.0	1.8	77.4
$\mathbf{f}_x^h$	23.0	57.2	0.3	21.3	16.7	1.1
Random	14.4	0	99.8	14.4	0	100

Underlined values indicate the lowest error.
[a] (i) Measured in information content
[b] (ii) the percentage of images for which a substantial decrease (-5%) or increase ($+5\%$) of the information content occurs. The experiment is performed with both 20 and 100 salient points per image.

The highest information content is obtained with $\mathbf{f}_x^{\tilde{o}}$, which is the color-saliency-boosted version of opponent derivatives. Boosting results in an 7% to 13% increase of the information content compared to the color-gradient-based detector. On the images of the Corel set, this resulted in a substantial increase on 22% to 63% of the images. The advantage of color boosting diminishes when increasing the number of salient points per image. This is caused by the limited number of color clues in many of the images. It is also noteworthy to see how small the difference is between luminance ($\lVert\mathbf{f}_x\rVert_1$) and *RGB*-based ($\mathbf{f}_x$) Harris detection. Since the intensity direction also dominates the *RGB* derivatives, using the *RGB* gradient instead of luminance-based Harris detection only results in a substantial increase in information content in 1% of the images.

13.2.6 Repeatability

We described two criteria for salient point detection, namely, distinctiveness and repeatability. The color boosting algorithm is designed to focus on color distinctiveness, while adopting the geometrical characteristics of the operator to which it is applied. Let us have a closer look at how color boosting influences repeatability. We identify two phenomena that influence the repeatability of $g(\mathbf{f}_x)$. Firstly, by boosting color saliency an anisotropic transformation is carried out. This will reduce the signal-to-noise ratio negatively, which would negatively influence the repeatability. Secondly, by boosting the photometric invariant directions more than the photometric variant directions, we improve robustness with respect to scene accidental changes (such as shadows) which improves repeatability. Let us analyze these two effects in more detail.

13.2.6.1 Signal-to-Noise For isotropic uncorrelated noise, ε, the measured derivative $\hat{\mathbf{f}}_x$ can be written as

$$\hat{\mathbf{f}}_x = \mathbf{f}_x + \varepsilon, \tag{13.57}$$

and after color saliency boosting

$$g\left(\hat{\mathbf{f}}_x\right) = g\left(\mathbf{f}_x\right) + \Lambda\varepsilon. \tag{13.58}$$

Note that isotropic noise remains unchanged under the orthogonal curvilinear transformations. Assume the worst case in which $\mathbf{f}_x$ only has signal in the photometric variant direction, then the noise can be written as

$$\frac{\left|g\left(\mathbf{f}_x\right)\right|}{\left|\Lambda\varepsilon\right|} \approx \frac{\Lambda_{33}\left|\mathbf{f}_x\right|}{\Lambda_{11}\left|\varepsilon\right|}. \tag{13.59}$$

Hence, the signal-to-noise ratio reduces by $\frac{\Lambda_{11}}{\Lambda_{33}}$, which will negatively influence repeatability to geometrical and photometrical changes.

The loss of repeatability caused by color saliency boosting is examined by adding uniform, uncorrelated Gaussian noise of $\sigma = 10$. This yields a good indication of loss in signal-to-noise, which in turn will influence results of repeatability under other variations, such as zooming, illumination changes, and geometrical changes. Repeatability is measured by comparing the Harris points detected in the noisy image to the points in the noise-free images. The results in Table 13.5 correspond to the expectation made by Equation 13.59, namely, the larger the difference between Λ_{11} and Λ_{33}, the poorer the repeatability.

Table 13.5 The percentage of harris points that remain detected after adding gaussian uncorrelated noise.

Method	20 points	100 points
$\mathbf{f}_x$	88	84
$\lvert\mathbf{f}_x\rvert_1$	88	83
$\mathbf{f}_x^{\tilde{s}}$	62	54
$\mathbf{f}_x^{\tilde{o}}$	51	41
$\mathbf{f}_x^{h}$	52	42

In Figure 13.16, the information content and repeatability as a function of the color boosting, determined by the α-parameter, are given (Eq. 13.53). The

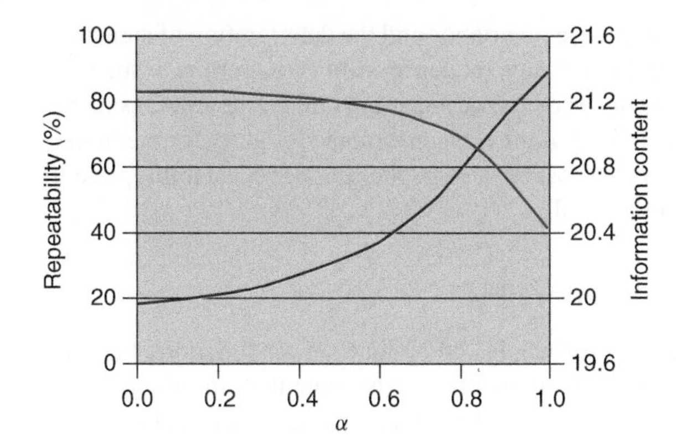

Figure 13.16 The information content (blue line) and the repeatability (red line) as a function of the amount of color saliency boosting.

experiment is performed by applying color boosting to the opponent color space. The results show that information content increases at the cost of stability. Depending on the application, a choice should be made about the amount of color saliency boosting.

13.2.6.2 Photometric Robustness The second phenomena that influences repeatability is the gain in photometric robustness. By boosting color saliency, the influence of the photometric variant direction diminishes, while the influence of the quasi-invariant directions increases. As a consequence, the repeatability under photometric changes, such as changing illumination and viewpoint, increases.

In Figure 13.17, the dependence of repeatability is tested on two image sequences with changing illumination conditions [57]. The experiment was performed by applying color boosting to the spherical color space, since changes due to shadow-shading will be along the photometric variant direction of the spherical system. For these experiments, two intertwining phenomena can be

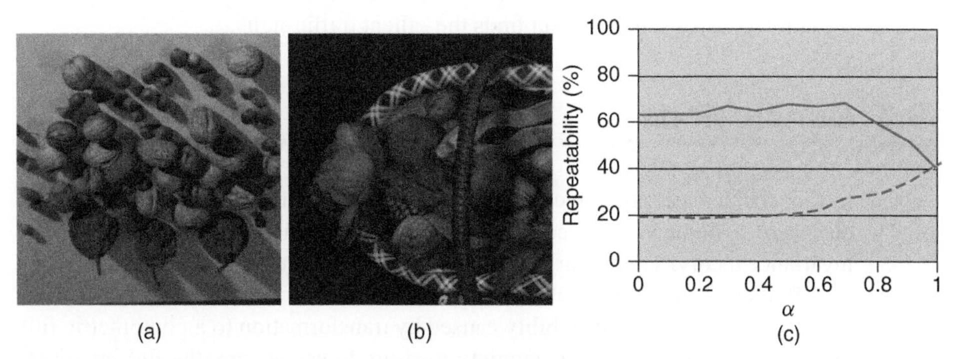

Figure 13.17 (a),(b) Two frames from two sequences with changing illumination conditions. (c) Repeatability as a function of the amount of color saliency boosting for the two sequences. Dotted line for the nuts sequence and the continuous line for the fruit basket sequence.

observed: the improved photometric invariance and the deterioration of signal-to-noise ratio with increasing α. For the nuts sequence, with very prominent shadows and shading, the photometric invariance is dominant, while for the fruit basket the gained photometric invariance only improves performance slightly for medium α values. For total color saliency boosting $\alpha = 1$, the loss of repeatability, due to loss of signal-to-noise, is substantial.

13.2.7 Illustrations of Generality

Color saliency boosting can, in principle, be applied on all functions that can be written as a function of the local derivatives. Also note that, in principle, the boosting theory can also be applied to higher order derivatives of images. Here, we show some additional examples. First, we apply saliency boosting to the focus point detector [254]. The detector focuses on the center of locally symmetric structures. In Figure 13.18b, the saliency map is shown. In Figure 13.18c, the result after saliency boosting is depicted. Although focus point detection is already an extension from luminance to color, black and white transition still dominate the result. Only after boosting the color saliency, the less interesting black-and-white structures in the image are ignored and most of the red Chinese signs are found. Similar difference in performance is obtained by applying color boosting to the linear symmetry detector [237]. This detector focuses on corner and junction-like structures. The *RGB*-gradient-based method focuses mainly on black and white events, while the more salient signboards are found only after color saliency boosting.

As a final illustration we illustrate color saliency boosting to gradient-based methods. In the third row of Figure 13.18, color boosting is applied to a gradient-based segmentation algorithm [257]. The algorithm finds globally optimal regions and boundaries. In Figure 13.18b and 13.18c, respectively, the *RGB* gradient and the color-boosted gradient are depicted. While the *RGB*-gradient-based segmentation is distracted by the many black and white events in the background, the color-boosted segmentation finds the salient traffic signs.

13.3 Conclusions

In this chapter we have investigated several aspects of color features. First, we discussed a framework to combine tensor-based features and the photometric invariance theory. The tensor basis of these features ensures that opposing vectors in different channels do not cancel out, but instead that they reinforce each other. To overcome the instability caused by transformation to a photometric full invariant, we introduce an uncertainty measure to accompany the full invariant. This uncertainty measure is incorporated in the color tensor to generate robust photometric invariant features.

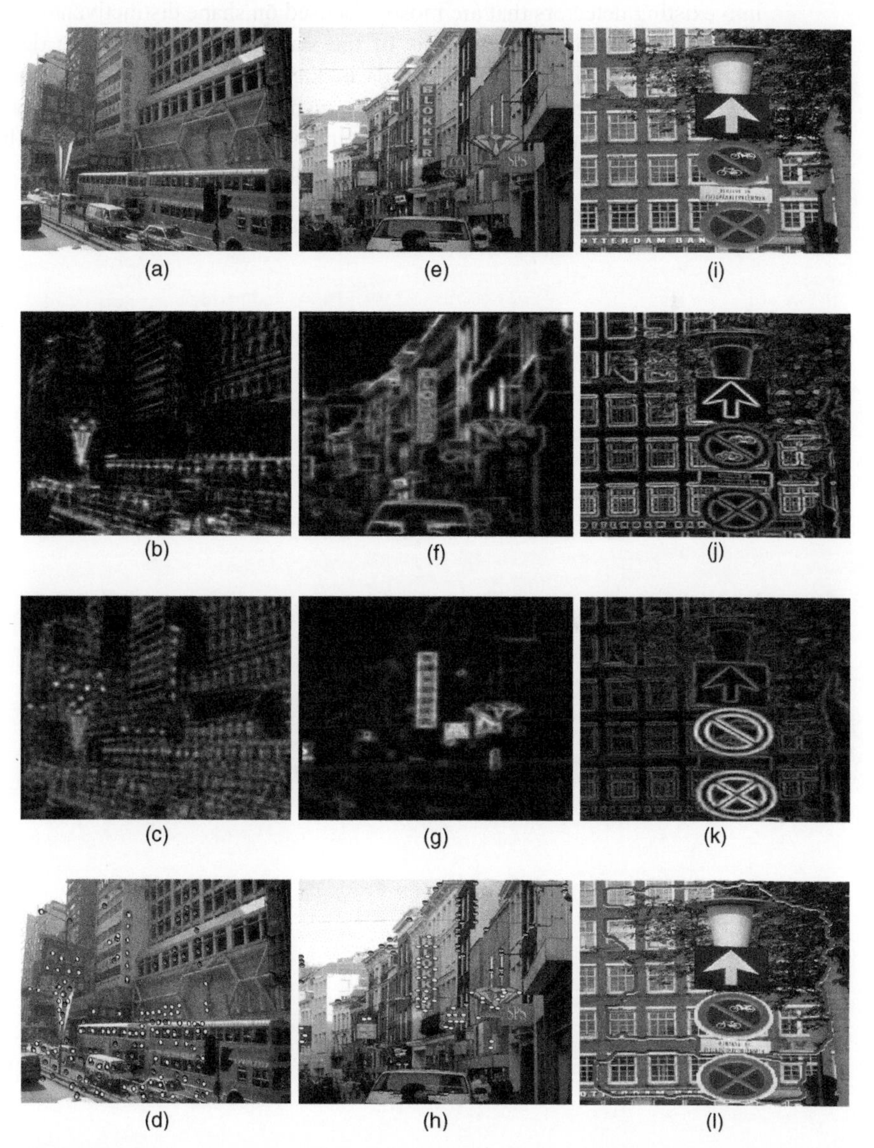

Figure 13.18 In sequence input image, *RGB*-gradient-based saliency map, color-boosted saliency map, and the results with red dots (lines) for gradient-based method and yellow dots (lines) for salient points after color saliency boosting. Results (a–d) for the focus points, (e–h) for the symmetry points, and (i–l) for the global optimal regions and boundary method.

Second, color distinctiveness is explicitly incorporated in the design of salient point detectors. The method, called *color saliency boosting*, can be incorporated into existing detectors that are mostly focused on shape distinctiveness. Saliency boosting is based on an analysis of the statistics of color image derivatives. Isosalient derivatives form ellipsoids in the color derivative histograms. This fact is exploited to adapt derivatives in such a way that equal saliency implies equal impact on the saliency map. Experiments show that color saliency boosting substantially increases the information content of the detected points.

14 Color Feature Description

With contributions by Gertjan J. Burghouts

In the previous chapters, we have outlined the theory of invariant feature extraction from color images. The advantage of the full invariants described in Chapter 6 is that they capture intrinsic scenes or object properties, robust to various arbitrary imaging conditions such as local illumination, shadows, and color of the light source. Hence, these invariant features are well suited to characterize the image content in the so-called image descriptors. In this chapter, we demonstrate the appropriateness of such invariant color descriptors. Much of the methodology described here is adopted from Burghouts and Geusebroek [258] and from van de Sande et al. [259].

Many computer vision tasks depend heavily on local feature extraction and matching. Object recognition is a typical case where local information is gathered to obtain evidence for recognition of previously learned objects. Recently, much emphasis has been placed on the detection and recognition of locally (weakly) affine invariant regions [55, 57, 260–262]. The rationale here is that planar regions transform according to well-known laws. Successful methods rely on fixing a local coordinate system to a salient image region, resulting in an ellipse describing local orientation and scale. After transforming the local region to its canonical form, image descriptors should be well able to capture the invariant region appearance. As pointed out by Mikolajczyk and Schmid [252], the detection of elliptic regions varies covariantly with the image (weak perspective) transformation,

Color in Computer Vision: Fundamentals and Applications, First Edition.
Theo Gevers, Arjan Gijsenij, Joost van de Weijer, and Jan-Mark Geusebroek.

while the normalized image pattern they cover and the image descriptors derived from them are typically invariant to the geometric transformation. Recognition performance is further enhanced by designing image descriptors to be photometric invariant such that local intensity transformations due to shading and variation in illumination have no or limited effect on the region description. State-of-the-art methods in object recognition normalize average and standard deviation of the intensity image [55, 252, 263]. Moreover, image measurements using a Gaussian filter and its derivatives are becoming increasingly popular as a way of detecting and characterizing image content in a geometric and photometric invariant way. Gaussian filters have interesting properties from an image processing point of view, among others, their robustness to noise [264], their rotational steerability [265], and their applicability in multiscale settings [54]. Many of the intensity-based descriptors proposed in the literature are based on Gaussian (derivative) measurements [53, 57, 253, 266, 267].

We consider the extension to color-based descriptors as color has high discriminative power. In many cases, objects can well be recognized merely by their color characteristics [43, 46, 47, 268–270]. However, photometric invariance is less trivial to achieve, as the accidental illumination and recording conditions affect the observed colors in a complicated way. Photometric invariance has been intensively studied for color features [46, 47, 50, 58, 164, 187]. The most successful local image descriptor so far is Lowe's SIFT descriptor [55]. The SIFT descriptor encodes the distribution of Gaussian gradients within an image region. The SIFT descriptor is a 128-bin histogram that summarizes local oriented gradients over 8 orientations and over 16 locations. This represents the spatial intensity pattern very well, while being robust to small deformations and localization errors. Nowadays, many modifications and improvements exist, among others, PCA-SIFT [271], GLOH [57], fast approximate SIFT [272], and SURF [273]. These region-based descriptors have achieved a high degree of invariance to overall illumination conditions for planar surfaces. Although designed to retrieve identical object patches, SIFT-like features turn out to be quite successful in bag-of-feature approaches to general scene and object categorization [274].

The important research question is whether color-based descriptors indeed improve on their gray-based counterparts in practice. The answer depends on the stability of the nonlinear combinations of Gaussian derivatives necessary to achieve a similar level of invariance as implemented in gray-value descriptors. For instance, the values of photometric invariants are distorted when the image is JPEG compressed, as the compression distorts the pixel values and spatial layout more for the color channels than for the intensity. Here we provide a study of local color descriptors in comparison with gray-value descriptors.

For (affine) region detection, many well-performing methods exist [149, 252, 254, 275–278]. Hence, we will concentrate on descriptor performance of the full photometric invariant derivatives, as well as their combination into color SIFT descriptors. Furthermore, to enable a fair comparison between intensity-based descriptors and color-based descriptors, we demand identical geometric invariance for both intensity-based and color-based features. This requirement is conveniently fulfilled by the Gaussian derivative framework.

For the evaluation of local gray-value and color invariants, we adopt the extensive methodology of Mikolajczyk and Schmid [57]. In this article, the authors propose the evaluation of descriptor performance by matching regions from one image to another image. Correct matches are determined using the homography between the two images. From Reference 57, we adopt the measures to evaluate discriminative power and invariance. Also, we adopt variety in recording conditions, being changes of illumination intensity, of the camera viewpoint, blurring of the image, and JPEG compression. We go beyond Reference 57 by extending this set with images recorded under different illumination colors and illumination directions. These conditions induce a significant variation in the image recording. For an illustration of images recorded under varying illumination directions, see Figures 14.1–14.4.

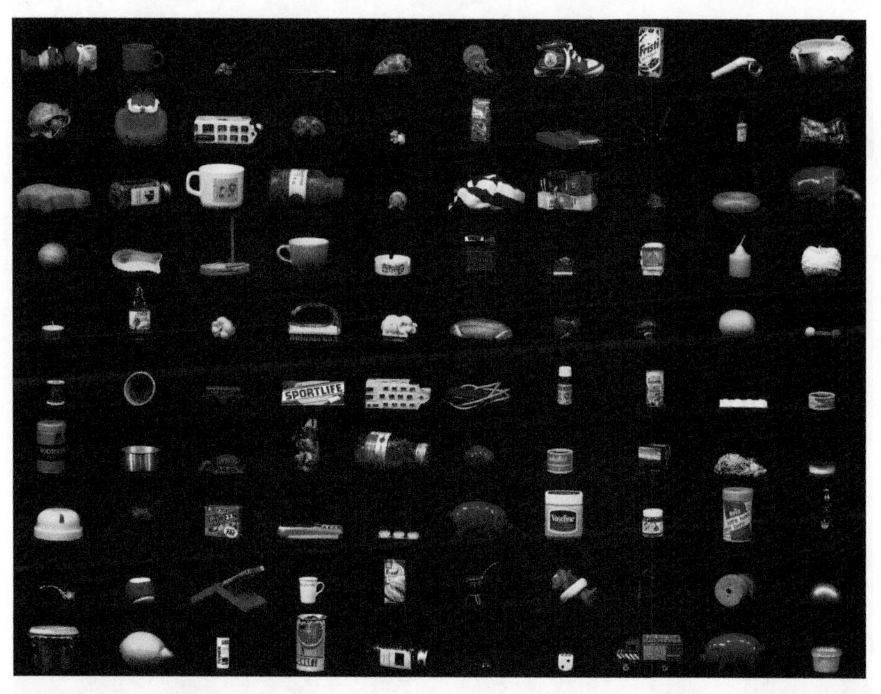

Figure 14.1 An illustration of the diverse objects from the ALOI collection [198]. A random sample of the objects in the collection is depicted.

We extend the number of images used in the evaluation framework [57] to 26,000, representing 1000 objects recorded under 26 imaging conditions. Moreover, we further decompose the evaluation framework in Reference 57 to the level of local gray-value invariants on which common region descriptors are

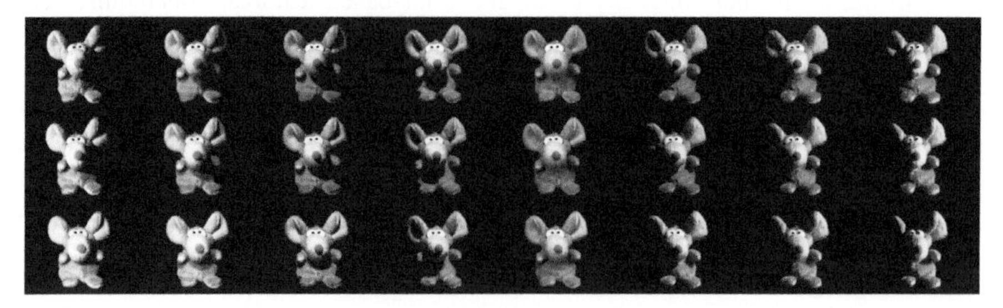

Figure 14.2 Example object from ALOI recorded under semihemispherical illumination, and images recorded under an illuminant at decreasing altitude angles. See Reference [198] for details.

Figure 14.3 Example object from ALOI recorded under varying illumination color.

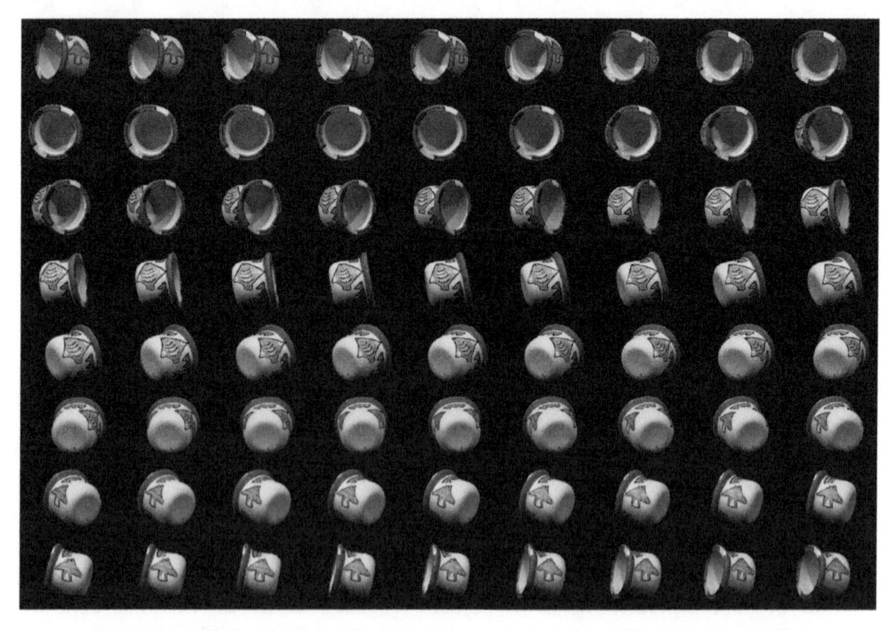

Figure 14.4 Example object from ALOI recorded under varying viewing angles.

based. We measure the performance of photometric invariants for the detection of color transitions only. Hence, we evaluate the performance of the Gaussian gray-value and color invariant derivatives. Finally, we establish performance criteria that are specific to color invariants, indicating the level of invariance with respect to photometric variation, and evaluating the ability to distinguish between various photometric effects.

14.1 Gaussian Derivative-Based Descriptors

We compare the local gray-value derivatives with the color invariant derivatives from Chapter 6 based on three evaluation criteria:

- *Discriminative power*. We establish the power of each invariant to discriminate between image regions. Discriminative power is measured by the quality of region matching, similar to Reference 57. The successful matching strategy as proposed by Lowe [55] is based on the rationale that for the recognition of an object it suffices to correctly match only a few regions of that object. In our experimental framework, we push this to the extreme and consider the matching of one region of an object against a database of 1000 regions: one noisy realization of the same object matched against 999 of other objects. Under noisy conditions we consider image deformations caused by blurring, JPEG compression and out-of-plane object rotation (viewpoint change), and photometric variation induced by changes in illumination direction and illumination color. Precision and recall characteristics reflect the discriminative power of the invariant under evaluation.
- *Invariance or robustness.* As above, but now we establish the degradation of the number of correct matches as a function of imaging condition or image transformation that increasingly deteriorates, similar to Reference 279. As with discriminative power, the conditions we test are blurring, JPEG compression, illumination direction, viewpoint change, and illumination color. The degradation in the recall reflects the constancy of the invariant under examination.
- *Information content.* We establish the power of each invariant to discriminate between true color transitions while remaining constant under nonobject-related transitions induced by shadow, shading, and highlights. Hence, we assess simultaneously for each invariant its power to discriminate between color transitions, and its invariance to photometric distortions. Note that this is different from the the two experiments above, as here we evaluate the property to discriminate between the variant and invariant aspects in the photometric condition, in isolation of a possible effect on recognition performance.

We consider for 1000 objects from the ALOI database [67] the following imaging conditions: JPEG Compression; blurring; and changes of the viewpoint, illumination direction, and illumination color. Figure 14.5 illustrates the imaging conditions for some of the objects.

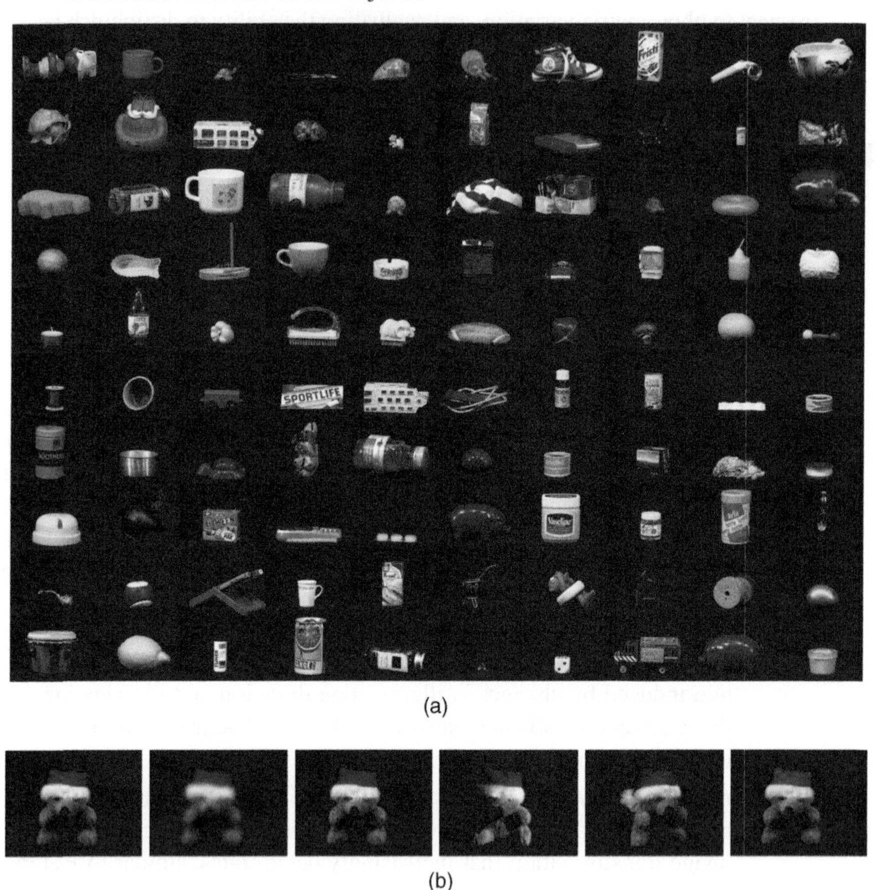

Figure 14.5 Randomly selected objects from the ALOI collection (100 example objects) are depicted in (a). Imaging and testing conditions are shown in (b): the reference image, blurring ($\sigma = 2.8$ pixels, image size 192×144), JPEG compression (50%), illumination direction change (to 30° altitude, from the right), viewpoint change (30°), illumination color change (3075K $\rightarrow$ 2175K).

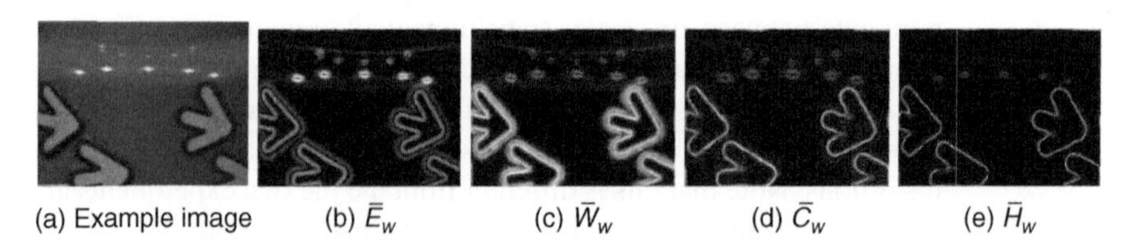

Figure 14.6 Photometric invariant gradients. $\overline{E}_w$ is not photometric invariant; $\overline{W}_w$ is invariant to illumination intensity; $\overline{C}_w$ is invariant to shadow and shading; and $\overline{H}_w$ is invariant to shadow, shading, and highlights.

For each object image, we determine its regions. For consistency with the literature, we determine Harris-affine regions [252]. As pointed out in Reference 57, to establish the correct matching of regions, one should either fix the camera viewpoint or consider the homography limiting oneself to more or less flat scenes. For 3D objects, the assertion of a flat scene fails. To overcome this problem, we consider images that have been recorded with a fixed camera viewpoint. However, the condition of viewpoint change has to be settled. Therefore, for each object, we manually selected the single region inside the object that is most consistent between the original and the image recorded under a viewpoint change. We copied the region from the original to all the remaining imaging conditions (see Fig. 14.7 for an example). Note that as we are dealing with regions inside objects only, the black background does not affect the experiments. Furthermore, trying to find 1 region from the 1000 selected regions could be seen as searching the 1 region in an image of 1000 cluttered objects for which all selected regions are visible. Together with the variation in image transformations and imaging conditions a total of 26,000 regions are available. The regions vary significantly in size and anisotropy (Fig. 14.7b and 14.7c). The ground truth of regions is publicly available on the website of the ALOI database [198].

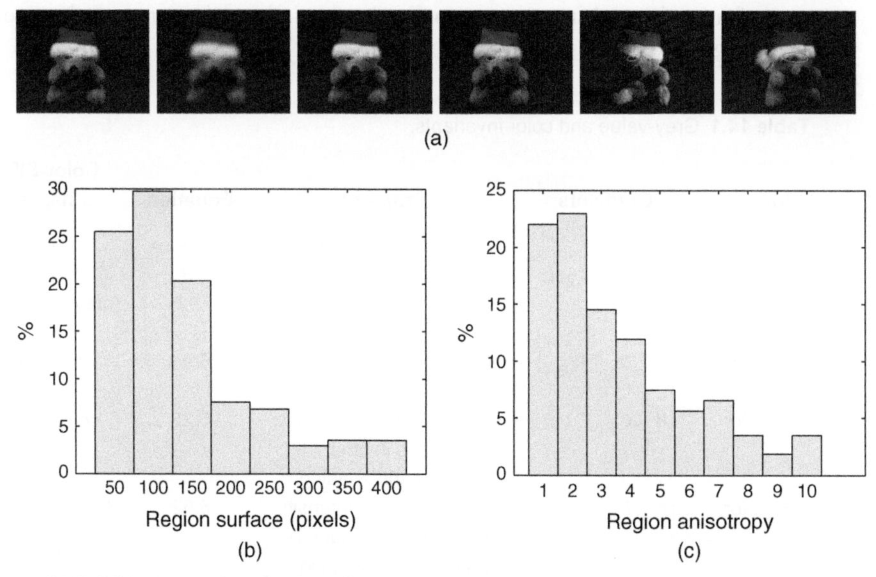

Figure 14.7 (a) Image regions for the reference image, blurring, JPEG compression, illumination color change, illumination direction change, and viewpoint change. For all imaging conditions except the change of viewpoint, the camera is fixed, so the regions are set identical. For the camera viewpoint change, we have manually selected the most stable region. Histogram of (b) the size of the region surfaces, and (c) of the anisotropy (where anisotropy = 1 indicates isotropy).

Next, we compute the invariants from each region. To be consistent with the literature, we normalize the regions as in Reference 252. We consider two experiments:

- *Single location computation* In the first experiment, we compute the invariant gradients from one location. We do so by computing them at a fixed scale (i.e., one-third of the region size). For each region, we determine the location in which the image gradient $\overline{E}_w$ is maximum. For all copied regions (see for region extraction the description above), this location is identical. From this location, we compute all invariants.
- *SIFT-based computation* In the second experiment, we compute the SIFT descriptor from the normalized region identical to Mikolayzcyk's computation [57], but with the gray-value gradient inside the SIFT descriptor replaced by one of the invariant color gradients.

For the performance evaluation, we consider the following sets of invariant gradients (Table 14.1). The extension ''SIFT'' to the name of the invariant implicates SIFT-based computation; otherwise, single-location Gaussian invariants are considered. Original SIFT is also included in the experiments and is equivalent to W-gray-SIFT. We include the intensity gradient W_w in the H and C color-based descriptors. Although this seems contradictory at first sight, the orthogonalization of intensity and intensity-normalized color information proves effective in matching.

Table 14.1 Grey-value and color invariants.

Invariant	Gradients	Property	Equation	Color-SIFT name
E-gray	$\{E_w\}$	Not photometric invariant	—	—
E-color	$\{E_w, E_{\lambda w}, E_{\lambda\lambda w}\}$	Not photometric invariant	6.9	—
W-gray	$\{W_w\}$	Invariant to local intensity level	—	(grey-) SIFT
W-color	$\{W_w, W_{\lambda w}, W_{\lambda\lambda w}\}$	Invariant to local intensity level	6.9	W-color- SIFT
C-color	$\{W_w, C_{\lambda w}, C_{\lambda\lambda w}\}$	Invariant to local intensity level, plus invariant to shadow and shading	6.28	C-color-SIFT
H-color	$\{W_w, H_w\}$	Invariant to local intensity level, plus invariant to shadow and shading, and highlights	6.52	H-color-SIFT

For fair comparison to the original SIFT descriptor, we reduce the dimensionality of all color SIFT descriptors to 128 numbers using PCA reduction (the covariances have been determined over 200 example regions computed from the reference images). Furthermore, we will evaluate the hue-based SIFT descriptor of Abdel-Hakim and Farag [280], termed *hue-color-SIFT*, and the HSV-based SIFT descriptor of Bosch and Zisserman [281], termed *hsv-color-SIFT*.

14.2 Discriminative Power

The objective of this experiment is to establish the distinctiveness of the invariants. To that end, we match image regions computed from a distorted image to regions computed from the reference images as in Reference 57. The discriminative power is measured by determining the recall of the regions that are to be matched and the precision of the matches:

$$\text{Recall} = \frac{\#\text{correct matches}}{\#\text{correspondences}}, \tag{14.1}$$

$$\text{Precision} = \frac{\#\text{correct matches}}{\#\text{correct matches} + \#\text{ false matches}}. \tag{14.2}$$

Here, recall indicates the number of correctly matched regions relative to the ground truth of corresponding regions in the dataset. Precision indicates the relative amount of correct matches in all the returned matches. The definition of recall is specific to the problem of matching based on a ground truth of one-to-one correspondences, and hence it deviates from the definition as used in information retrieval. The aim in our experiment is to correctly match all regions (recall of one) with ideally no mismatches (precision of one).

We consider the nearest-neighbor matching as employed in Reference 57. Distances between values of photometric invariants are computed from the Mahalanobis distance (the covariances have been determined over 200 examples computed from reference images). Over various thresholds, the number of correct and false matches are evaluated to obtain a recall versus precision curve. A good descriptor would produce a small decay in this curve, reflecting the maintenance of a high precision while matching more image regions.

We randomly draw a test set of regions and use 1000-fold cross-validation to measure performance over our dataset. To end up with graphs that allow a comparison between various levels of color invariance, we vary the number of regions to match per experiment. The number of regions to which a single region is compared is set to 20 for the invariants computed from one location. We consider a successful distinction between 20 image points to be the minimal requirement of a point-based descriptor. For the SIFT-based computation of invariants, we increase this number, as the region-based description is more distinctive. The number of regions to which one region is compared is between 100 or 500, depending on the difficulty of the imaging condition. We consider a successful distinction between 100 regions to be the minimal requirement of a region-based descriptor. We consider a successful distinction between 500 regions to be sufficient for realistic computer vision tasks; this is in line with the validation in References 57, 279.

The results of the SIFT region matching for invariant gradients are shown in Figure 14.8. All photometric invariants are plotted using solid lines. All color-based invariants are plotted using red lines, opposed to gray-value invariants that are plotted in black lines.

Overall, all color invariants have better performance than gray-value-based features. Gray-value derivatives E-gray and W-gray are outperformed by color-based descriptors, except when illumination color is changed (Fig. 14.8e). In that case, normalized intensity W-gray performs reasonably, but is still outperformed by many color-based invariants, as expected.

The performance of H-color is a bit disappointing compared to the other color invariants. Two effects play a role here. First, this descriptor misses one color channel of information, and better discriminative power could be achieved when adding a saturation channel. However, in that case one would, at best, expect a performance similar to W-color. We will see a comparison later on when establishing performance for the color SIFT descriptors. A second issue affecting the H-color feature is the instabilities caused by the normalization in the denominator of Equation 6.52. The expression becomes unstable for colors that are unsaturated, and hence is grayish. Blurring by the Gaussian filter enhances this effect, as color at boundaries—which we are evaluating in this setup—are mixed. Hence, H-color seems unsuitable for region descriptors based on Gaussian derivatives.

The effect of blurring, shown in Figure 14.8a, causes the image values to be smoothed. Hence, details are lost, but no photometric variation is introduced. The color gradient with no photometric invariant properties, E-color, performs best. Besides the decay in performance due to additional blur, the graph clearly illustrates the gain in discriminative power when using color information.

The compression of images by JPEG, shown in Figure 14.8b, causes the color values to be distorted more than the intensity channel. Still, color information is distinctive, as the color gradient that is invariant to the intensity level, W-color, performs best. At the beginning of the recall-precision curves, one clearly sees the advantage of orthogonalizing intensity and color information, as W-color, C-color, and H-color perform significantly better than E-color, for which all channels are correlated with intensity. In the latter case, all values of the SIFT descriptor will be severely corrupted by the JPEG compression. For the invariant color descriptors, the intensity channel will be relatively mildly corrupted by the compression, whereas the color channels still add extra discriminative power. Compression effects become more influential at the tail of the recall-precision curves, where one sees H-color drop off quite early because of the instability of the descriptor, followed by C-color. Although W-color had a slower start, it ends up doing quite well because of the more stable calculation of the nonlinear derivative combination.

For changes in the illumination direction (Fig. 14.8c), the main imaging effects are darker and lighter image patches, and shadow and shading changes. However,

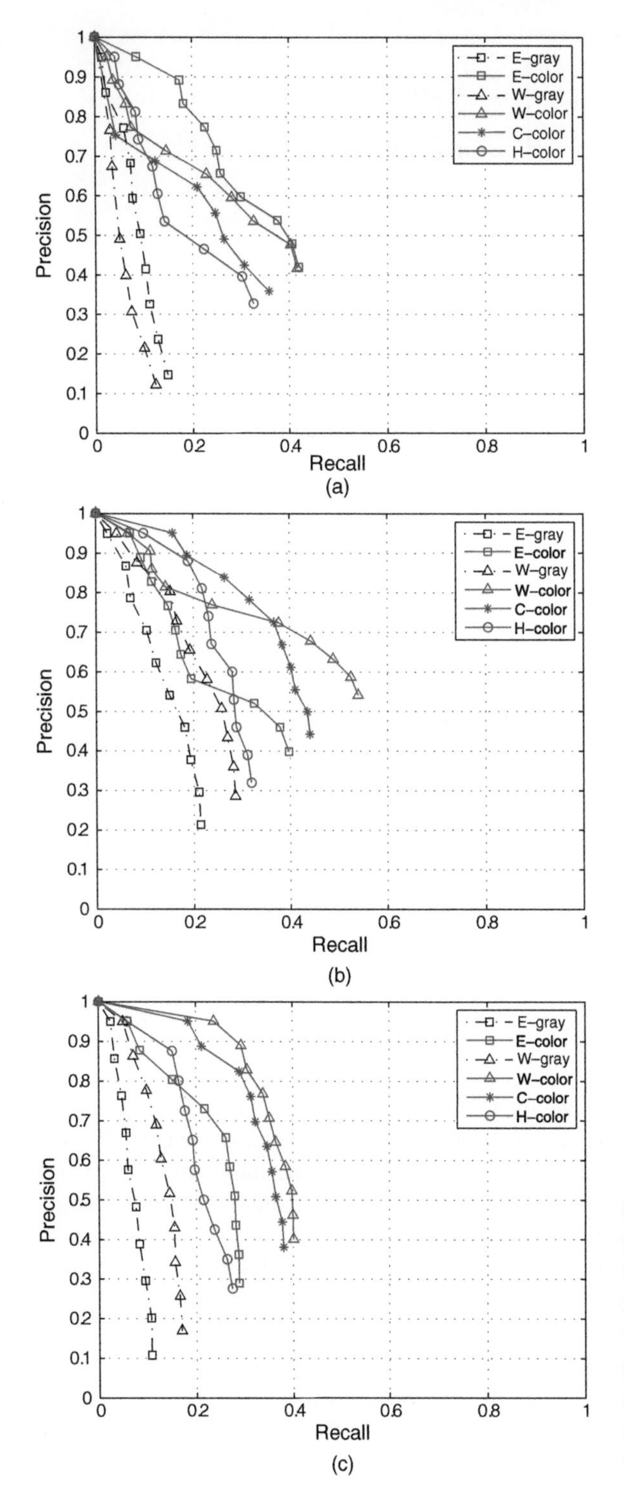

Figure 14.8 Discriminative power of photometric invariant gradients. (a) Blurring ($\sigma = 1$ pixel), 1 versus 20, (b) JPEG compression (50%), 1 versus 20, (c) Illumination direction (30°), 1 versus 20, (d) viewpoint change (30°), 1 versus 20, (e) illumination color (2100K), 1 versus 20.

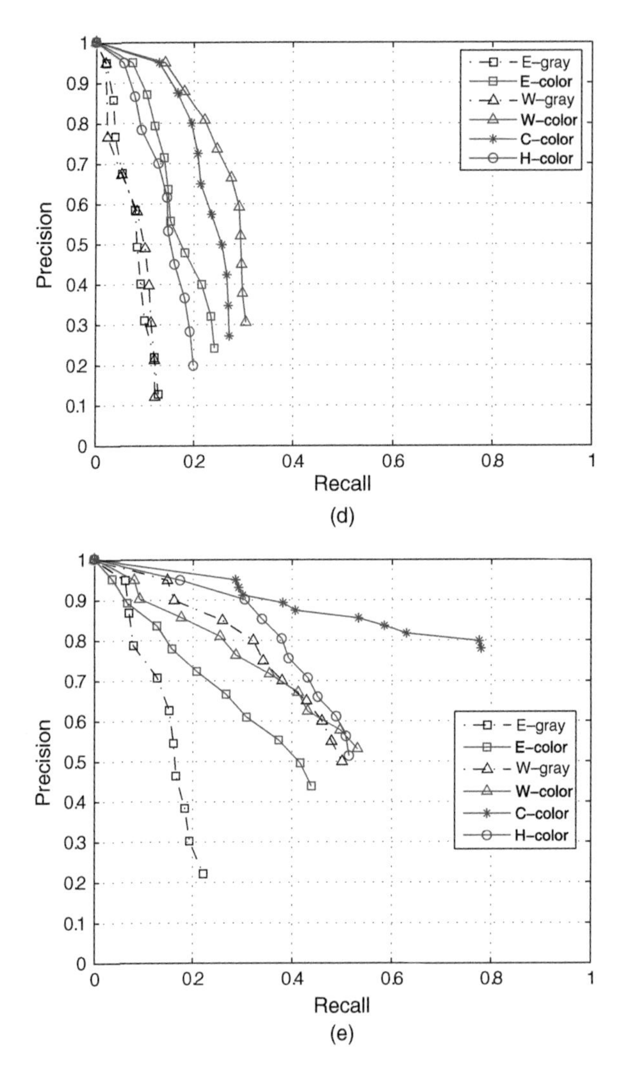

Figure 14.8 (*Continued*)

for the small scale at which we measure the Gaussian derivative descriptors, we expect intensity changes to dominate over shadow and shading edges. Shadow and shading (geometry) edges are expected to become more important when assessing SIFT-based descriptors, which capture information over a much larger region. Hence, both color gradients that are invariant to intensity changes, W-color and C-color, perform well. Clearly, the color invariant descriptors outperform gray-value descriptors and noninvariant color descriptors.

The results of a change in viewpoint (Fig. 14.8d), clearly demonstrate the advantage of adding color information. The patches, manually indicated to be stable, merely contain a change in information content due to a projective transformation and small errors in the affine region detection. Furthermore, the light field will be distributed somewhat differently over the image, causing W-color and C-color to perform superior to gray-value descriptors, noninvariant color descriptors, and the H-color descriptor.

For varying illumination color (Fig. 14.8e), obviously the color values become distorted. The color gradient invariant to shadow, C-color, is very robust here. Although C-color is based on color, its gradients are computed in such a way that can be shown to be reasonably color constant [67]. Furthermore, one would expect the gray-value descriptors not to be affected by illumination color changes. However, a change in overall intensity is also present, making direct use of E-gray infeasible. The intensity-normalized invariant W-gray performs reasonably, but lacks the discriminative power that comes with the use of color.

Figure 14.9 shows the discriminative power of the invariants when they are plugged into the SIFT descriptor. The figure has an identical organization as Figure 14.8. The only exception in the experimental setup is that the number of regions, to which a single region is matched, is increased. This number varies over the imaging conditions, and is either 100 or 500, to obtain suitable resolution in the performance graphs. Furthermore, note that two extra methods from the literature have been added, the hue-color-SIFT descriptor [280] and the hsv-color-SIFT descriptor [281].

Overall, the relative performance of SIFT-based computation of invariants corresponds largely to the relative performance of invariants from single points. Color-based SIFT invariant to shadow and shading effects, C-color-SIFT, performs best.

Generally, the SIFT-based computation significantly improves the discriminative power compared to single-point computation. Almost all color and gray-value descriptors perform well under blurring (Fig. 14.9a), JPEG compression (Fig. 14.9b), and illumination color changes (Fig. 14.9e). Note that the C-color-SIFT descriptor performs equally well as the intensity-based SIFT descriptor in the case of illumination color changes, implying a high degree of color constancy for this descriptor.

Discriminative power drops when considering illumination direction or viewpoint changes (Fig. 14.9a,b). These cases are much harder to distinguish using a SIFT descriptor. In these cases, the gray-value-based SIFT is outperformed by the color-based SIFT descriptors. In particular, the color-based SIFT invariant to shadow and shading effects, C-color-SIFT, is very discriminative in these cases. This can be explained by the large spatial area over which the SIFT descriptor captures image structure. Hence, shadow and shading (object geometry) effects are more likely to be captured by the SIFT descriptor, but the effects are cancelled by the C invariant.

The shadow and highlight invariant H-color-SIFT is generally not very distinctive compared to W-color-SIFT and C-color-SIFT. Lack of discriminative power affects the performance of hue-color-SIFT, H-color-SIFT, and SIFT under blurring. Furthermore, the hue-based descriptors, hue-color-SIFT, and H-color-SIFT are affected by JPEG compression and by illumination color changes. The distinctiveness of hue-color-SIFT is generally much less than that of H-color-SIFT.

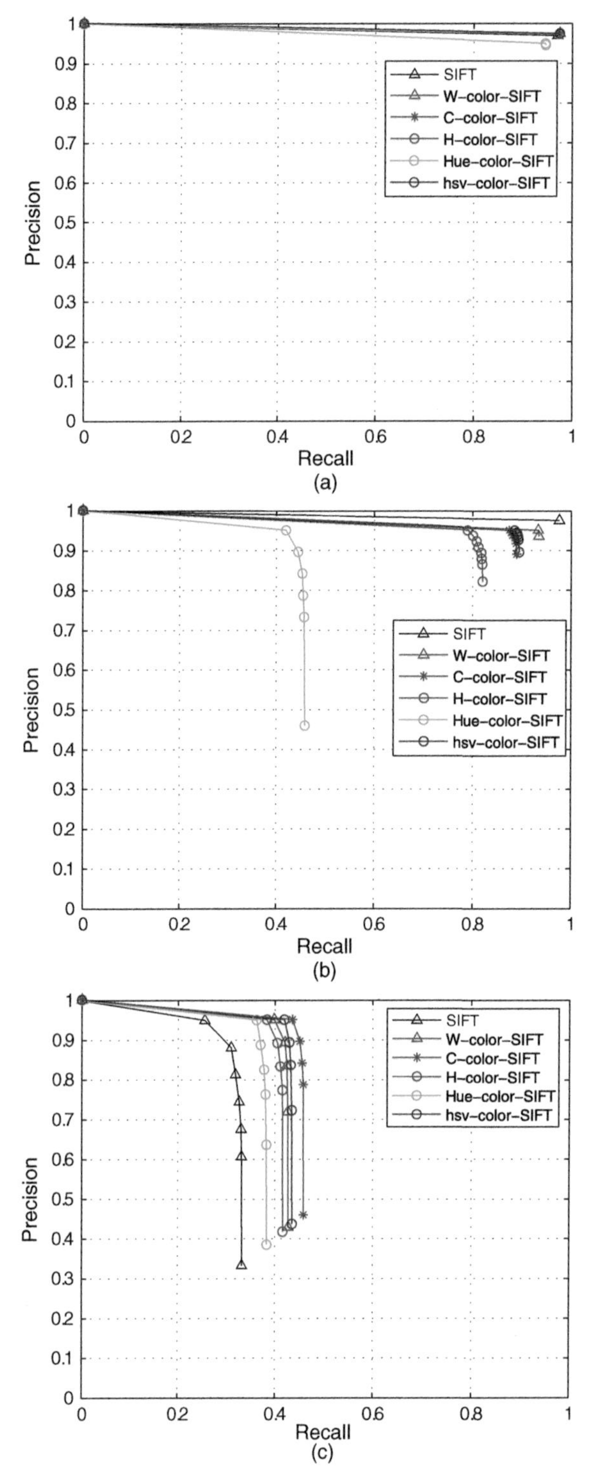

Figure 14.9 Discriminative power of photometric invariant gradients when plugged into the SIFT descriptor. (a) Blurring ($\sigma = 1$ pixel), 1 versus 500, (b) JPEG compression (50%), 1 versus 500, (c) Illumination direction (30°), 1 versus 100, (d) Viewpoint change (30°), 1 versus 100, (e) Illumination color (2100K), 1 versus 500.

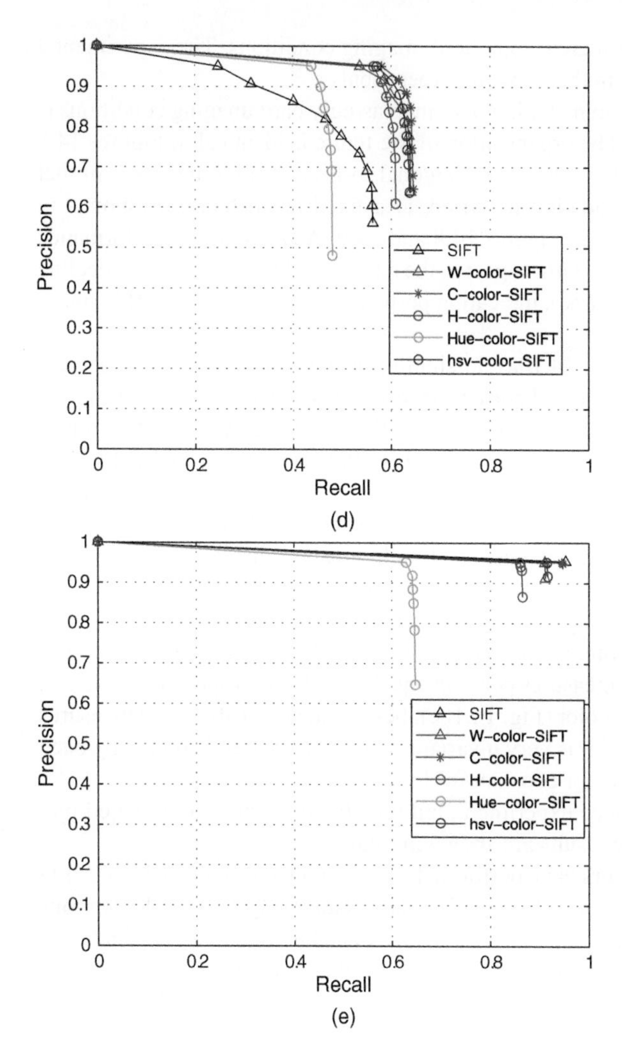

Figure 14.9 (*Continued*)

Hence, using the hue alone is not a distinctive region property. The distinctiveness of hsv-color-SIFT is generally somewhat higher than that of H-color-SIFT. Thus, the saturation s in the hsv color space is a distinctive property. However, the distinctiveness of hsv-color-SIFT is generally less than that of W-color-SIFT and C-color-SIFT because of Instability, as argued before.

14.3 Level of Invariance

The objective of this experiment is to establish the constancy of the invariants against varying imaging conditions. Likewise [279], we measure the degradation of recall (Eq. 14.1) over increasingly hard imaging conditions. The experimental setup is identical to that in the previous experiment.

The results of the region matching over increasingly hard imaging conditions is shown in Figure 14.10. The organization of the figure is identical to Figures 14.8 and 14.9. The presented graphs are orthogonal to Figures 14.8 and 14.9, in that now the amount of degradation is varied, at a fixed recall that corresponds to the endpoint of the curves in Figures 14.8 and 14.9. Any decline in performance indicates lack of constancy with respect to the tested condition. Ideally, the decline would be zero (horizontal line), indicating perfect invariance to the set of imaging conditions.

For image blurring (Fig. 14.10a) no significant imaging effects are observed. Hence, all descriptors have equal performance with respect to constancy, although initial discriminative power varies from a recall of 0.2 for gray-value derivatives to more than 0.7 for color-based derivatives. For JPEG compression (Fig. 14.10b), the gray-value invariants E-gray and W-gray are slightly more constant than the color invariants, as the image intensity is less affected by JPEG compression than the image chromaticity. For changes in the illumination direction (Fig. 14.10c) due to the small scale of the derivative descriptors, the main imaging effect is the change in region intensity. Hence, W-gray, W-color, C-color, and H-color are very stable. For a viewpoint change (Fig. 14.10d), only marginal imaging effects are observed. Hence, all measures perform equally well with respect to constancy. For varying illumination color (Fig. 14.10e), besides the intensity-based measures E-gray and W-gray, C-color is very invariant. This measure has theoretically been shown to be reasonably color constant [67].

We repeat the invariance experiment, but now the invariants are plugged into the SIFT descriptor. The results are shown in Figure 14.11.

Overall, most descriptors have performed well for blurring (Fig. 14.11a), JPEG compression (Fig. 14.11b), and illumination color change (Fig.14.11e). Exceptions again are the hue-based descriptors H-color-SIFT and hue-color-SIFT, which lack discriminative power, and are more affected by these conditions. A change in illumination direction or viewpoint is much harder for the SIFT descriptor to deal with, even with the color invariance built in. Overall, the C-color-SIFT seems the best choice, for which shadow and shading edges are discounted. This descriptor has invariance comparable to the intensity-based SIFT descriptor, but gains considerably in discriminative power.

14.4 Information Content

The objective of this final experiment is to establish the information content of the photometric invariants. Information content refers to the ability of an invariant to distinguish between color transitions and photometric events such as shadow, shading, and highlights. Ideally, the invariant's values covary with color transitions and its value is constant to photometric events to which it is designed

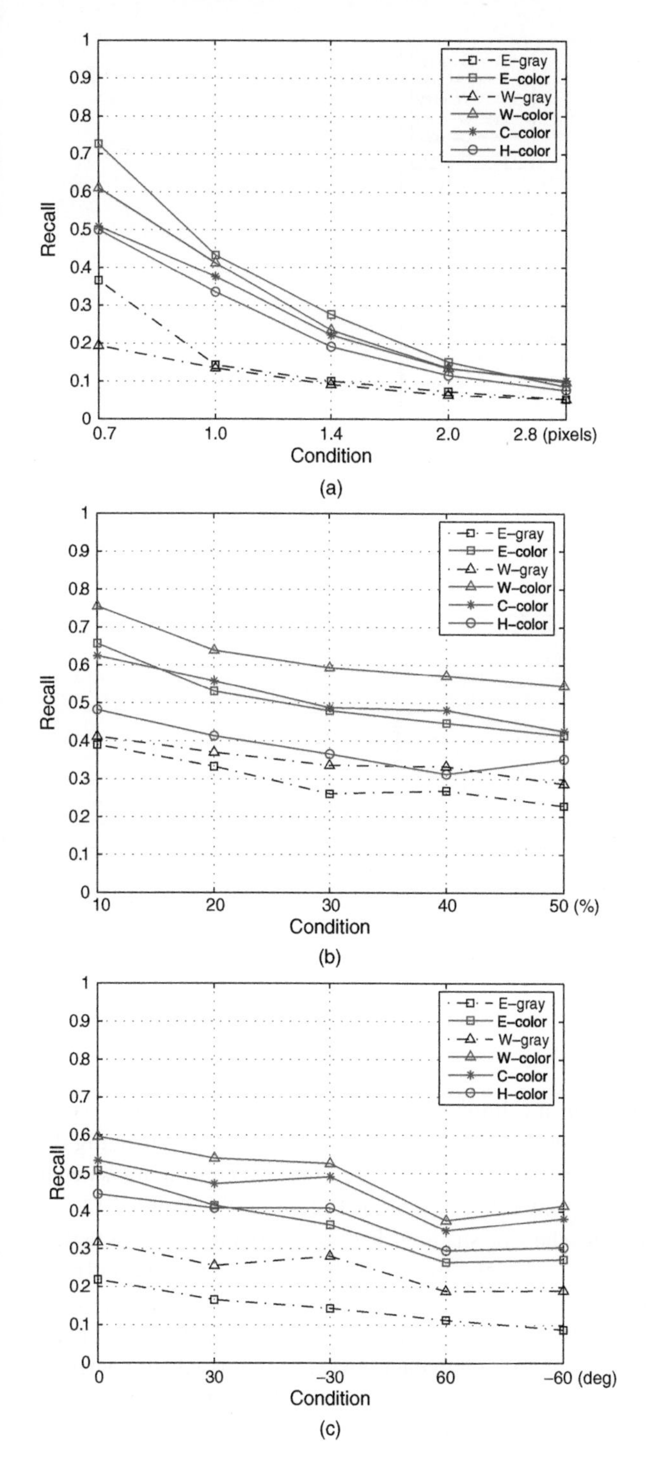

Figure 14.10 Invariance of photometric invariant gradients over increasingly hard imaging conditions. (a) Blurring, 1 versus 20, (b) JPEG compression, 1 versus 20, (c) Illumination direction, 1 versus 20, (d) Viewpoint change, 1 versus 20, (e)Illumination color, 1 versus 20.

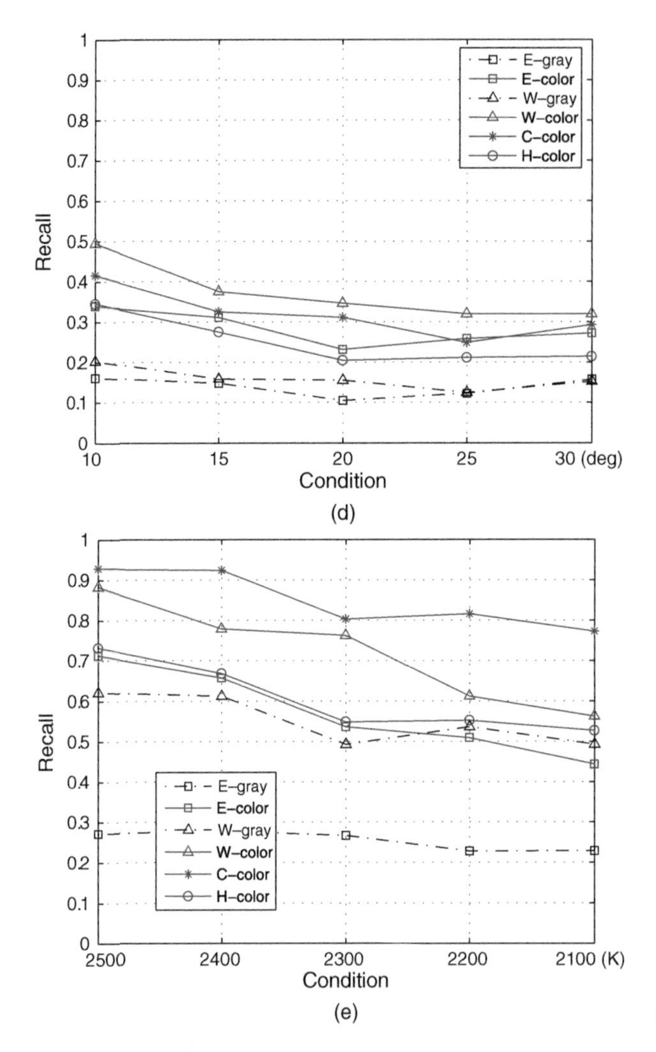

Figure 14.10 (*Continued*)

to be invariant. We illustrate the information content of W-color and C-color (Fig. 14.12). For the first object, new image edges are introduced by changing the illumination direction in Figures 14.12b and 14.12c. Hence, the matching is better with the shadow and shading invariant descriptor C-color-SIFT. Figures 14.12e and 14.12f show an example where no shadow/shading invariance performs better. Here, no new edges are introduced by the change in illumination direction, and only the local intensity is affected because of the relatively large-scale shading effects.

To establish the information content, we measure the discriminative power and invariance over individual image regions. Each image region is labeled whether it contains a color transition, or a shadow, shading, or highlight transition. In this way, the information content evaluates the invariant's discriminative power and

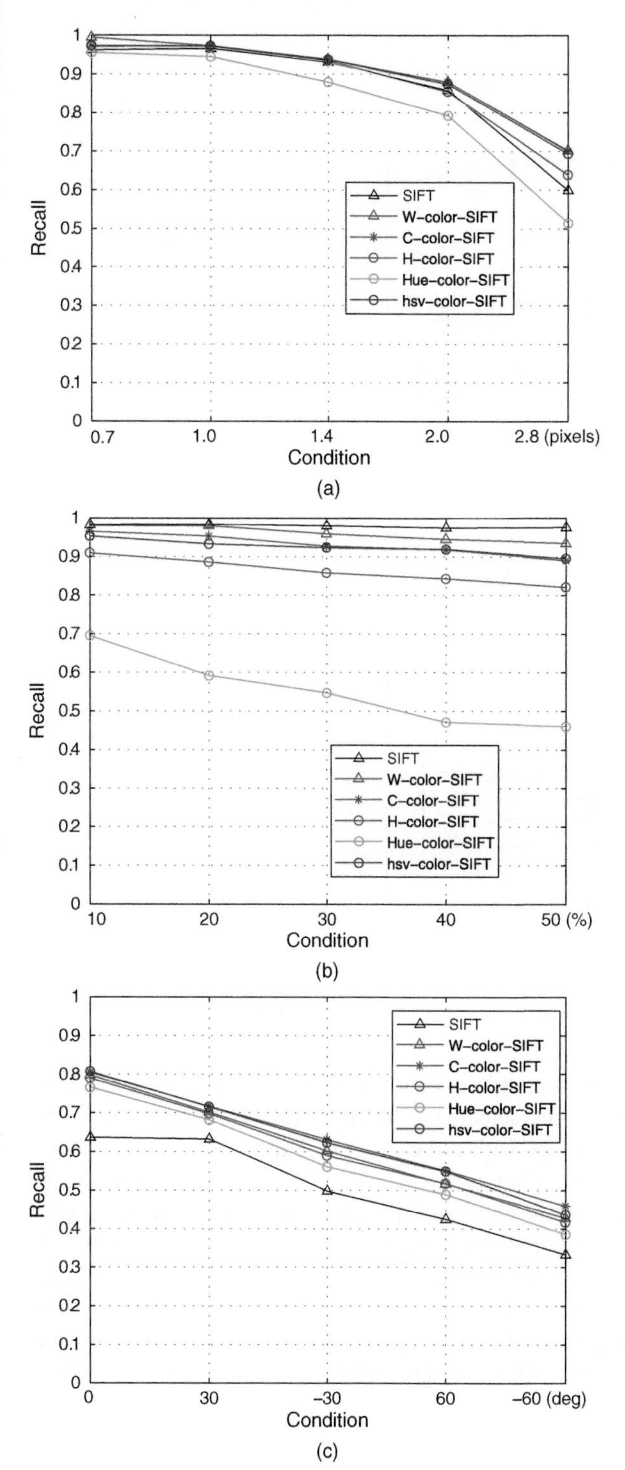

Figure 14.11 Invariance of photometric invariant gradients over increasingly hard imaging conditions when plugged into the SIFT descriptor. (a) Blurring, 1 versus 500, (b) JPEG compression, 1 versus 500, (c) Illumination direction, 1 versus 100, (d) Viewpoint change, 1 versus 100, (e) Illumination color, 1 versus 500.

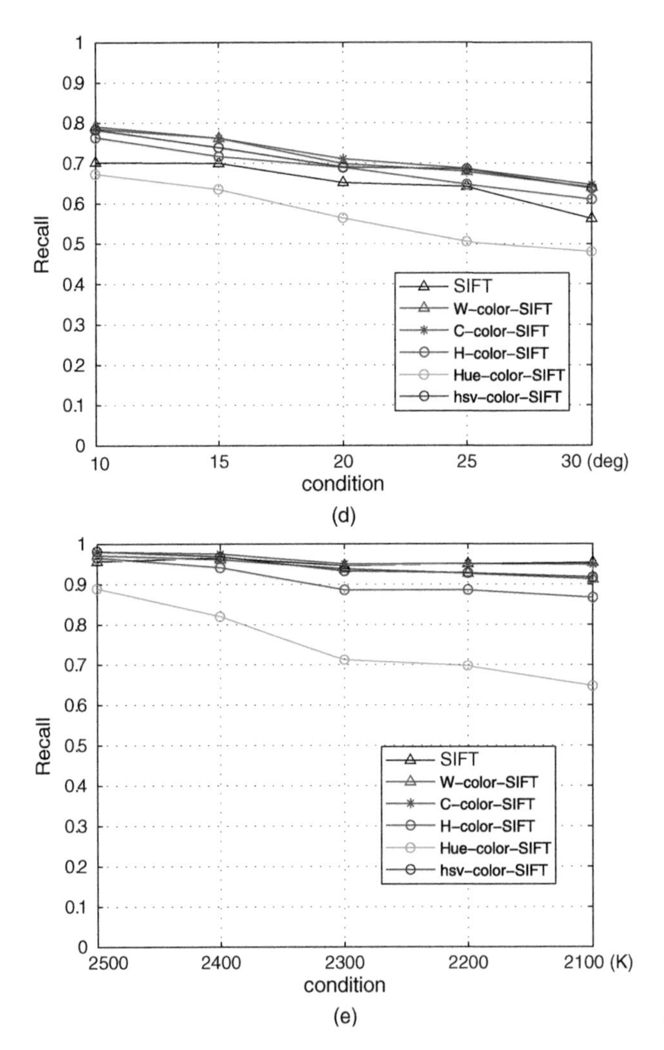

Figure 14.11 (*Continued*)

invariance over various photometric events. To that end, we construct a large annotated dataset from images selected from the CURET dataset [282]. This dataset contains tens of images within the order of hundreds of labeled image points located at the various photometric events. The selected texture images contain many edges, where we annotated for each image whether the texture was generated mainly by either shadow/shading (*sponge, cracker b, lambswool, quarry tile, wood b*, and *rabbit fur*) or highlight effects (*aluminum foil, rug a*, and *styrofoam*). From these images, regions have been detected by applying a Harris corner detector [53]. Figure 14.13a,b illustrates, for two fragments of texture images, shadow/shading and highlight edges, respectively. In addition, we have

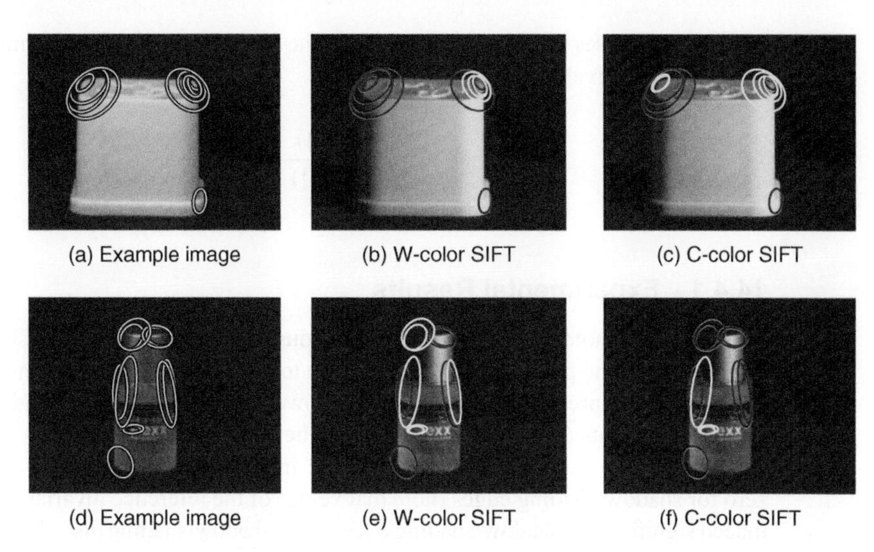

(a) Example image (b) W-color SIFT (c) C-color SIFT

(d) Example image (e) W-color SIFT (f) C-color SIFT

Figure 14.12 Illustration of matching for two objects. One is better matched with C-color-SIFT, the other with W-color-SIFT, respectively. Correct matches are shown in yellow, false matches are shown in blue. *Source*: Reprinted with permission, © 2009 Elsevier.

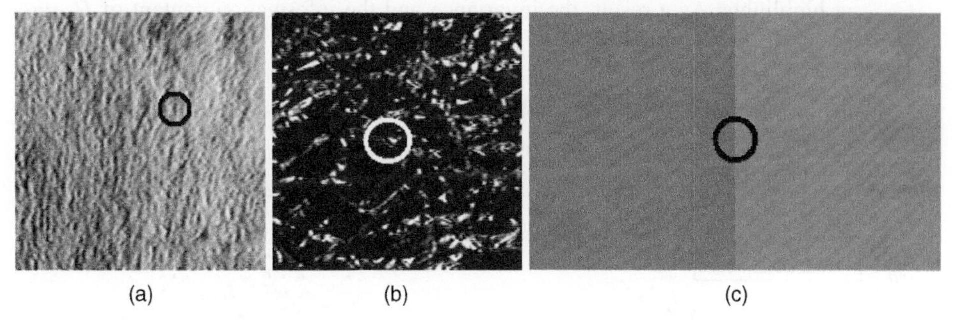

(a) (b) (c)

Figure 14.13 Examples of the photometric events dataset. Detected points are given a label whether the point is located on a (a) shadow/shading edge, (b) highlight edge, or (c) color edge.

collected image points located at color transitions. To that end, images have been taken from PANTONE color patches [70] (see Fig. 14.13c for an illustration). From the PANTONE patch combinations, we have selected the 100 combinations that have the largest hue difference, that is, patches that reflect true changes in object color rather than intensity or saturation differences.

We measure an invariant's power to distinguish between color transitions and disturbing photometric events by the Fisher criterion. From many color transitions, we compute a first cloud of points; from transitions of a particular disturbing photometric event, we compute a second point cloud. The Fisher criterion expresses the separation between the two clouds of points, termed $\{x_1\}$ and $\{x_2\}$ respectively:

$$\text{Information} = \frac{|\mu(\{x_1\}) - \mu(\{x_2\})|^2}{\sigma^2(\{x_1\}) + \sigma^2(\{x_2\})}. \tag{14.3}$$

14.4.1 Experimental Results

The values of photometric invariants to various photometric events are shown in Figure 14.14. The plots show values relative to the total color edge strength $\overline{W}_w$. We do so to express simultaneously the power of $\overline{W}_w$ and of the shadow and shading invariants $\overline{C}_w$ and $\overline{H}_w$ to distinguish between photometric events and true color edges. As expected, the values of the invariants $\overline{C}_w$ and $\overline{H}_w$ are close to zero for shadow/shading edges (note that values of the reference invariant $\overline{W}_w$ are indeed significant to shadow/shading edges). For shadow/shading disturbances, we obtain information($\overline{C}_w$) = 2.6, and information($\overline{H}_w$) = 4.9. Thus, the invariant $\overline{H}_w$ separates shadow/shading from object transitions much better than $\overline{C}_w$. Furthermore, the value of $\overline{H}_w$ is also low for highlights (Fig. 14.14b). However, as expected, not all of the values are close to zero because of pixel saturations at highlights. As a result, the invariance and the information content of $\overline{H}_w$ are

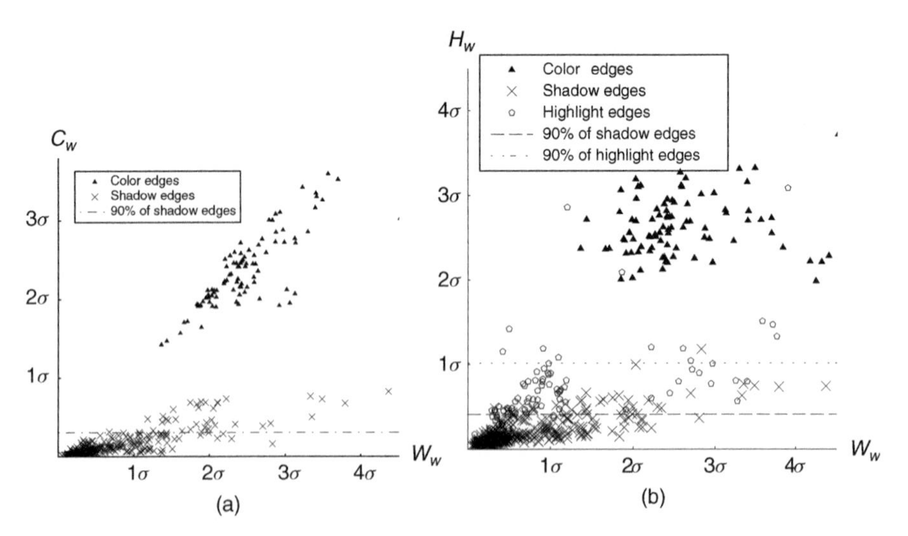

Figure 14.14 Scatter plots of invariant values to photometric events. The figures depict (a) $\overline{C}_w$ versus $\overline{W}_w$ and (b) $\overline{H}_w$ versus $\overline{W}_w$. All invariants are sensitive to color edges. $\overline{C}_w$ and $\overline{H}_w$ are invariant to shadow and shading, where $\overline{H}_w$ is additionally invariant to highlights. The horizontal lines describe a 90% interval of the invariant values. This gives an indication of the invariant's ability to distinguish between values to color edges and to disturbing photometric events. *Source*: Reprinted with permission, © 2009 Elsevier.

somewhat lower for highlight disturbances than for shadow/shading disturbances, $information(\overline{H}_w) = 2.9$.

Overall, the photometric invariant H-color is more constant to shadow and shading than C-color. Both perform well when separating color transitions from shadow and shading transitions. The separation of color transitions and highlights by H-color is harder because of saturated highlights. As a consequence, most of the highlights are separated well, but some highlights are misclassified as color transitions.

14.5 Summary

In this chapter, we have discussed color invariant descriptors for image description and recognition. We evaluated the descriptive power of local derivative-based color invariants and the descriptive power of color invariant SIFT descriptors. In Chapter 16 we will show the application of these descriptors in image and video retrieval.

15 Color Image Segmentation

With contributions by Gertjan J. Burghouts

In this chapter, we consider the invariant assessment of color and texture in combination with applications in image segmentation and material classification. For texture segmentation, we consider the work on Gabor filters [283] and Gaussian derivative filters as the most important [284, 285]. For the modeling of materials, the mapping of image features onto a codebook of feature representatives receives extensive treatment. For reason of generality and simplicity, filterbank outputs are commonly used as features. These methods are often referred to as texton-based methods [286, 287], or nowadays bag-of-word approaches. The combination of color and texture has attracted attention in the recent literature. In Mirmehdi and Petrou [288], color-textured images are roughly segmented based on a spatial color model [289]. The assumption underlying their approach implies that texture can be characterized by its color histogram over a region. The drawback here is that the spatial structure of the texture is not considered since only first-order statistics, the histogram, is taken into account. Thai et al. [290] propose measuring color–texture by embedding the Gabor filters into an opponent color representation. The method provides a useful structural representation for color–texture.

Color in Computer Vision: Fundamentals and Applications, First Edition.
Theo Gevers, Arjan Gijsenij, Joost van de Weijer, and Jan-Mark Geusebroek.

We show the extension of the Gaussian color model presented in Chapter 6 to the domain of texture by extending the Gaussian color into the spectral Fourier domain, closely following the work presented by Hoang et al. [291]. By doing so, we extend the Gaussian framework with a family of color Gabor filters, suitable to capture texture and color. Following the methodology outlined in Chapter 6 to obtain full photometric invariant features, we arrive at texture descriptors invariant under the Lambertian reflection model. Application of these color–texture features in the area of image segmentation result in robust methods for color and texture segmentation.

15.1 Color Gabor Filtering

Recall from Chapter 6 that a color image is observed by integrating over some spatial extent and over a spectral bandwidth. Before observation, a color image may be regarded as a three-dimensional energy density function $E(x, y, \lambda)$, where (x, y) denotes the spatial coordinate and λ denotes the wavelength. Observation of the energy density $E(x, y, \lambda)$ boils down to correlation of the incoming signal with a Gaussian measurement probe $G(x, y, \lambda)$. In Section 6.1 we have shown that three Gaussian derivative functions over the visual spectrum are appropriate to measure color. The Gaussian measurement function $G(x, y, \lambda)$ estimates quantities of the energy density $E(x, y, \lambda)$.

In the case of texture, we are interested in the local spatial frequency characteristics of $E(x, y, \lambda)$. These properties are better investigated in the domain of spatial frequency. Thus, it is appropriate to represent the joint color–texture properties in a combined *wavelength-Fourier* domain $\mathcal{E}(u, v, \lambda)$, where λ remains the wavelength of the light energy, and (u, v) denotes the spatial frequency. Probing this Fourier domain with a Gaussian function now yields the appropriate measurements to assess the image frequency content. The measurement of the signal $\mathcal{E}(u, v, \lambda)$ at a given spatial frequency (u_0, v_0) and wavelength λ_0 is obtained by a 3D Gaussian probe centered at (u_0, v_0, λ_0) at a frequency scale σ_f and wavelength scale σ_λ,

$$\hat{\mathcal{M}}(u, v, \lambda) = \int \mathcal{E}(u, v, \lambda) G(u - u_0, v - v_0, \lambda - \lambda_0; \sigma_f, \sigma_\lambda)\, d\lambda. \tag{15.1}$$

Frequency selection is achieved by tuning the parameters u_0, v_0, and σ_f, and color information is captured by the Gaussian specified by λ_0 and σ_λ. The central wavelength λ_0 and spectral bandwidth σ_λ is fixed, as detailed in Section 6.1.1. However, the choice of the central frequencies (u_0, v_0) and frequency bandwidth σ_f is free. Centring the Gaussian at the origin of the Fourier domain yields our spatial Gaussian color model. Any other choice of (u_0, v_0) leads to the well-known Gabor functions in the spatial domain, but is now calculated over each of the three

Gaussian (opponent) color channels $E(x,y)$, $E_{\lambda}(x,y)$, and $E_{\lambda\lambda}(x,y)$,

$$M_{\lambda^{(n)}}(x,y) = h(x,y) * \hat{E}_{\lambda^{(n)}}(x,y), \qquad (15.2)$$

where

$$h(x,y) = \frac{1}{2\pi\sigma_s^2}\,\mathrm{e}^{-\frac{x^2+y^2}{2\sigma_s^2}}\,\mathrm{e}^{2\pi j(Ux+Vy)} \qquad (15.3)$$

is the 2D Gabor function at the radial center frequency $F = \sqrt{U^2+V^2}$ (cycles/pixel) and the filter orientation $\tan(\theta) = V/U$, and $j^2 = -1$. The Gabor filters in 15.3) are illustrated in Figure 15.1.

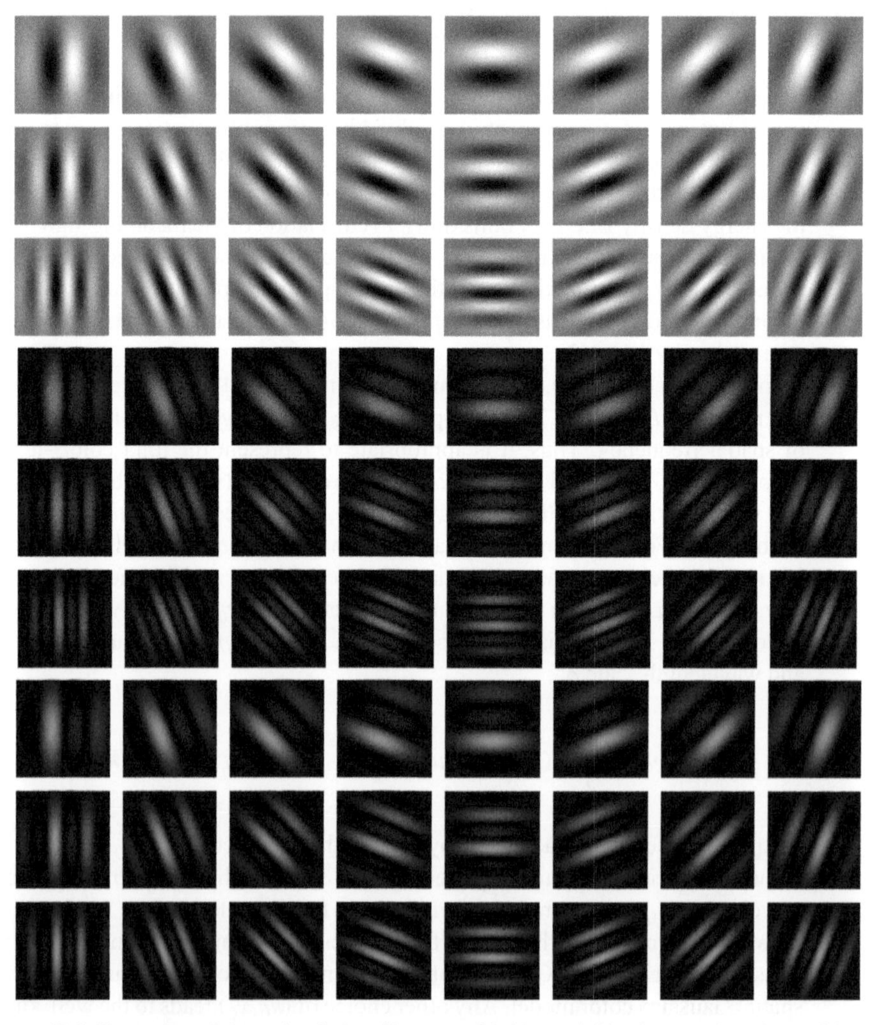

Figure 15.1 Illustration of the color Gabor filter sets M, M_{λ}, and $M_{\lambda\lambda}$ for certain values of (u_0, v_0) and σ_f.

15.2 Invariant Gabor Filters Under Lambertian Reflection

Within a single texture patch, the value of the Gabor filter response varies proportionally to the local intensity of the texture. The darker region has a response value smaller than the value of the brighter one. Therefore, illumination intensity, shadow, and shading effects may compromise the segmentation process. Hence, we aim to correct the effect of intensity variations on the Gabor filter responses. Similar to the derivation of the invariant set C in Chapter 6, we may directly extend the results to Gabor filtering, as all derivations are equivalent for the Gabor filter with respect to the Gaussian filter. In that case, the expressions for set $\tilde{C}$ (the *tilde* indicating frequency tuning by Gabor filters) become

$$\tilde{C}_\lambda = \frac{\tilde{E}_\lambda}{E}, \tag{15.4}$$

which is the Gabor filtered yellow-blue opponent color channel, pixel-wise normalized by the Gaussian smoothed intensity channel. Similarly,

$$\tilde{C}_{\lambda\lambda} = \frac{\tilde{E}_{\lambda\lambda}}{E}. \tag{15.5}$$

These frequency responses are independent of the local intensity, shadow and shading, assuming Lambertian reflection.

Besides the invariant C, the set N has been derived in Section 6.1.4. This set is color constant, that is, invariant for the illumination color. For the case of Gabor filtering, the expressions can be summarized as

$$\tilde{N}_\lambda = \frac{\tilde{E}_\lambda E - E_\lambda \tilde{E}}{E^2}, \tag{15.6}$$

$$\tilde{N}_{\lambda\lambda} = \frac{\tilde{E}_{\lambda\lambda} E^2 - E_{\lambda\lambda} \tilde{E} E - 2\tilde{E}_\lambda E_\lambda E + 2E_\lambda^2 \tilde{E}}{E^3}, \tag{15.7}$$

where the high frequency filtering effect of the Gabor filters cancel the low frequency effects of illumination variations. For further details, see Reference 291.

15.3 Color-Based Texture Segmentation

Here we combine the Gaussian color model with the Gabor filtering results to illustrate combined color–texture segmentation. We employ a simple segmentation algorithm with a scheme similar to [292]. The overall scheme is depicted in Figure 15.2.

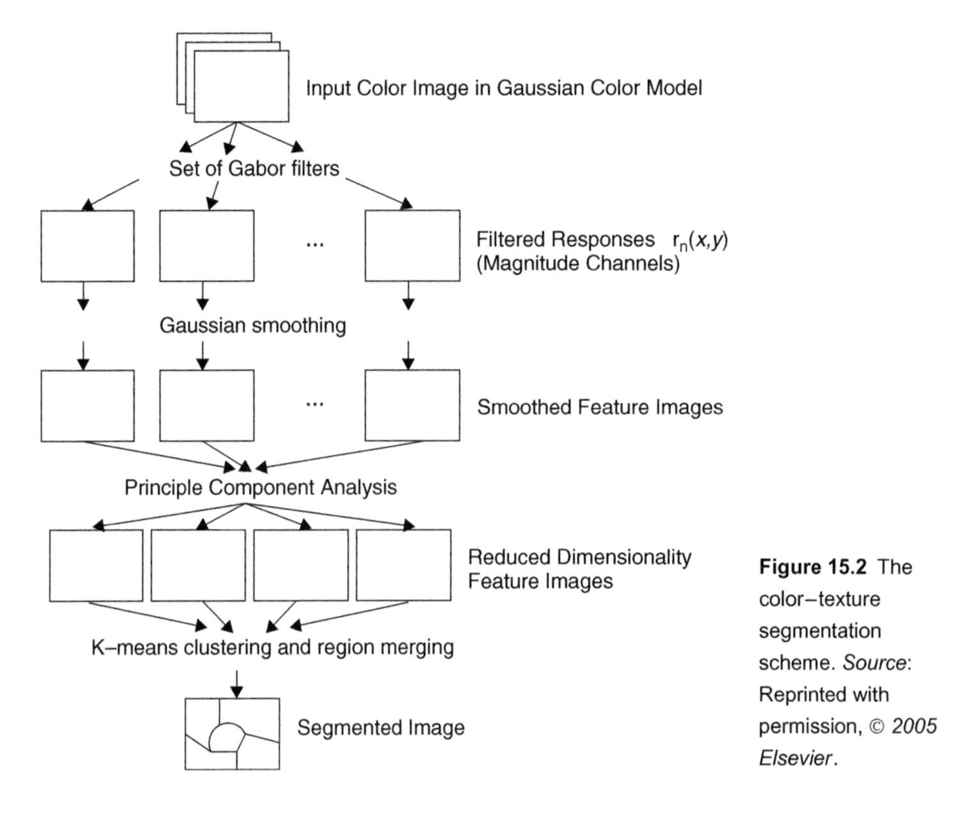

Figure 15.2 The color–texture segmentation scheme. *Source*: Reprinted with permission, © *2005 Elsevier*.

The magnitude of the Gabor filter responses emphasize texture regions, which are in tune with the chosen frequencies of the filter. Methods for designing an efficient set of Gabor filters can be found in References 292–294. In our setup, we use 20 Gabor filters built from five scales $\sigma_s = 4, 3.5, 2.95, 2.35, 1.75$, corresponding to five center frequencies $F = 0.05, 0.08, 0.14, 0.22, 0.33$ (cycles/pixel) and four orientations $\theta = 0, \pm\pi/4, \pi/2$. These values of scale and center frequency are calculated based on the method proposed by Manjunath in Reference 294. We therefore obtain 60 filtered response images from which we consider the magnitude, $r_n(x, y)$, $n = 1, \ldots, 60$. Each image pixel (x_i, y_j) is now represented by a 60-dimensional feature vector whose nth component is denoted by $r_n(x_i, y_j)$. Pixels in one color–texture homogeneous region will form a cluster in the feature space, which is compact and may be discriminated from clusters corresponding to other regions.

The segmentation algorithm is based on clustering pixels using their associated feature vectors. For preprocessing, every filtered magnitude image $r_n(x, y)$ is smoothed by a Gaussian kernel to suppress the variation of the feature vectors within the same color–texture region. Since the feature vectors are highly correlated, we apply the principal components analysis (PCA) to reduce the feature space dimensionality to only four principal dimensions. The four-dimensional feature vectors are used as the input for clustering. The clustering algorithm has two steps. The first step calculates ''super-pixels'' by k-means clustering with a high number of k applied to the feature space. In a second step, a region merging method is used to combine adjacent clusters, which are statistically similar (Fig. 15.2).

The region merging is done in an agglomerative manner where in each iteration the two most similar regions are merged. We employ a region similarity measure analogous to the one proposed in Reference 295. The similarity between regions R_i and R_j is given by

$$S_{i,j} = (\mu_i - \mu_j)^{\mathrm{T}}[\Sigma_i + \Sigma_j]^{-1}(\mu_i - \mu_j), \tag{15.8}$$

where μ_i, μ_j are the mean vectors and Σ_i, Σ_j are the covariance matrices computed from feature vectors of regions R_i and R_j, respectively. Here, $S_{i,j}$ measures the distance between two sets. If one of the two reduces to a single point, $S_{i,j}$ becomes the Mahalanobis distance. The advantage of this measure is that the uncertainty of the vectors μ_i and μ_j as expressed by their respective covariances $\Sigma_{i,j}$ is taken into account. The two regions R_i and R_j are merged if the value of $S_{i,j}$ is under a threshold. In our experiment, the similarity threshold t in the range of $[6 \ldots 9]$ produces almost the same result for every test image. Therefore, we fix the similarity threshold at $t = 7.5$ for all our experiments. Finally, a simple postprocessing technique is utilized to remove small-sized isolated regions.

The segmentation results are illustrated in Figure 15.3. The input image is created by collaging five subimages of natural and artificial color–texture. In this image, two patches on top are chosen to be similar in texture but different in color. The two patches on the left are chosen to be similar in color but different in texture. The results in Figure 15.3 show that five regions are correctly discriminated when using the presented measurement.

Segmentations of real images using the presented method are illustrated in Figure 15.4. Furthermore, segmentation results obtained by using invariant features are shown in Figure 15.5.

15.4 Material Recognition Using Invariant Anisotropic Filtering

The appearance of materials change significantly under different imaging settings, depending on the settings themselves [282] and also on the physical properties of a material [296]. Hence, materials-specific image representations may improve on the recognition performance, as they capture properties that are distinctive to the material and are balanced with the variation of imaging settings. For instance, for one material, the local intensity variation is a distinctive property, while the other is distinguished best from other materials based on its color properties. Figure 15.6 depicts some materials from the ALOT dataset [297] and the testing conditions (Fig. 15.7). The first and second materials are distinguished best when comparing their colors, more specifically, the red channel. For the third and fourth materials,

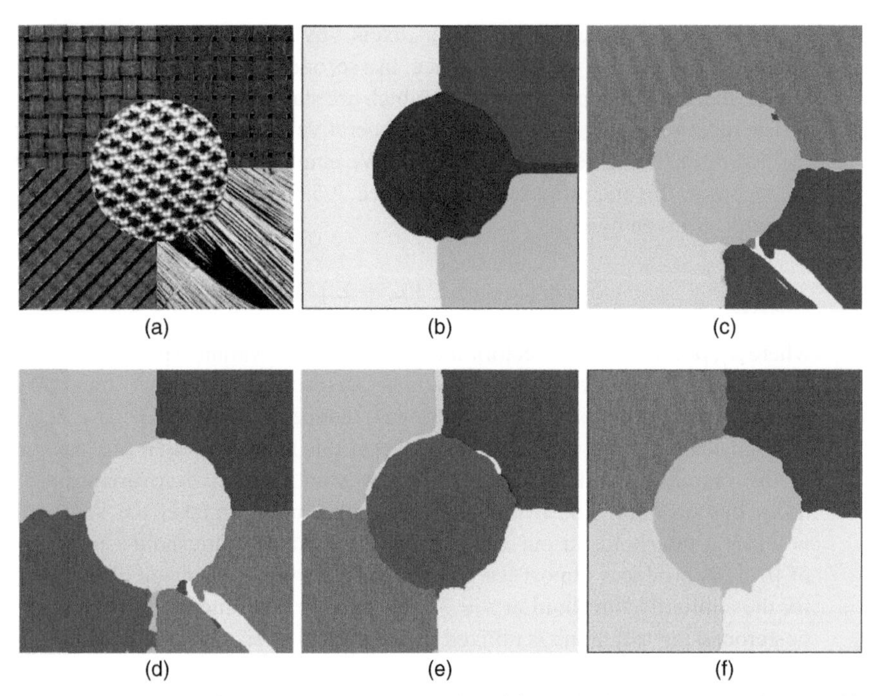

Figure 15.3 Illustration of color–texture segmentation. (a) Synthetic color–texture image with five different regions. (b) The segmentation result using only color features. The original color image is smoothed by a set of Gaussian filters at different scales as in Reference 288. Here, two regions with identical color are merged. (c) Segmentation result using only gray-value texture. Note that the regions with identical texture but different colors are merged. (d) The segmentation result using the color–texture features without shadow invariance. The regions are correctly segmented, but affected by shadow. (e) The segmentation result using the shadow invariant color–texture feature. In this case, all regions are correctly segmented. (f) Postprocessing of the invariant segmentation result to remove small isolated regions. *Source*: Reprinted with permission, © *2005 Elsevier*.

the most discriminative feature is the amount of intensity edges, while the fifth image in the first row and the second image in the third row are distinguished best when comparing the information in the green channel. These examples illustrate the advantage of material-specific representations.

Not only for material recognition [285, 298] and classification [299] but also for object and scene classification [300] the mapping of image features onto a codebook of feature representatives [274, 301] has received extensive treatment. Commonly used features are the class of SIFT-based features [55, 57], see, for example, Reference 263. Alternatively, filterbank outputs are in use as features. Promising methods that use filterbanks to model object and scenes have been proposed by Winn et al. [302] and by Shotton et al. [303].

Here we adapt the approach put forward by Varma and Zisserman [285] and include color invariant properties in their proposed MR8 anisotropic filterbank,

Figure 15.4 Segmentation of a number of example images. *Source*: Reprinted with permission, © *2005 Elsevier*.

Figure 15.5 Segmentations of example images using invariant features. Note that the backgrounds with cast shadows are well segmented, that is, disregarding the cast shadows. *Source*: Reprinted with permission, © *2005 Elsevier*.

Figure 15.6 Example materials from the ALOT dataset [297].

Figure 15.7 Test images for the ALOT material depicted above on third row, first column.

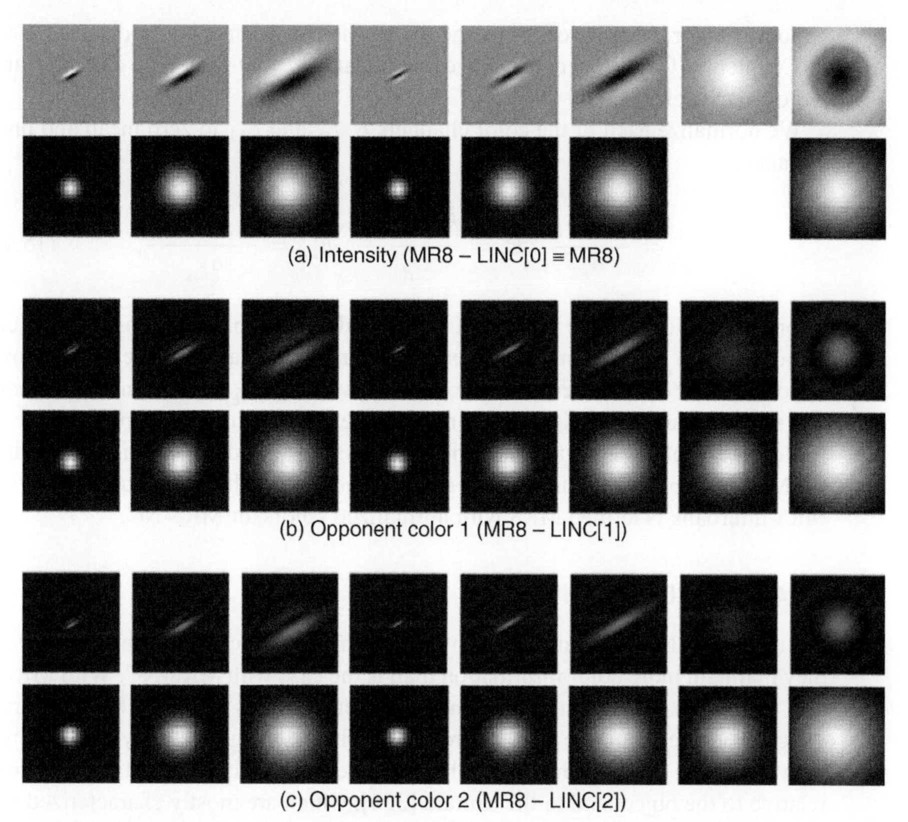

Figure 15.8 MR8-LINC: a color invariant filterbank. The original MR8-filterbank (a, top row) is convolved with each of the image's opponent colors channels (a–c, upper rows), to yield 24 responses per pixel. Each of the 24 filter outputs is normalized by the local intensity as is measured by a Gaussian kernel of the same size of the MR8 filter (a–c, lower rows). The only MR8 filter that is not normalized is the Gaussian kernel that measures intensity (otherwise it would yield a constant output). The normalization achieves invariance to local intensity changes.

closely following [297]. The MR8 filterbank is shown in Figure 15.8a. Typically, before the image is convolved with the MR8 filterbank, the image is normalized to zero mean and unit variance, to achieve invariance to imaging conditions; see Reference 285.

15.4.1 MR8-NC Filterbank

In a first modification of the MR8 filterbank to extend it to use color information, we apply the filterbank to the image's color channels directly. This is a straightforward extension, which is also employed by Winn et al. [302], who have applied the MR8 filterbank to *Lab* color values. We largely follow Reference 302 here. However, we restrain to a linear subspace of RGB and apply the filterbank to the three opponent color channels of the image, being the Gaussian color model presented in Chapter 6. Opponent colors have the advantage that the color channels are largely decorrelated.

We normalize each of the color channels $\hat{E}$, $\hat{E}_\lambda$ and $\hat{E}_{\lambda\lambda}$ to zero mean and unit variance,

$$\hat{E}' = \frac{\hat{E} - \mu_{\hat{E}}}{\sigma_{\hat{E}}}, \quad \hat{E}'_\lambda = \frac{\hat{E}_\lambda - \mu_{\hat{E}_\lambda}}{\sigma_{\hat{E}_\lambda}}, \quad \hat{E}'_{\lambda\lambda} = \frac{\hat{E}_{\lambda\lambda} - \mu_{\hat{E}_{\lambda\lambda}}}{\sigma_{\hat{E}_{\lambda\lambda}}}, \tag{15.9}$$

where $\mu_{\hat{E}}$, $\sigma_{\hat{E}}$ denote the mean and standard deviation of the intensity channel, respectively; $\mu_{\hat{E}_\lambda}$, $\sigma_{\hat{E}_\lambda}$ denote the mean and standard deviation of the blue-yellow opponent color channel, respectively; and, equivalently, $\mu_{\hat{E}_{\lambda\lambda}}$, $\sigma_{\hat{E}_{\lambda\lambda}}$ denote the mean and standard deviation of the green-red opponent color channel, respectively.

Next, each of the normalized color channels $\hat{E}'$, $\hat{E}'_\lambda$, and $\hat{E}'_{\lambda\lambda}$ is convolved with the MR8 filterbank, yielding 24 filter outputs per pixel. This first extension of the MR8 filterbank is termed MR8 with normalized colors, or MR8-NC.

15.4.2 MR8-INC Filterbank

For the MR8-INC invariant filterbank, we normalize the color channels such that they maintain more color information than is the case with MR8-NC. With MR8-NC, the means of the yellow-blue and red-green channels are normalized to zero, effectively discarding the actual chromaticity in the image and only considering the variation. The color channels will be affected mainly by the lighting direction relative to the object and to the camera [304], which are mostly characterized by intensity fluctuations. Hence, we propose to normalize the three opponent color channels only by the standard deviation of the intensity. Normalizing the intensity channel by the standard deviation of intensity,

$$\hat{E}' = \frac{\hat{E} - \mu_{\hat{E}}}{\sigma_{\hat{E}}}, \tag{15.10}$$

sets the variance of this channel to unity. Here, $\mu_{\hat{E}}$ and $\sigma_{\hat{E}}$ indicate the mean and standard deviation of the intensity channel as before. Normalizing the yellow-blue and red-green channels also by the intensity standard deviation,

$$\hat{E}'_\lambda = \frac{\hat{E}_\lambda}{\sigma_{\hat{E}}}, \quad \hat{E}'_{\lambda\lambda} = \frac{\hat{E}_{\lambda\lambda}}{\sigma_{\hat{E}}}, \tag{15.11}$$

yields a more stable response when the intensity variation fluctuates as a consequence of lighting or viewpoint changes. At the same time, it maintains information about the chromaticity in the image. Likewise MR8-NC, each of the normalized color channels is convolved with the MR8 filterbank, yielding 24 filter outputs per pixel. We refer to this filterbank as MR8 with intensity-normalized colors, or MR8-INC.

15.4.3 MR8-LINC Filterbank

In a third modification, we modify the MR8-filterbank to achieve invariance to local intensity changes by a local color normalization rather than a global one. We follow closely the invariant Gaussian features outlined in Chapter 6.

For each pixel, we obtain the Gaussian filtered non-normalized opponent color values using the MR8-filterbank, to obtain 24 filter outputs per pixel. Also, for each pixel, we measure the local intensity with a Gaussian kernel at the same scale as the MR8 filter under consideration. Per pixel, we normalize each output of the MR8 filterbank by the local intensity as measured by that Gaussian filter, yielding the transformed filter responses MR8′,

$$\mathrm{MR8}'\left(\hat{E}\right) = \frac{\mathrm{MR8}(\hat{E})}{\hat{E}^{\sigma}}, \quad \mathrm{MR8}'\left(\hat{E}_{\lambda}\right) = \frac{\mathrm{MR8}(\hat{E}_{\lambda})}{\hat{E}^{\sigma}},$$
$$\mathrm{MR8}'\left(\hat{E_{\lambda\lambda}}\right) = \frac{\mathrm{MR8}(\hat{E_{\lambda\lambda}})}{\hat{E}^{\sigma}}, \tag{15.12}$$

where MR8(.) indicates the successive application of a filter from the filterbank, and $\hat{E}^{\sigma}$ represents the intensity image smoothed at the same spatial scale as the filter of MR8 under consideration, see Figure 15.8. Obviously, the zeroth order Gaussian filter from the MR8-filterbank is not normalized by the local intensity; otherwise, its output would be constant. We refer to this color filterbank as MR8 with local intensity-normalized colors, or MR8-LINC.

15.4.4 MR8-SLINC Filterbank

Finally, we construct a shadow and shading invariant filterbank, termed MR8-SLINC. Similar to MR8-LINC, the invariance is achieved locally. With MR8-LINC, first the filterbank outputs are computed before normalization by the local intensity. Alternatively, the color values $\hat{E}_{\lambda}(x,y)$ and $\hat{E}_{\lambda\lambda}(x,y)$ can be normalized locally first before filtering the thus obtained images,

$$\mathrm{MR8}'\left(\hat{E}\right) = \frac{\mathrm{MR8}(\hat{E})}{\hat{E}^{\sigma}}, \quad \mathrm{MR8}'\left(\hat{E}_{\lambda}\right) = \mathrm{MR8}\left(\frac{\hat{E}_{\lambda}}{\hat{E}}\right),$$
$$\mathrm{MR8}'\left(\hat{E_{\lambda\lambda}}\right) = \mathrm{MR8}\left(\frac{\hat{E_{\lambda\lambda}}}{\hat{E}}\right). \tag{15.13}$$

Under Lambertian reflection, the normalization of color values by the local intensity results in color values independent of the intensity distribution. Hence, the filterbank outputs of MR8-SLINC are invariant to shadow and shading.

15.4.5 Summary of Filterbank Properties

Similar to MR8, the color-based filterbanks MR8-NC and MR8-INC involve a global color normalization. In other words, the normalization is dependent on the contents of the image. Hence, clutter will affect the normalization. This makes the output of MR8-NC and MR8-INC *scene dependent*. In contrast, the local normalizations that are employed in MR8-LINC and MR8-SLINC are not scene dependent, but only *locally dependent* on the actual color values.

Furthermore, the filterbanks can be ordered by their degree of invariance. MR8-SLINC is the most invariant, as its color channels aim to discard intensity variation. MR8 and MR8-NC, respectively, are the intensity and color variations, but they discard their mean and variance. MR8-LINC retains more of the intensity and color variations, as it discards locally the variance due to intensity fluctuations. Finally, MR8-INC is less invariant than MR8-LINC, as it discards only the global variance due to intensity fluctuations.

15.5 Color Invariant Codebooks and Material-Specific Adaptation

In this section, we consider the construction of color invariant codebooks from the several filterbanks, and the methodology to apply the codebooks in a material-specific setting. First, we formalize the color invariant filterbanks as follows: MR8-X = { MR8-X, MR8-$X[1]$, MR8-$X[2]$}, where $X \in$ {NC, INC, LINC, SLINC}. We learn one codebook for each color channel MR8-$X[i]$, with $i \in \{0, 1, 2\}$. For codebook construction, we follow the common scheme of learning textons by k-means clustering of filterbank outputs [262, 285, 298, 305]. We consider a single set of 20 images randomly drawn from the learning set of material images. Each is filtered by one of the filterbanks MR8-$X[i]$, and from each filtered image we store 10 cluster centers. As a result, for each filterbank MR8-$X[i]$, we obtain a codebook of 200 textons. For the filterbank MR8-X, we have obtained 3 codebooks of length 200. For fair comparison with the single-channel MR8 filterbank, the length of the MR8 codebook is increased to 600 by storing 30 instead of 10 cluster centers per learning image.

To represent an image in terms of codebooks, it is filtered by each of the color channel filterbanks MR8-$X[i]$ first, before mapping the filter outputs onto the corresponding codebook and counting the most similar occurrences. For each MR8-$X[i]$, a histogram of length 200 is obtained; hence for MR8-X three histograms are obtained. After concatenation of the histograms per color channel, a histogram of length 600, which corresponds to the filterbank MR8-X is obtained. The codebook representation is outlined in Figure 15.9.

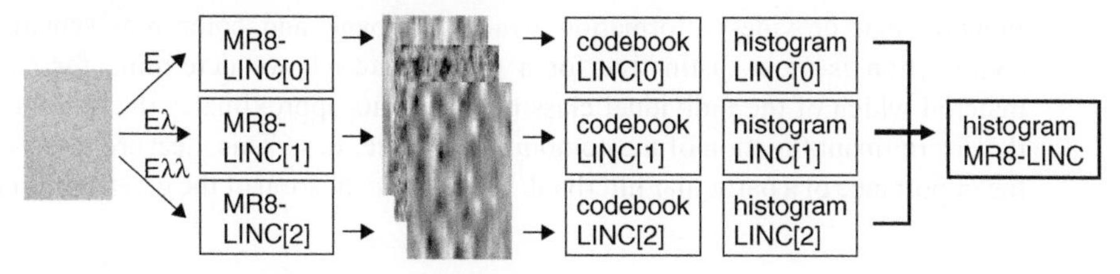

Figure 15.9 Color codebook approach where the three color channels are separately filtered and represented by a histogram. Subsequently, the histograms are combined into one. *Source*: Reprinted with permission, © *2009 Elsevier*.

The limitation of the color codebook representation as presented above is that the discriminative power of the color channels is averaged by using a single histogram comparison measure. For instance, the intensity information may be less distinctive for a given material than is the color information. The averaging of the information in the color channels may lead to incorrect classification of materials. The misclassification of an image of the bluish material, mistakenly considered to be more similar to the pink material, is illustrated in Figure 15.10a.

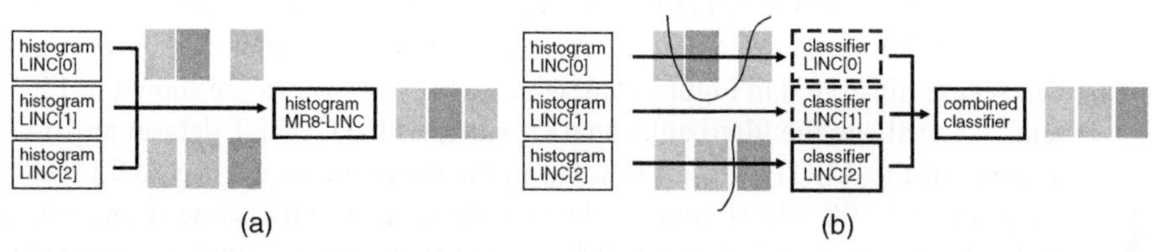

Figure 15.10 Separation of two images of the same material from one image of another material. The fixed representation in (a) is not able to distinguish correctly between the two, while the material-specific representation is able to distinguish between the two (third color channel). *Source*: Reprinted with permission, © *2009 Elsevier*.

To overcome the limited resolving power of the direct combination of the three color channels, we start with classification of a material at the level of individual color channels and to give preference to a distinctive combination thereof. Figure 15.10b illustrates that the bluish material is well separated from the pink material using the information in the third color channel.

We propose to train one classifier per color channel per filterbank to discriminate one material from all other materials. Hence, with I filterbanks, $F_{1\ldots I}$, and J color channels, $c_1 \ldots J$, we obtain $I \times J$ classifiers. With N materials, each classifier outputs N posterior probabilities. With this procedure, $I \times J \times N$ values are produced by the first classifier stage.

In the combination stage, one classifier is trained using the $I \times J \times N$ values obtained for each material image. This one versus all classifier learns per material the discriminant function from the posterior probabilities assigned to each material by the individual classifiers. As a result, the combined classifier learns *implicitly* the filterbank and color channel that is most distinctive for the specific material. To infer *explicitly* from the material-specific discriminant function that provides information which filterbank and color representation combination is most distinctive for a given material, we determine for each material which of the individual classifier's outputs approximates the normal to the discriminant function of the combining classifier best. This measure indicates the importance of a particular filterbank for the classification of the given material.

15.6 Experiments

In the experiments, we evaluate the color filterbanks and their combination. We take two datasets into account to cover a wide range of real-world materials and imaging conditions under which they can be viewed. First, we consider the well-known CURET dataset [282]. This dataset enables us to test the robustness under varying imaging conditions, that is, changes of the illumination direction and of the camera viewpoint. For color-based methods, a critical issue is whether the method is robust to color transformations in the image as a consequence of varying illumination color. Second, we consider the ALOT dataset [297] to also include variations of the illumination color. In addition, this dataset contains more color and 3D variation. Some of the materials that are included in the ALOT dataset are illustrated in Figure 15.6, while some test images are shown in 15.7. In total, we evaluate the filterbanks on 61 textures of the CURET dataset and on 200 textures of the ALOT dataset. In total, in the experiments we use 5,612 CURET images and 7,200 ALOT images. For CURET, we use the same train, test, and texton learn sets as in Reference 285; for ALOT the sets are publicly available on the website of the ALOT database.

In the experiments, the number of textons is always set to 200 (as in Reference 285). For the individual and combined classifiers, we prefer, respectively, the nearest mean classifier (Euclidean distance) and the linear Bayes-normal classifier, as these are performing best.

15.6.1 Material Classification by Color Invariant Codebooks

We start the performance evaluation by establishing the classification accuracy when randomly selecting the learning images. This experiment gives an indication of the discriminative power and robustness of each of the color filterbanks. We include the original MR8 as a baseline comparison. We consider the mean and standard deviation of classification accuracy over 1000 repetitions (random selections).

In Figure 15.11a and b the recognition results for the CURET and ALOT datasets are shown, respectively. First, we discuss the results for the CURET dataset. The filterbanks with most invariant properties, MR8, MR8-NC, and MR8-SLINC filterbanks performance is degraded compared to the less invariant MR8-INC and MR8-LINC filterbanks. MR8 performs somewhat better than MR8-NC and MR8-SLINC, as the nearest mean classifier puts all emphasis on

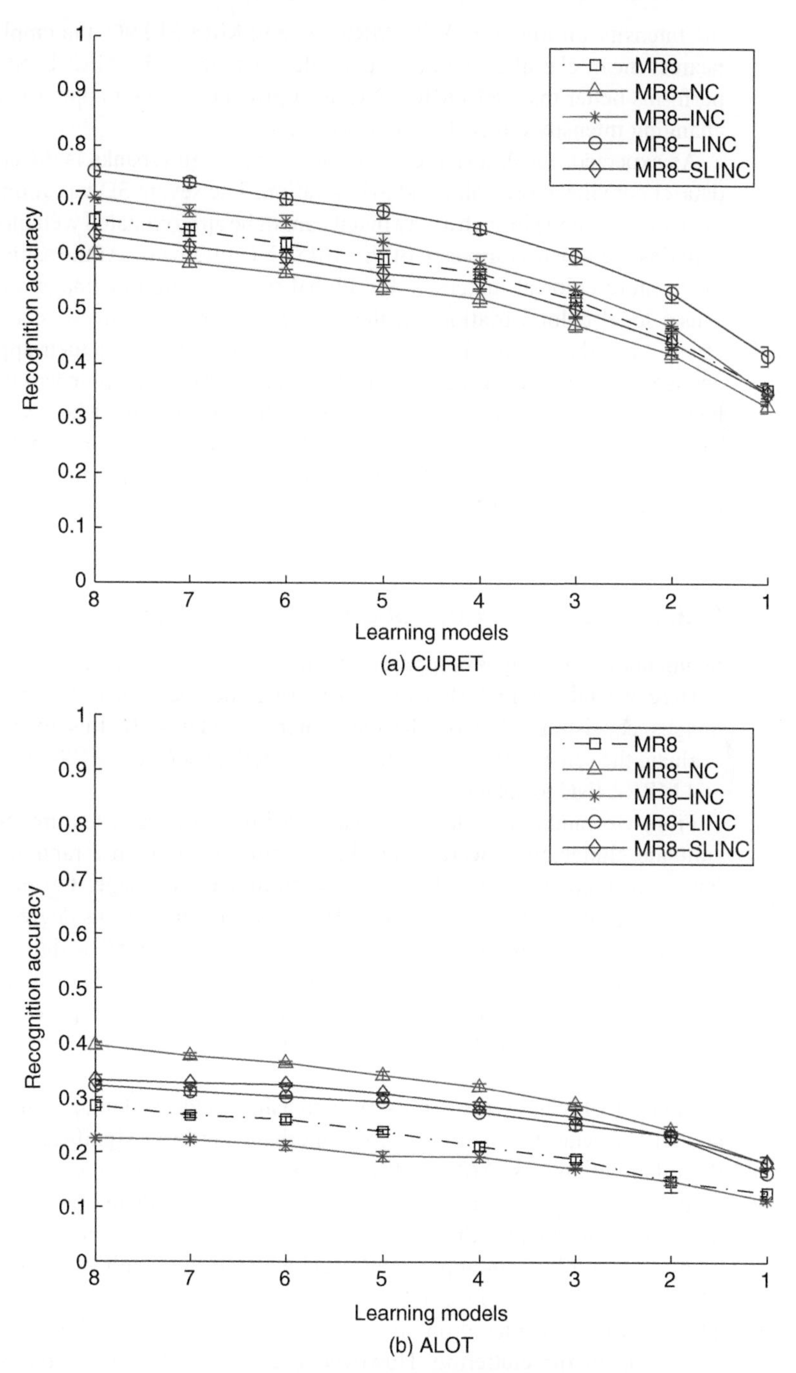

Figure 15.11 Accuracy of material recognition for various filterbanks with randomly selected images of (a) the CURET dataset and (b) the ALOT dataset. The vertical bars indicate standard deviation over 1000 repetitions.

the intensity information. With MR8-NC and MR8-SLINC, the emphasis of the nearest mean classifier is put on the color channels. The MR8-LINC filterbank performs better than does MR8-INC, as it provides a better approximation of the changing intensity effects by doing so locally.

As expected, for ALOT the performance of the filterbanks is different, as this dataset contains more color and 3D variation. The severe 3D variations cause the intensity to change in such a way that it cannot be approximated well globally. This explains the low performance of the MR8-INC filterbank. At the same time, with much more colorful materials, the global normalization of image colors makes sense: local color variations in the image are now kept albeit relative to each other. Also, the severe 3D variations across materials causes their appearance to change significantly with different illumination. Keeping color variations, while being very invariant, explains the good performance of the MR8-NC filterbank. The MR8-INC and MR8-LINC filterbanks are less invariant, hence they perform somewhat less than MR8-NC. The distinctive color information maintained by MR8-INC and MR8-LINC explains their better performance compared to the MR8 filterbank.

15.6.2 Color–Texture Segmentation of Material Images

Segmentation of images composed of various materials is a challenging problem.

Here we take a first step by considering the segmentation performance of images consisting of two adjacent material textures. In this experiment, we evaluate the sensitivity of the color-based filterbanks MR8, MR8-NC, MR8-INC, and MR8-LINC to such cluttered images.

First, we randomly select one learning image for each texture. Second, we simulate clutter by concatenating the learning image with a randomly selected image of another texture. For the first cluttered test image, the percentage of original versus clutter is 90% versus 10%. To simulate various degrees of clutter, we increase the clutter percentage, up to 40% (note: with 50%, the segmentation would become chance). The cluttered images are publicly available on the website of the ALOT [297] database. For generalization purposes, we use the texton dictionary from the previous experiment (i.e., we do not learn new textons from cluttered images).

Figure 15.12 shows the results for increasingly cluttered images of the CURET and ALOT datasets. The MR8-LINC filterbank performs significantly better than the other filterbanks, MR8, MR8-NC, and MR8-INC, over various degrees of clutter. The low performance of MR8, MR8-NC, and MR8-INC is due to the global normalization schemes that they employ. A global normalization is distorted by clutter, so the filterbank input is different when dealing with variations of clutter. The local normalization employed in MR8-LINC is not distorted by clutter. The small performance drop here is due to ambiguity in the images themselves as a result of the cluttering. However, even with 40% clutter, the MR8-LINC filterbank achieves a classification accuracy of 75.5% on the ALOT dataset, while the runner-up (MR8-LINC) has an accuracy of 39.0% only.

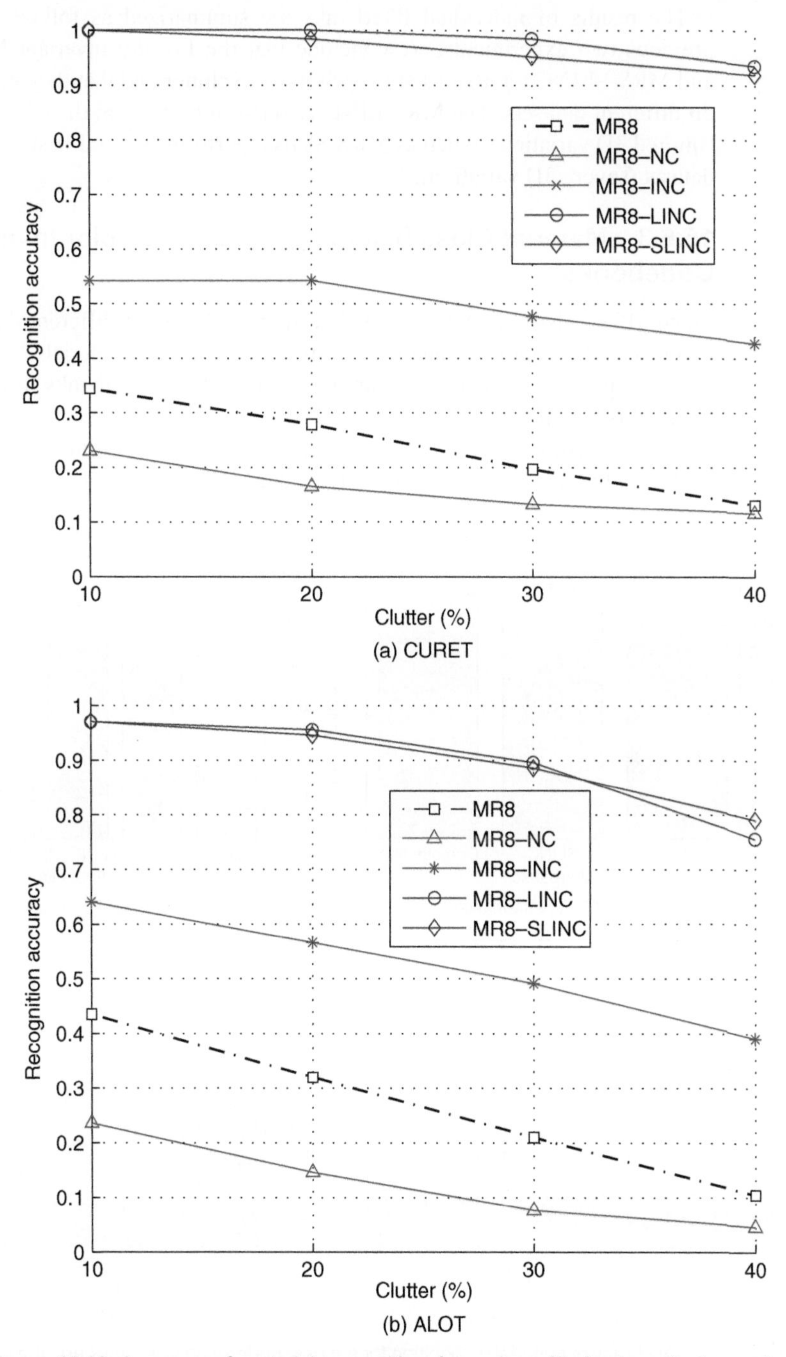

Figure 15.12 Accuracy of material recognition for various filterbanks with increasingly cluttered images of (a) the CURET dataset and (b) the ALOT dataset.

The results of individual filterbanks are summarized as follows. From the previous two experiments, we conclude that the Locally invariant MR8-LINC and MR8-SLINC filterbanks are very robust to clutter and that they perform well on different datasets. The MR8-LINC is performing best on the CURET dataset (limited 3D variation), whereas MR8-SLINC performs second-best on the ALOT dataset (severe 3D variation).

15.6.3 Material Classification by Adaptive Color Invariant Codebooks

Since MR8-LINC and MR8-SLINC perform well but on different datasets, and given that the datasets contain very different types of materials, we establish in this experiment whether the tuning of each of the filterbanks to a particular material is beneficial.

As expected, Figure 15.13a and c indicates that the classification accuracy is increased by combining the MR8-LINC and MR8-SLINC filterbanks. While the

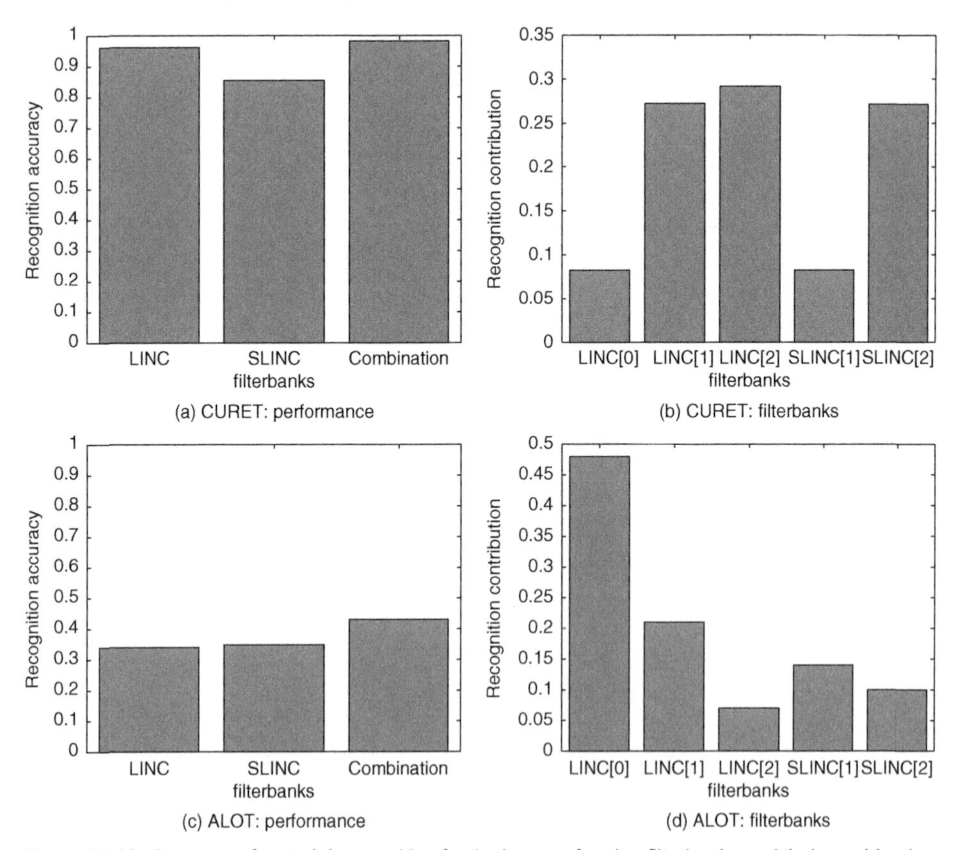

Figure 15.13 Accuracy of material recognition for the best performing filterbanks and their combination for the CURET dataset (a) and the ALOT dataset (c). Percentages indicate how often a particular filterbank is most distinctive (b,d).

classification accuracy of MR8-LINC is almost saturated for the CURET dataset, 0.96, the combination achieves a marginal improvement, 2%. For the ALOT dataset, the performance is increased from 0.35 to 0.42, achieving an improvement of 19.8% .

Indeed, as laid down in Figure 15.13b and d, the most distinctive filterbank per material varies significantly across the datasets, and also across the individual materials. The CURET dataset contains many materials of which the structure is similar. Hence, the intensity variation, although very discriminative (see previous experiments), is not *most* discriminative. Rather, color information is most discriminative, as the color channels of the filterbanks are often most distinctive. The information in the filterbanks that are not invariant to shadow and shading, MR8-LINC, is in 56% most distinctive. Most CURET materials are uni-colored, hence the color information is distinctive. With unicolored materials, too much information is lost when discarding shadow and shading variation. Hence, the shadow and shading invariant filterbank MR8-SLINC is in a few cases, 27%, most distinctive.

For the ALOT dataset, the performance improvement due to filterbank tuning is significant. As this dataset contains more variation of the material properties, and because more materials are included, the results generalize better. For ALOT the most distinctive filterbanks corresponds to intensity information. This can be explained from the fact that intensity variation rather than color variation is the dominating factor in material appearance [296]. The information in the filterbanks that are not invariant to shadow and shading, MR8-LINC, is in 28% most distinctive. The shadow and shading invariant filterbank MR8-SLINC is in 25% most distinctive. We conclude that MR8-LINC and MR8-SLINC are discriminative for large but different sets of materials, respectively.

Finally, we stress that the recognition of materials from the ALOT dataset is obviously a far-from-solved problem. Here, we have demonstrated the merit of automatically tuning filterbanks with different invariant properties to individual materials with different physical properties.

15.7 Image Segmentation by Delaunay Triangulation

The methods discussed so far are based on local properties, as can be estimated by localized filters. Alternatively, a more regional segmentation can be devised by integrating edge evidence over longer contours. In this section, an adaptive image segmentation scheme is discussed employing the Delaunay triangulation for image splitting. The tessellation grid of the Delaunay triangulation is adapted to the structure of the image data by combining photometric invariant region and edge information. To achieve robustness against imaging conditions (*e.g.*, shading, shadows, illumination and highlights), photometric invariant similarity measures and edge computation is used. We consider the Delaunay triangulation as the geometric data structure for image segmentation. The Delaunay triangulation maximizes the minimum angle, minimizes the maximum circumscribing circle, and minimizes the maximum smallest enclosing circle for each triangle.

The adaptive image segmentation method is as follows:

Initialization: Let $\mathcal{D}^j$ denote the incremental Delaunay triangulation after j insertions of points in $\mathcal{R}^2$. Let d_i^j be the ith triangle of the j triangulation. Furthermore, consider the function $g : \mathcal{R}^2 \rightarrow \mathcal{R}$ defining an image surface $g(x, y)$. $g_i^j(x, y)$ is a compact area of g, which is bounded by the vertices of triangle d_i^j. Because it is assumed that the image data points are limited to a rectangular image domain, the image segmentation method starts with the construction of the initial triangulation $\mathcal{D}^0$ consisting of two triangles d_i^0 for $i = 1, 2$ whose vertices are the corners of g.

Splitting: After the construction of $\mathcal{D}^0$, the algorithm successively examines triangles d_i^j by computing the similarity predicate $H()$. The similarity predicate is defined on g_i^j denoting the underlying image data of triangle d_i^j. If the similarity predicate is false, edge pixels in g_i^j are classified topographically based on their local neighborhood by the difference function $D()$. Then, the splitting function $S()$ assigns a *transition error* to every edge point. The goal is to adapt the image tessellation grid properly to the underlying structure of the image data. As a consequence, the edge point with the lowest transition error is taken and entered into $\mathcal{D}^j$ to generate the next triangulation $\mathcal{D}^{j+1}$. The splitting phase continues until all triangles satisfy $H()$.

Merging: Let R_i be a point set in $\mathcal{R}^2$ forming the ith polygon with corresponding $r_i \subset \mathcal{R}^2$, which is a compact area of the plane by merging triangular areas of the final Delaunay triangulation ($\mathcal{D}^N$). In fact, $r_i = g_1^N \cup g_2^N \cup \ldots g_n^N$, where all n triangular image regions are adjacent. The merging phase starts with the triangulation produced by the splitting phase $R_i = d_i^N$ for all i. Function $H()$ provides the criterion by which two adjacent polygons are merged into one.

The algorithm is determined by functions $H()$, $D()$ and $S()$. They are discussed in the next section.

15.7.1 Homogeneity Based on Photometric Color Invariance

To provide robustness against imaging conditions (e.g., illumination, shading, highlights, and inter-reflections), photometric color invariants are used, which have been discussed in Section 4.4.1:

$$c_1(R, G, B) = \arctan\left(\frac{R}{G}\right), \tag{15.14}$$

$$c_2(R, G, B) = \arctan\left(\frac{R}{B}\right), \tag{15.15}$$

$$c_3(R, G, B) = \arctan\left(\frac{G}{B}\right), \tag{15.16}$$

where R, G, and B are the red, green, and blue channels of a color camera.

$c_1c_2c_3$ is insensitive to a large extent to a change in camera viewpoint, object pose, and for the direction and intensity of the incident light. Furthermore, when shadows correspond to a change in intensity, which is often the case, $c_1c_2c_3$ is also insensitive to shadows. When shadows are strongly colored, $c_1c_2c_3$ is not shadow invariant.

Furthermore, we focus on (see Section 4.4.2)

$$l_1(R, G, B) = \frac{|R - G|}{|R - G| + |B - R| + |G - B|}, \tag{15.17}$$

$$l_2(R, G, B) = \frac{|R - B|}{|R - G| + |B - R| + |G - B|}, \tag{15.18}$$

$$l_3(R, G, B) = \frac{|G - B|}{|R - G| + |B - R| + |G - B|}, \tag{15.19}$$

also insensitive to highlights under the restriction of white illumination or a white-balanced camera.

15.7.2 Homogeneity Based on a Similarity Predicate

We define region $\mathcal{R}$ to be homogeneous when the observed color invariant values of the region can be approximated in the color invariant space by a Gaussian distribution with mean and standard deviation due to noise. If the standard deviation is below a predefined threshold, region $\mathcal{R}$ is considered to be homogeneous. Again, the mean noise standard deviation $\hat{\sigma}$ in the image is estimated by applying a least-squares fit to a uniformly colored region (e.g., derived from a 5×5 mask). Then, the similarity predicate $H()$, returning a Boolean value, is given by

$$H(\mathcal{R}) = \begin{cases} \textit{true}, & \text{if } \epsilon \leq \hat{\sigma} \\ \textit{false}, & \text{otherwise} \end{cases}, \tag{15.20}$$

where region $\mathcal{R}$ is considered to be homogeneous if the color invariant values of $\mathcal{R}$ form a Gaussian distribution, which falls within the limit of the noise standard deviation.

15.7.3 Difference Measure

In this section, the principled way is taken to compute gradients in vector images as described by di Zenzo [234] and discussed in Chapter 13, which is summarized as follows.

Let $\Theta(x_1,x_2) : \Re^2 \rightarrow \Re^m$ be a m-band image with components $\Theta_i(x_1,x_2) : \Re^2 \rightarrow \Re$ for $i = 1,2,\ldots,m$. For color images we have $m = 3$. Hence, at a given image location the image value is a vector in $\Re^m$. The difference at two nearby points $P = (x_1^0, x_2^0)$ and $Q = (x_1^1, x_2^1)$ is given by $\Delta\Theta = \Theta(P) - \Theta(Q)$. Considering an infinitesimal displacement, the difference becomes the differential $\mathrm{d}\Theta = \sum_{i=1}^{2} \frac{\partial\Theta}{\partial x_i}\mathrm{d}x_i$ and its squared norm is given by

$$\mathrm{d}\Theta^2 = \sum_{i=1}^{2}\sum_{k=1}^{2} \frac{\partial\Theta}{\partial x_i}\frac{\partial\Theta}{\partial x_k}\mathrm{d}x_i\mathrm{d}x_k = \begin{bmatrix} \mathrm{d}x_1 \\ \mathrm{d}x_2 \end{bmatrix}^T \begin{bmatrix} g_{11} & g_{12} \\ g_{21} & g_{22} \end{bmatrix} \begin{bmatrix} \mathrm{d}x_1 \\ \mathrm{d}x_2 \end{bmatrix}, \tag{15.21}$$

where $g_{ik} := \frac{\partial\Theta}{\partial x_i} \cdot \frac{\partial\Theta}{\partial x_k}$ and the extrema of the quadratic form are obtained in the direction of the eigenvectors of the matrix $[g_{ik}]$ and the values at these locations correspond with the eigenvalues given by

$$\lambda_{\pm} = \frac{g_{11} + g_{22} \pm \sqrt{(g_{11} - g_{22})^2 + 4g_{12}^2}}{2}, \tag{15.22}$$

with corresponding eigenvectors given by $(\cos\theta_{\pm}, \sin\theta_{\pm})$, where $\theta_+ = \frac{1}{2}\arctan\frac{2g_{12}}{g_{11}-g_{22}}$ and $\theta_- = \theta_+ + \frac{\pi}{2}$. Hence, the direction of the minimal and maximal changes at a given image location is expressed by the eigenvectors θ_- and θ_+, respectively, and the corresponding magnitude is given by the eigenvalues λ_- and λ_+, respectively. Note that λ_- may be different than zero and that the strength of a multivalued edge should be expressed by how λ_+ compares to λ_-, for example, by subtraction $\lambda_+ - \lambda_-$ as proposed by Sapiro and Ringach [241].

Then, the color gradient for RGB is as follows:

$$\nabla C_{RGB} = \sqrt{\lambda_+^{RGB} - \lambda_-^{RGB}}, \tag{15.23}$$

for $\lambda_{\pm} = \frac{g_{11}^{RGB} + g_{22}^{RGB} \pm \sqrt{(g_{11}^{RGB} - g_{22}^{RGB})^2 + 4(g_{12}^{RGB})^2}}{2}$, where $g_{11}^{RGB} = |\frac{\partial R}{\partial x}|^2 + |\frac{\partial G}{\partial x}|^2 + |\frac{\partial B}{\partial x}|^2$, $g_{22}^{RGB} = |\frac{\partial R}{\partial y}|^2 + |\frac{\partial G}{\partial y}|^2 + |\frac{\partial B}{\partial y}|^2$, $g_{12}^{RGB} = \frac{\partial R}{\partial x}\frac{\partial R}{\partial y} + \frac{\partial G}{\partial x}\frac{\partial G}{\partial y} + \frac{\partial B}{\partial x}\frac{\partial B}{\partial y}$, where the partial derivatives are computed through Gaussian smoothed derivatives.

Then, the color invariant gradient (based on $c_1c_2c_3$) for matte objects is given by

$$\nabla C_{c_1c_2c_3} = \sqrt{\lambda_+^{c_1c_2c_3} - \lambda_-^{c_1c_2c_3}}, \tag{15.24}$$

for $\lambda_{\pm} = \frac{g_{11}^{c_1c_2c_3} + g_{22}^{c_1c_2c_3} \pm \sqrt{(g_{11}^{c_1c_2c_3} - g_{22}^{c_1c_2c_3})^2 + 4(g_{12}^{c_1c_2c_3})^2}}{2}$, where $g_{11}^{c_1c_2c_3} = |\frac{\partial c_1}{\partial x} 5|^2 + |\frac{\partial c_2}{\partial x}|^2 + |\frac{\partial c_3}{\partial x}|^2$, $g_{22}^{c_1c_2c_3} = |\frac{\partial c_1}{\partial y}|^2 + |\frac{\partial c_2}{\partial y}|^2 + |\frac{\partial c_3}{\partial y}|^2$, $g_{12}^{c_1c_2c_3} = \frac{\partial c_1}{\partial x}\frac{\partial c_1}{\partial y} + \frac{\partial c_2}{\partial x}\frac{\partial c_2}{\partial y} + \frac{\partial c_3}{\partial x}\frac{\partial c_3}{\partial y}$.

In a similar way, we propose that the color invariant gradient (based on $l_1l_2l_3$) for shiny objects is given by

$$\nabla \mathcal{C}_{l_1l_2l_3} = \sqrt{\lambda_+^{l_1l_2l_3} - \lambda_-^{l_1l_2l_3}}, \tag{15.25}$$

for $\lambda_\pm = \frac{g_{11}^{l_1l_2l_3}+g_{22}^{l_1l_2l_3}\pm\sqrt{(g_{11}^{l_1l_2l_3}-g_{22}^{l_1l_2l_3})^2+4(g_{12}^{l_1l_2l_3})^2}}{2}$, where $g_{11}^{l_1l_2l_3} = |\frac{\partial l_1}{\partial x}|^2 + |\frac{\partial l_2}{\partial x}|^2 + |\frac{\partial l_3}{\partial x}|^2$, $g_{22}^{l_1l_2l_3} = |\frac{\partial l_1}{\partial y}|^2 + |\frac{\partial l_2}{\partial y}|^2 + |\frac{\partial l_3}{\partial y}|^2$, $g_{12}^{l_1l_2l_3} = \frac{\partial l_1}{\partial x}\frac{\partial l_1}{\partial y} + \frac{\partial l_2}{\partial x}\frac{\partial l_2}{\partial y} + \frac{\partial l_3}{\partial x}\frac{\partial l_3}{\partial y}$.

15.7.4 Segmentation Results

Figure 15.15a shows an image of several objects against a background consisting of four squares. The size of the image is 256 × 256. The image has been recorded by the SONY XC-003P and the Matrox Magic Color frame grabber. The digitization was done in 8 bits per color. Two light sources of average daylight color are used to illuminate the objects in the scene. The image is clearly contaminated by shadows, shading, highlights, and inter-reflections. Inter-reflections occur when an object receives the reflected light from other objects. In 15.14a, edges are shown obtained from the *RGB* image with nonmaximum suppression with $\sigma_g = 1.0$ used for the Gaussian-based fuzzy derivatives. Clearly, edges are introduced by abrupt surface orientations, shadows, inter-Reflections, and highlights. In contrast, computed edges for $c_1c_2c_3$ and $l_1l_2l_3$ defined by $\nabla \mathcal{C}_{c_1c_2c_3}$ and $\nabla \mathcal{C}_{l_1l_2l_3}$, respectively, shown in Figure 15.14b and c, are insensitive for shadows, surface orientation changes, and highlights (only for $\nabla \mathcal{C}_{l_1l_2l_3}$).

To avoid edge grouping, to obtain proper region outlines with closed contours, the $l_1l_2l_3$ edge map is used as the input of the region-based segmentation method. Again, in Figure 15.15a, the recorded color image is shown. The mean noise standard deviation is estimated by applying a least-squares fit to a uniformly colored region (5 × 5 mask). The measured mean noise standard deviation is

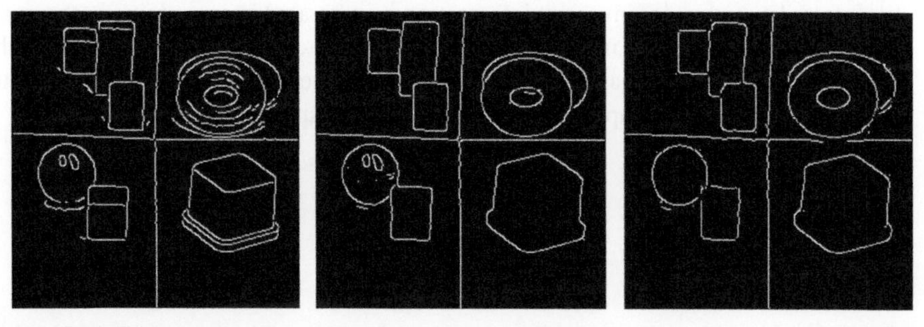

Figure 15.14 Edge maps of the various color models computed from the first recorded color image shown in Figure 15.15a. (a) Edge map based on *RGB* gradient field $\nabla \mathcal{C}_{RGB}$ with nonmaximum suppression. (b) Edge map based on $c_1c_2c_3$ gradient field $\nabla \mathcal{C}_{c_1c_2c_3}$ with nonmaximum suppression. (c) Edge map based on $l_1l_2l_3$ gradient field $\nabla \mathcal{C}_{l_1l_2l_3}$ with nonmaximum suppression.

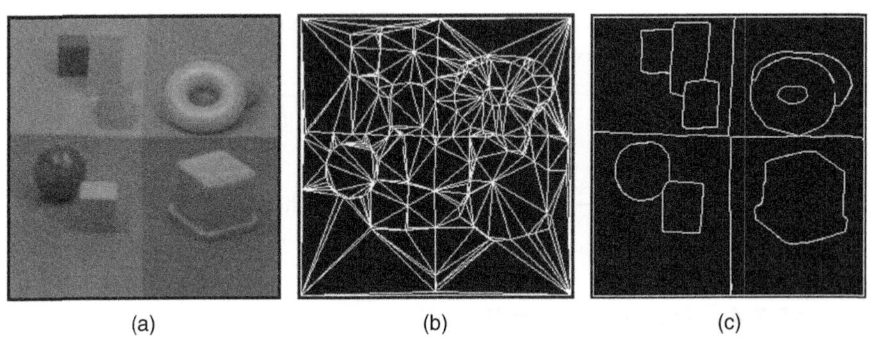

(a) (b) (c)

Figure 15.15 (a) First recorded color image. (b) Splitting result based on Delaunay splitting. (c) The final segmentation result of the region-based segmentation method. *Source*: Reprinted with permission, *© 2002 IEEE*.

$\hat{\sigma} = 3.1$ and used as the threshold for the similarity predicate $H_C()$. The splitting result is shown in Figure 15.15b. The final segmentation result is shown in Figure 15.15c. Despite the various radiometrical and geometrical variations caused by the imaging process, region outlines correspond neatly to material boundaries.

15.8 Summary

In this chapter, we have discussed schemes for image segmentation and material recognition. The chapter extends the theory of Gaussian color measurement from Chapter 6 to Gabor filtering, effective in texture segmentation and categorization. Furthermore, we applied anisotropic invariant color filter banks to recognize materials from single images. Finally, a more regional strategy has been demonstrated, where color invariant information is integrated over contours in the image by a Delaunay-based image triangulation. The latter method demonstrates global analysis of color image edge content for segmentation, rather than the fully localized filter bank approaches from the former approaches.

PART V

APPLICATIONS

16 Object and Scene Recognition

With contributions by Koen E. A. van de Sande and Cees G. M. Snoek

Image category recognition is important to access visual information on the level of objects (buildings, cars, etc.) and scene types (outdoor, vegetation, *etc.*). In general, systems for category recognition on images [300, 306–308] and video [309, 310] use machine learning based on image descriptions to distinguish object and scene categories. In Chapter 13, methods have been discussed to detect salient points that are invariant to translation, rotation, and scale. Further, different color descriptors are presented in Chapter 14. Because there are many different descriptors, a structured overview is required of color invariant descriptors in the context of image category recognition.

Therefore, this chapter gives an overview of the invariance properties and the distinctiveness of the different color descriptors. The analytical invariance properties of color descriptors are explored using a taxonomy based on invariance properties with respect to photometric transformations, and evaluated experimentally using two benchmarks from the image domain [311] and the video domain [312]. The benchmarks are very different in nature: the image benchmark consists of consumer photographs and the video benchmark consists of key frames from broadcast news videos.

This chapter is organized as follows. In Section 16.1, the diagonal model is revisited to provide a taxonomy of photometric invariance. Then, in Section 16.2,

Color in Computer Vision: Fundamentals and Applications, First Edition.
Theo Gevers, Arjan Gijsenij, Joost van de Weijer, and Jan-Mark Geusebroek.

different color descriptors and their invariance properties are given. Then, the color descriptors are applied in the context of object and scene classification.

16.1 Diagonal Model

In Chapter 3, the diagonal model has been introduced, which models the changes of the camera responses under variations of the illuminant:

$$\begin{pmatrix} R_c \\ G_c \\ B_c \end{pmatrix} = \begin{pmatrix} a & 0 & 0 \\ 0 & b & 0 \\ 0 & 0 & c \end{pmatrix} \begin{pmatrix} R_u \\ G_u \\ B_u \end{pmatrix}, \tag{16.1}$$

and can be written as a brief notation as follows:

$$\mathbf{f}_c = \mathbf{D}_{u,c}\mathbf{f}_u, \tag{16.2}$$

in which $\mathbf{f}_u$ is the image taken under an unknown light source, $\mathbf{f}_c$ is the same image as it appears when it is taken under the canonical illuminant, and $\mathbf{D}_{u,c}$ is a diagonal matrix that maps colors taken under an unknown light source u to their corresponding colors under the canonical illuminant c.

To include the diffuse light term, the diagonal model is extended with an offset $(o1, o2, o3)$, resulting in the diagonal-offset model:

$$\begin{pmatrix} R_c \\ G_c \\ B_c \end{pmatrix} = \begin{pmatrix} a & 0 & 0 \\ 0 & b & 0 \\ 0 & 0 & c \end{pmatrix} \begin{pmatrix} R_u \\ G_u \\ B_u \end{pmatrix} + \begin{pmatrix} o1 \\ o2 \\ o3 \end{pmatrix}. \tag{16.3}$$

On the basis of the diagonal model and the diagonal-offset model (Eq. 16.3), five types of common changes in the image values $\mathbf{f}(\mathbf{x})$ can be categorized.

Firstly, for Equation 16.3, when the image values change by a constant factor in all channels $(a = b = c)$, then this is equal to a *light intensity change*:

$$\begin{pmatrix} R^c \\ G^c \\ B^c \end{pmatrix} = \begin{pmatrix} a & 0 & 0 \\ 0 & a & 0 \\ 0 & 0 & a \end{pmatrix} \begin{pmatrix} R^u \\ G^u \\ B^u \end{pmatrix}. \tag{16.4}$$

In addition to differences in the intensity of the light source, light intensity changes also include (no-colored) shadows and shading. Hence, when a descriptor is invariant to light intensity changes, it is *scale invariant* with respect to (light) intensity.

Secondly, an equal shift in image intensity values in all channels, *light intensity shift*, where $(o_1 = o_2 = o_3)$ and $(a = b = c = 1)$ will yield:

$$\begin{pmatrix} R^c \\ G^c \\ B^c \end{pmatrix} = \begin{pmatrix} R^u \\ G^u \\ B^u \end{pmatrix} + \begin{pmatrix} o_1 \\ o_1 \\ o_1 \end{pmatrix}. \tag{16.5}$$

Light intensity shifts are due to diffuse lighting including scattering of a white light source, object highlights (specular component of the surface) under a white light source. When a descriptor is invariant to a light intensity shift, it is *shift invariant* with respect to light intensity.

Thirdly, the above classes of changes can be combined to model both intensity changes and shifts:

$$\begin{pmatrix} R^c \\ G^c \\ B^c \end{pmatrix} = \begin{pmatrix} a & 0 & 0 \\ 0 & a & 0 \\ 0 & 0 & a \end{pmatrix} \begin{pmatrix} R^u \\ G^u \\ B^u \end{pmatrix} + \begin{pmatrix} o_1 \\ o_1 \\ o_1 \end{pmatrix}, \tag{16.6}$$

an image descriptor robust to these changes is scale invariant and shift invariant with respect to light intensity.

Fourthly, in the full diagonal model (allowing $a \neq b \neq c$), the image channels scale independently:

$$\begin{pmatrix} a & 0 & 0 \\ 0 & b & 0 \\ 0 & 0 & c \end{pmatrix} \begin{pmatrix} R \\ G \\ B \end{pmatrix}. \tag{16.7}$$

This allows for *light color changes* in the image. Hence, this class of changes can model a change in the illuminant color and light scattering, amongst others.

Finally, the full diagonal-offset model models arbitrary offsets ($o_1 \neq o_2 \neq o_3$), besides the light color changes ($a \neq b \neq c$) offered by the full diagonal model:

$$\begin{pmatrix} a & 0 & 0 \\ 0 & b & 0 \\ 0 & 0 & c \end{pmatrix} \begin{pmatrix} R \\ G \\ B \end{pmatrix} + \begin{pmatrix} o_1 \\ o_2 \\ o_3 \end{pmatrix}. \tag{16.8}$$

This type of change is called *light color change and shift.*

In conclusion, five types of common changes are identified on the basis of the diagonal-offset model of illumination change, that is, variations to light intensity changes, light intensity shifts, light intensity changes and shifts, and light color changes and light color changes and shifts.

16.2 Color SIFT Descriptors

In this section, color descriptors are presented and their invariance properties are summarized. As color descriptors based on histograms have been discussed in the previous chapters, this section focuses on color descriptors based on SIFT. See Table 16.1 for an overview of the descriptors and their invariance properties. More information can be found in Reference 313.

Table 16.1 Invariance of descriptors (Section 16.2) against types of changes in the diagonal-offset model and its specializations (Section 16.1).

	Light intensity change $\begin{pmatrix} a & 0 & 0 \\ 0 & a & 0 \\ 0 & 0 & a \end{pmatrix}\begin{pmatrix} R \\ G \\ B \end{pmatrix}$	**Light intensity shift** $\begin{pmatrix} R \\ G \\ B \end{pmatrix} + \begin{pmatrix} o_1 \\ o_1 \\ o_1 \end{pmatrix}$	**Light intensity change and shift** $\begin{pmatrix} a & 0 & 0 \\ 0 & a & 0 \\ 0 & 0 & a \end{pmatrix}\begin{pmatrix} R \\ G \\ B \end{pmatrix} + \begin{pmatrix} o_1 \\ o_1 \\ o_1 \end{pmatrix}$	**Light color change** $\begin{pmatrix} a & 0 & 0 \\ 0 & b & 0 \\ 0 & 0 & c \end{pmatrix}\begin{pmatrix} R \\ G \\ B \end{pmatrix}$	**Light color change and shift** $\begin{pmatrix} a & 0 & 0 \\ 0 & b & 0 \\ 0 & 0 & c \end{pmatrix}\begin{pmatrix} R \\ G \\ B \end{pmatrix} + \begin{pmatrix} o_1 \\ o_2 \\ o_3 \end{pmatrix}$
RGB histogram	−	−	−	−	−
O_1, O_2	−	+	−	−	−
O_3, intensity	−	−	−	−	−
Hue	+	+	+	−	−
Saturation	−	−	−	−	−
r, g	+	−	−	−	−
Transformed color	+	+	+	+	+
SIFT (∇I)	+	+	+	−	−
HSV−SIFT	−	−	−	−	−
HueSIFT	+	+	+	−	−
OpponentSIFT	+	+	+	−	−
C−SIFT	+	−	−	−	−
rgSIFT	+	−	−	−	−
Transformed color SIFT	+	+	+	+	+
RGB-SIFT	+	+	+	+	+

Invariance is indicated with "+" and lack of invariance is indicated with "-". The invariance of a descriptor to condition A is defined as follows: under a condition A, the descriptor is independent of changes in condition A. The independence is derived analytically under the assumption that no color clipping occurs.

SIFT The SIFT descriptor proposed by Lowe [55] describes the local shape of a region using gradient orientation histograms. The gradient of an image is shift invariant: taking the derivative cancels out offsets (Section 16.1). Under light intensity changes (scaling of the intensity channel) the gradient direction and the relative gradient magnitude remain the same. Because the SIFT descriptor is normalized, the gradient magnitude changes have no effect on the final descriptor. The SIFT descriptor is not invariant to illumination color changes because the intensity channel is a combination of the R, G, and B channels. To compute SIFT descriptors, the version described by Lowe [55] is used.

HSV-SIFT Bosch et al. [314] compute SIFT descriptors over all three channels of the HSV color model. This gives 3×128 dimensions per descriptor, 128 per channel. As stated earlier, the H color model is scale invariant and shift invariant with respect to light intensity. However, owing to the combination of the HSV channels, the complete descriptor has no invariance properties. Further, the instability of the hue for low saturation is not addressed here.

HueSIFT Van de Weijer et al. [149] introduce a concatenation of the hue histogram with the SIFT descriptor. When compared to HSV–SIFT, the usage of the weighed hue histogram addresses the instability of the hue near the gray axis. Because the bins of the hue histogram are independent, the periodicity of the hue channel for HueSIFT is addressed. Similar to the hue histogram, the HueSIFT descriptor is scale invariant and shift invariant.

OpponentSIFT OpponentSIFT describes all the channels in the opponent color space using SIFT descriptors. The opponent color space has been defined in Chapter 3 by Equations 3.48–3.50. The information in the O_3 channel is equal to the intensity information, while the other channels describe the color information in the image. These other channels do contain some intensity information, but because of the normalization of the SIFT descriptor they are invariant to changes in light intensity.

C-SIFT In the opponent color space, the O_1 and O_2 channels still contain some intensity information. To add invariance to intensity changes, in Sections 6.1.4 and 14.1, the C-invariant has been discussed, which eliminates the remaining intensity information from these channels. The C-SIFT descriptor [258] uses the C-invariant, which can be intuitively seen as the normalized opponent color space $\frac{O_1}{O_3}$ and $\frac{O_2}{O_3}$. Because of the division by intensity, the scaling in the diagonal model will cancel out, making C-SIFT scale invariant with respect to light intensity. Owing to the definition of the color space, the offset does not cancel out when taking the derivative: it is not shift invariant.

***rg*SIFT** For the rgSIFT descriptor, descriptors are added for the r and g chromaticity components of the normalized RGB color model from Equations 4.1–4.3, which is already scale invariant.

Transformed color SIFT For the transformed color SIFT, the same normalization is applied to the *RGB* channels as for the transformed color histogram. For every normalized channel, the SIFT descriptor is computed. The descriptor is scale invariant, shift invariant and invariant to light color changes and shift.

RGB-SIFT For the RGB-SIFT descriptor, SIFT descriptors are computed for every *RGB* channel independently. An interesting property of this descriptor is that its descriptor values are invariant to different photometric transformations. Because the SIFT descriptor operates on derivatives only, the subtraction of the means in the transformed color model is redundant, as this offset is already cancelled out by taking derivatives. Similarly, the division by the standard deviation is already implicitly performed by the normalization of the vector length of SIFT descriptors.

16.3 Object and Scene Recognition

In this section, the *distinctiveness* of the color descriptors is assessed experimentally through their discriminative power on two different datasets: an image benchmark and a video benchmark.

16.3.1 Feature Extraction Pipelines

To empirically test the different color descriptors, the descriptors are computed at scale-invariant points [55, 300]. See Figure 16.1 for an overview of the processing pipeline. In the pipeline shown, scale-invariant points are obtained with the Harris–Laplace point detector on the intensity channel. Other region detectors, such as the dense sampling detector, maximally stable extremal regions [315] and maximally stable color regions [316], can be plugged in. For the experiments, the Harris–Laplace point detector is used because it has shown good performance for category recognition [300]. This detector uses the Harris corner detector to find potential scale-invariant points. It then selects a subset of these points for which

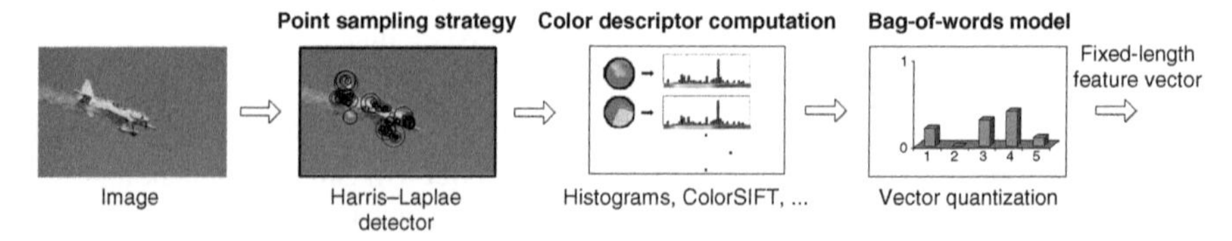

Figure 16.1 The stages of the primary feature extraction pipeline used in this chapter. First, the Harris–Laplace salient point detector is applied to the image. Then, for every point a color descriptor is computed over the area around the point. All the color descriptors of an image are subsequently vector quantized against a codebook of prototypical color descriptors. This results in a fixed-length feature vector representing the image. *Source*: Reprinted with permission, © *2010 IEEE*.

the Laplacian-of-Gaussians reaches a maximum over scale. The color descriptors from Section 16.2 are computed over the area around the points. The size of this area depends on the maximum scale of the Laplacian-of-Gaussians.

To obtain fixed-length feature vectors per image, the bag-of-words model is used [262]. The bag-of-words model is also known as *textons* [298], *object parts* [317] and *codebooks* [274, 318]. The bag-of-words model performs vector quantization of the color descriptors in an image against a visual codebook. A descriptor is assigned to the codebook element that is closest in Euclidean space. To be independent of the total number of descriptors in an image, the feature vector is normalized to sum to 1. The visual codebook is constructed by applying k-means clustering to 200,000 randomly sampled descriptors from the set of images available for training. In this chapter, visual codebooks with 4000 elements are used. After performing point sampling, color descriptor computation, and vector quantization, an image is represented by a fixed-length feature vector.

16.3.2 Classification

The (support vector machine (SVM) classifier is used for image categorization. The decision function of an SVM classifier for a test sample with feature vector $\vec{F}'$ has the form

$$g(\vec{F}') = \sum_{\vec{F} \in trainset} \alpha_{\vec{F}} y_{\vec{F}} k(\vec{F}, \vec{F}') - \beta, \tag{16.9}$$

where $y_{\vec{F}}$ is the class label of $\vec{F}$ $(-1$ or $+1)$, $\alpha_{\vec{F}}$ is the learned weight of train sample $\vec{F}$, β is a learned threshold and $k(\vec{F}, \vec{F}')$ is the value of a kernel function based on the χ^2 distance, which has shown good results in object recognition [300]:

$$k(\vec{F}, \vec{F}') = e^{-\frac{1}{D} dist_{\chi^2}(\vec{F}, \vec{F}')}, \tag{16.10}$$

where D is a scalar that normalizes the distances. We set D to the average χ^2 distance between all elements of the train set.

The LibSVM implementation is used to train the classifier. As parameters for the training phase, the weight of the positive class is set to $\frac{\#pos+\#neg}{\#pos}$ and the weight of the negative class is set to $\frac{\#pos+\#neg}{\#neg}$, with $\#pos$ the number of positive instances in the train set and $\#neg$ the number of negative instances. The cost parameter is optimized using threefold cross-validation with a parameter range of 2^{-4} through 2^4.

To use multiple features, instead of relying on a single feature, the kernel function is extended in a weighted fashion for m features:

$$\begin{aligned} k(\{\vec{F}_{(1)}, ldots, \vec{F}_{(m)}\}, \{\vec{F'}_{(1)}, ldots, \vec{F'}_{(m)}\}) \\ = e^{-\frac{1}{\sum_{j=1}^{m} w_j}\left(\sum_{j=1}^{m} \frac{w_j}{D_j} dist(\vec{F}_{(j)}, \vec{F'}_{(j)})\right)}, \end{aligned} \tag{16.11}$$

with w_j the weight of the jth feature, D_j the normalization factor for the jth feature, and $\vec{F}_{(j)}$ the jth feature vector.

An example of the use of multiple features is the spatial pyramid [308]; it is illustrated in Figure 16.2. When using the spatial pyramid, additional features are extracted for specific parts of the image. For example, in a 2×2 subdivision of the image, feature vectors are extracted for each image quarter with a weight of $\frac{1}{4}$ for each quarter. Similarly, a 1×3 subdivision consisting of three horizontal bars, which introduces three new features (each with a weight of $\frac{1}{3}$). In this setting, the feature vector for the entire image has a weight of 1.

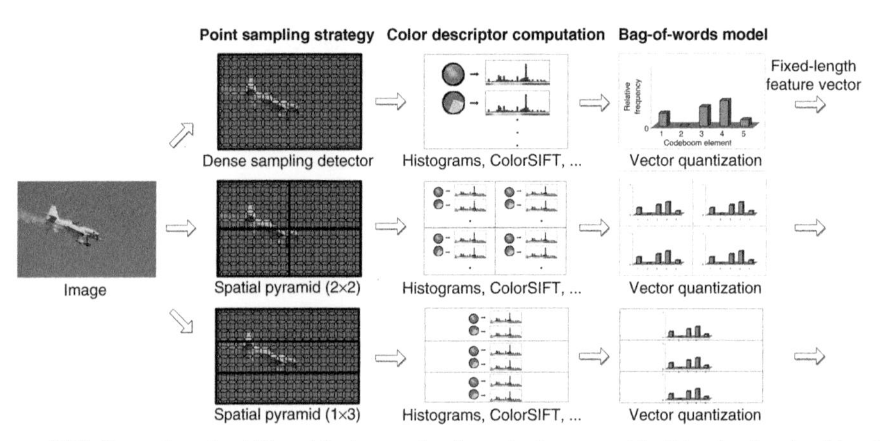

Figure 16.2 Examples of additional feature extraction pipelines used in this chapter, besides the primary pipeline shown in Figure 16.1. The pipelines shown are examples of using a different point sampling strategy or a spatial pyramid [308]. The spatial pyramid constructs feature vectors for specific parts of the image. For every pipeline, first, a point sampling method is applied to the image. Then, for every point a color descriptor is computed over the area around the point. All the color descriptors of an image are subsequently vector quantized against a codebook of prototypical color descriptors. This results in a fixed-length feature vector representing the image. *Source*: Reprinted with permission, © *2010 IEEE.*

16.3.3 Image Benchmark: PASCAL Visual Object Classes Challenge

The PASCAL (Visual Object Classes ((VOC) Challenge [311] provides a yearly benchmark for comparison of object classification systems. The PASCAL VOC Challenge 2007 dataset contains nearly 10,000 images of 20 different object categories, for example, bird, bottle, car, dining table, motorbike, and people. The dataset is divided into a predefined train set (5011 images) and test set (4952 images).

16.3.4 Video Benchmark: Mediamill Challenge

The Mediamill Challenge by Snoek et al. [312] provides an annotated video dataset, based on the training set of the NIST TRECVID 2005 benchmark [310]. Over this dataset, repeatable experiments have been defined. The experiments decompose automatic category recognition into a number of components, for which they provide a standard implementation. This provides an environment to analyze which components affect the performance most.

The dataset of 86 hours is divided into a Challenge training set (70% of the data or 30,993 shots) and a Challenge test set (30% of the data or 12,914 shots). For every shot, the Challenge provides a single representative keyframe image. So, the complete dataset consists of 43,907 images, one for every video shot. The dataset consists of television news from November 2004 broadcast on six different TV channels in three different languages: English, Chinese, and Arabic. On this dataset, the 39 LSCOM-Lite categories [319] are employed. These include object categories such as aircraft, animal, car, and faces, and scene categories such as desert, mountain, sky, urban, and vegetation.

16.3.5 Evaluation Criteria

For our benchmark results, the average precision is taken as the performance metric for determining the accuracy of ranked category recognition results. The average precision is a single-valued measure that is proportional to the area under a precision-recall curve. This value is the average of the precision over all images/keyframes judged to be relevant. Hence, it combines both precision and recall into a single performance value. For the PASCAL VOC Challenge 2007, the official standard is the 11-point interpolated average precision, and for TRECVID, the official standard is the noninterpolated average precision. The interpolated average precision is an approximation of the noninterpolated average precision. As the difference between the two is generally very small, we will follow the official standard for each dataset and refer to them as average precision scores. When performing experiments over multiple object and scene categories, the average precisions of the individual categories are aggregated. This aggregation, mean average precision, is calculated by taking the mean of the average precisions. As average precision depends on the number of correct object and scene categories present in the test set, the mean average precision depends on the dataset used.

To obtain an indication of significance, the bootstrap method [320, 321] is used to estimate confidence intervals for mean average precision. In bootstrap, multiple test sets T_B are created by selecting images at random from the original test set T, with replacement, until $|T| = |T_B|$. This has the effect that some images are replicated in T_B, whereas other images may be absent. This process is repeated

1000 times to generate 1000 test sets, each obtained by sampling from the original test set T. The statistical accuracy of the mean average precision score can then be evaluated by looking at the standard deviation of the mean average precision scores over the different bootstrap test sets.

16.4 Results

16.4.1 Image Benchmark: PASCAL VOC Challenge

From the results shown in Figure 16.3, it is observed that for object category recognition the SIFT variants perform significantly better than color histograms (see Reference 313 for a detailed explanation on these descriptors). Histograms are not very distinctive when compared to SIFT-based descriptors: they contain too little relevant information to be competitive with SIFT.

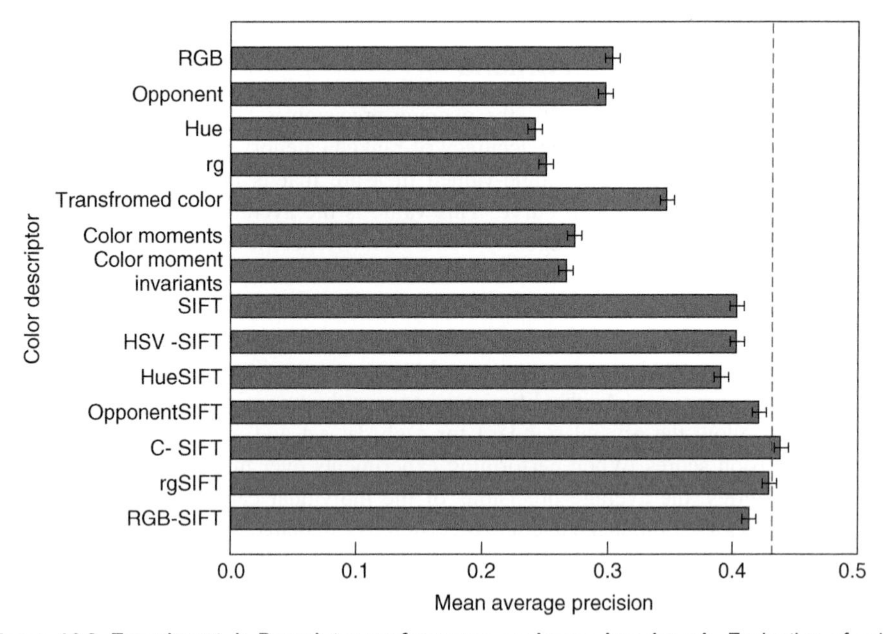

Figure 16.3 Experiment 1: Descriptor performance on image benchmark. Evaluation of color descriptors on an image benchmark, the PASCAL VOC Challenge 2007 [311], averaged over the 20 object categories. Error bars indicate the standard deviation in mean average precision, obtained using bootstrap. The dashed lines indicate the lower bound of the C-SIFT confidence interval. *Source*: Reprinted with permission, © *2010 IEEE*.

For SIFT and the four best color SIFT descriptors from Figure 16.3 (OpponentSIFT, C-SIFT, *rg*SIFT, and RGB-SIFT), the results per object category are shown in Figure 16.4. For bird, boat, horse, motorbike, person, potted plant, and sheep, it can be observed that the descriptors that perform best have scale invariance for light intensity (C-SIFT and *rg*SIFT). Of these two scale-invariant descriptors, C-SIFT has the highest overall performance. The performance of the OpponentSIFT descriptor, which is also shift invariant compared to C-SIFT, indicates that only scale invariance, that is, invariance to light intensity changes, is important for these object categories. RGB-SIFT includes additional invariance against light intensity shifts and light color changes and shifts when compared to C-SIFT. However, this additional invariance makes the descriptor less discriminative for these object categories because a reduction in performance is observed.

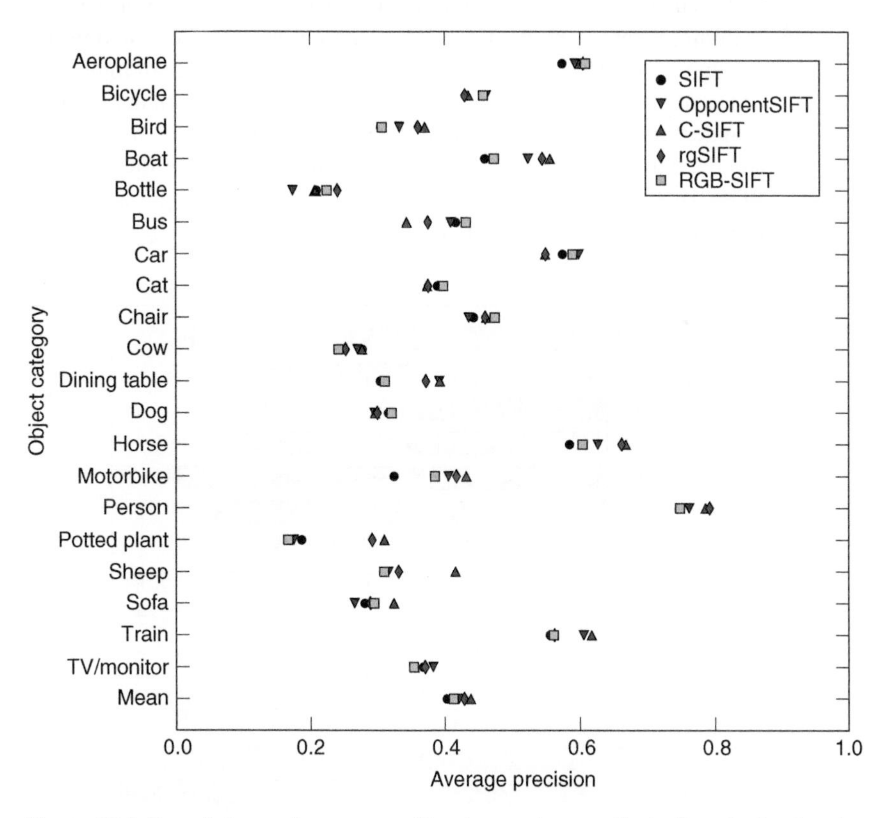

Figure 16.4 Descriptor performance split out per category. Evaluation of color descriptors on an image benchmark, the PASCAL VOC Challenge 2007, split out per object category. SIFT and the best four color SIFT variants from Figure 16.3 are shown. *Source*: Reprinted with permission, © *2010 IEEE*.

In conclusion, C-SIFT is significantly better than all other descriptors except *rg*SIFT (Fig. 16.3) on the image benchmark. The corresponding invariant property of both of these descriptors is given by Equation 16.4. However, the difference between the *rg*SIFT descriptor and OpponentSIFT, which corresponds to Equation 16.6, is not significant. Therefore, the best choice for this dataset is C-SIFT.

16.4.2 Video Benchmark: Mediamill Challenge

From the visual categorization results shown in Figure 16.5, the same overall pattern as for the image benchmark is observed: SIFT and color SIFT variants perform significantly better than the other descriptors. The shift-invariant OpponentSIFT has left C-SIFT behind and is now the only descriptor that is significantly better than all other descriptors. An analysis on the individual object and scene categories shows that the OpponentSIFT descriptor performs best for building, meeting, mountain, office, outdoor, sky, studio, walking/running, and weather news. All these concepts occur under a wide range of light intensities and different amounts of diffuse lighting. Therefore, its invariance to light intensity changes and shifts makes OpponentSIFT a good feature for these categories, and explains why it is better than C-SIFT and *rg*SIFT for the video benchmark. RGB-SIFT, with additional invariance to light color changes and shifts, does not differ significantly from C-SIFT and *rg*SIFT. For some categories, there is a small

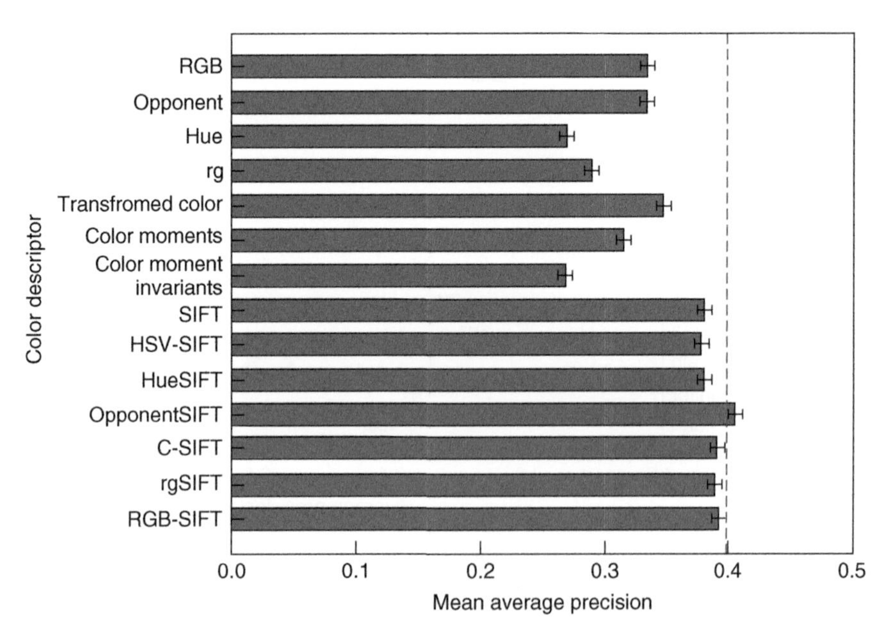

Figure 16.5 Descriptor performance on video benchmark. Evaluation of color descriptors on a video benchmark, the Mediamill Challenge [312], averaged over 39 object and scene categories. Error bars indicate the standard deviation in mean average precision, obtained using bootstrap. The dashed line indicates the lower bound of the OpponentSIFT confidence interval. *Source*: Reprinted with permission, © *2010 IEEE.*

performance gain, for others there is a small loss. This contrasts with the results on the image benchmark, where a performance reduction was observed.

In conclusion, OpponentSIFT is significantly better than all other descriptors on the video benchmark (Fig. 16.5). The corresponding invariant property is given by Equation 16.6.

16.4.3 Comparison

So far, the performance of single descriptors has been analyzed. It is worthwhile to investigate combinations of several descriptors, since they are not completely redundant. State-of-the-art results on the PASCAL VOC Challenge 2007 also employ combinations of several methods. Table 16.2 gives an overview of combinations on this dataset. For example, the best entry in the PASCAL VOC Challenge 2007, by Marszałek et al. [322], has achieved a mean average precision of 0.594 using SIFT and HueSIFT descriptors, the spatial pyramid [308], additional point sampling strategies besides Harris–Laplace such as Laplacian point sampling and dense sampling, and a feature selection

Table 16.2 Combinations on image benchmark.[a]

Author	Point sampling	Descriptor	Spatial pyramid	Mean average precision
This chapter	Harris–Laplace, dense sampling	SIFT	$1 \times 1 + 2 \times 2 + 1 \times 3$	0.558
This chapter	Harris–Laplace, dense sampling	C-SIFT	$1 \times 1 + 2 \times 2 + 1 \times 3$	0.566
Marszałek et al. [322]	Harris–Laplace, dense sampling, Laplacian	SIFT, HueSIFT, other	$1 \times 1 + 2 \times 2 + 1 \times 3$	0.575
Marszałek et al. [322]	Harris–Laplace, dense sampling, Laplacian	SIFT, HueSIFT, other; with feature selection	$1 \times 1 + 2 \times 2 + 1 \times 3$	0.594
This chapter	Harris–Laplace, dense sampling	SIFT, OpponentSIFT, *rg*SIFT, C-SIFT, RGB-SIFT	$1 \times 1 + 2 \times 2 + 1 \times 3$	0.605

[a] In this table, combinations of descriptors on the image benchmark are compared to Marszałek et al. [322], who obtains state-of-the-art results on this dataset. Adding color descriptors improves over intensity-based SIFT alone by 8%.

scheme. When the feature selection scheme is excluded and simple flat fusion is used, Marszałek reports a mean average precision of 0.575.

To illustrate the potential of the color descriptors from Table 16.1, a simple fusion experiment has been performed with SIFT and the best four color SIFT variants (Section 16.3.2 details how the combination is constructed). To be comparable, a setting similar to Marszałek is used: both Harris–Laplace point sampling and dense sampling are employed and the same spatial pyramid is used (see Figure 16.1 for an overview of the feature extraction pipelines used). In this setting, the best single-color descriptor achieves a mean average precision of 0.566. The combination gives a mean average precision of 0.605. This convincing gain of 7% suggests that the color descriptors are not entirely redundant. Compared to the intensity-based SIFT descriptor, the gain is 8%. Further gains should be possible, if the descriptors with the right amount of invariance are fused, preferably using an automatic selection strategy.

As shown in Table 16.3, similar gains are observed on the Mediamill Challenge: mean average precision increases by 7% when combinations of color descriptors are used, instead of intensity-based SIFT only. Relative to the best single-color descriptor, an increase of 3% is observed. Furthermore, when the descriptors of this chapter are compared to the baseline provided by the Mediamill Challenge, there is a relative improvement of 104%.

Table 16.3 Combinations on video benchmark.[a]

Author	**Point sampling**	**Descriptor**	**Spatial pyramid**	**Mean average precision**
Snoek et al. [312]	Grid	Weibull [200]	1×1	0.250
This chapter	Harris–Laplace, dense sampling	SIFT	$1 \times 1 + 2 \times 2 + 1 \times 3$	0.476
This chapter	Harris–Laplace, dense sampling	OpponentSIFT	$1 \times 1 + 2 \times 2 + 1 \times 3$	0.494
This chapter	Harris–Laplace, dense sampling	SIFT, OpponentSIFT, *rg*SIFT, C-SIFT, RGB-SIFT	$1 \times 1 + 2 \times 2 + 1 \times 3$	0.510

[a]In this table, combinations of descriptors on the video benchmark are compared to the baseline set by the Mediamill Challenge [312] for the 39 LSCOM-Lite categories [319]. Adding color descriptors improves over intensity-based SIFT alone by 7%.

For reference, combinations of color descriptors from this chapter were submitted to the PASCAL VOC 2008 benchmark [323] and the TRECVID 2008 evaluation campaign [310]. In both cases, top performance was achieved. The color descriptors as presented in this chapter were the foundation of these submissions. For additional details, see Tables 16.4 [324, 325] and 16.5 [326].

Table 16.4 PASCAL VOC 2008 evaluation: best overall performance.[a]

Author	Point sampling	Descriptor	Spatial pyramid	Mean average precision
This chapter and Tahir et al. [324]	Harris–Laplace, dense sampling	SIFT, OpponentSIFT, *rg*SIFT, C-SIFT, RGB-SIFT	$1 \times 1 + 2 \times 2 + 1 \times 3$	0.549

[a]In this table, results of descriptor combinations from this chapter as submitted to the classification task of the PASCAL VOC Challenge 2008 [323] are shown.

Table 16.5 NIST TRECVID 2008 evaluation: best overall performance.[a]

Author	Point sampling	Descriptor	Spatial pyramid	Inferred mean average precision
This chapter and Snoek et al. [326]	Harris–Laplace, dense sampling	SIFT, OpponentSIFT, *rg*SIFT, C-SIFT, RGB-SIFT	1x1+2x2+1x3	0.194

[a]In this table, results of descriptor combinations from this chapter as submitted to the NIST TRECVID 2008 video benchmark [310] are shown.

16.5 Summary

From the results, it can be noticed that invariance to light color changes and shifts is domain specific. For the image dataset, a significant reduction in performance was observed, whereas for the video dataset there was no performance difference. However, there are specific samples where invariance to light color changes provides a benefit. The overall performance is not improved by light color invariance, presumably because light color changes are quite rare in both benchmarks because of the white balancing performed during data recording.

Overall, when choosing a single descriptor and no prior knowledge about the dataset and object and scene categories is available, the best choice is OpponentSIFT. The corresponding invariance property is scale and shift invariance, given by Equation 16.6. The second best is C-SIFT for which the corresponding invariance property is scale invariance, given by Equation 16.4. Table 16.6 summarizes the recommendations for the datasets from this chapter and datasets where no prior knowledge is available.

Table 16.6 Recommended color descriptors per dataset.[a]

PASCAL VOC 2007	Mediamill challenge	Unknown data
1. C-SIFT	1. OpponentSIFT	1. OpponentSIFT
2. OpponentSIFT	2. RGB-SIFT	2. C-SIFT
3. RGB-SIFT	3. C-SIFT	3. RGB-SIFT
4. SIFT	4. SIFT	4. SIFT

[a]The recommended choice of descriptors for different datasets: the PASCAL VOC 2007, Mediamill Challenge and datasets where no prior knowledge about the lighting conditions or the object and scene categories is available. Without such prior knowledge, OpponentSIFT is the best choice.

To obtain state-of-the-art performance on real-world datasets with large variations in lighting conditions, multiple color descriptors should be chosen, each one with a different amount of invariance. As shown earlier, even a simple combination of color descriptors improves over the individual descriptors, suggesting that they are not completely redundant. Results on the two categorization benchmarks have shown that the choice of a single descriptor for all categories is suboptimal (Fig. 16.4). While the addition of color improves category recognition by 8–10% over intensity-based SIFT only, further gains should be possible if the descriptor with the appropriate amount of invariance is selected per category, using either a feature selection strategy or domain knowledge.

17 Color Naming

With contributions by Robert Benavente, Maria Vanrell, Cordelia Schmid, Ramon Baldrich, Jakob Verbeek, and Diane Larlus

Within a computer vision context, color naming is the action of assigning linguistic color labels to pixels, regions, or objects in images. Humans use color names routinely and seemingly without effort to describe the world around us. They have been primarily studied in the fields of visual psychology, anthropology, and linguistics [327]. Color names are, for example, used in the context of image retrieval. A user might query an image search engine for ''red cars''. The system recognizes the color name ''red'', and orders the retrieved results on ''car'' based on their resemblance to the human usage of ''red''. Furthermore, knowledge of visual attributes can be used to assist object recognition methods. For example, for an image annotated with the text ''Orange stapler on table'', knowledge of the color name orange would greatly simplify the task of discovering where (or what) the stapler is. Color names are further applicable in automatic content labeling of images, colorblind assistance, and linguistic human–computer interaction [328].

In this chapter, we first discuss the influential linguistic study on color names by Berlin and Kay [329] in Section 17.1. In their work they define the concept of

Portions reprinted, with permission, from ''Learning Color Names for Real-World Applications.'', by J. van de Weijer, Cordelia Schmid, Jakob Verbeek, Diane Larlus, in IEEE Transaction in Image Processing, Volume 18 (7), © 2009 IEEE, and from ''Parametric fuzzy sets for automatic color naming,'' R. Benavente, M. and Vanrell, and R. Baldrich, Journal of the Optical Society of America A, Volume 25. (10).

Color in Computer Vision: Fundamentals and Applications, First Edition.
Theo Gevers, Arjan Gijsenij, Joost van de Weijer, and Jan-Mark Geusebroek.

basic color terms. As we will see, the basic color terms of the English language are black, blue, brown, gray, green, orange, pink, purple, red, white, and yellow. Next, we discuss two different approaches to computational color naming. The main difference between the two methods is the data from which the methods learn the color names. The first method, discussed in Section 17.2, is based on *calibrated data* acquired from a psychophysical experiment. With calibrated we mean that the color samples are presented in a controlled laboratory environment on stable viewing conditions with a known illumination setting. The second method, discussed in Section 17.3, is instead based on *uncalibrated data* obtained from Google Image. These images are uncalibrated in the worst sense. They have unknown camera settings, unknown illuminant, and unknown compression. However, the advantage of uncalibrated data is that it is much easier to collect. At the end of the chapter, in Section 17.4, we compare both computational color-naming algorithms on both calibrated and uncalibrated data.

17.1 Basic Color Terms

Color naming, and all the semantic fields in general, have been involved for many years in a discussion between two points of view in linguistics. On the one hand, relativists support the idea that semantic categories are conditioned by experience and culture, and, therefore, each language builds its own semantic structures in a quite arbitrary form. On the other hand, universalists defend the existence of semantic universals shared across languages. These linguistic universals would be based on the human biology and directly linked to neurophysiological mechanisms. Color has been presented as a clear example of relativism since each language has a different set of terms to describe color.

Although some works had investigated the use of color terms in English [330], the anthropological study by Berlin and Kay [329] about color naming in different languages was the starting point of many works about this topic in the subsequent years.

Berlin and Kay studied the use of color names in speakers of a total of 98 different languages (20 experimentally and 78 through literature review). With their work, Berlin and Kay wanted to support the hypothesis of semantic universals by demonstrating the existence of a set of color categories shared across different languages. To this end, they first defined the concept of "basic color term" by setting the properties that any basic color term should fulfill. These properties are

- It is monolexemic, that is, its meaning cannot be obtained from the meaning of its parts.
- It has a meaning that is not included in that of other color terms.
- It can be applied to any type of objects.

- It is psychologically salient, that is, it appears at the beginning of elicited lists of color terms, it is consistently used along time by speakers and across different speakers, and it is used by all the speakers of the language.

In addition, they defined a second set of properties for the terms that might be doubtful according to the previous rules. These properties are

- The doubtful form should have the same distributional potential as the previously established basic terms.
- Basic color terms should not be also the name of an object that has that color.
- Foreign words that have recently been incorporated to the language are suspect.
- If the monolexemic criterion is difficult to decide, the morphological complexity can be used as a secondary criterion.

The work with informants from the different languages was divided in two parts. In the first part, the list of basic color names in each informant's language, according to the previous rules, was verbally elicited. This part was done in the absence of any color stimuli and using as little as possible of any other language. In the second part, subjects were asked to perform two different tasks. First, they had to indicate on the Munsell color array all the chips that they would name under any condition with each of their basic terms, that is, the area of each color category. Second, they had to point out the best example (focus) of each basic color term in their language.

Data obtained from the 20 informants was completed with information from published works in the other 78 languages. After the study of these data, Berlin and Kay extracted three main conclusions from their work:

1. *Existence of Basic Color Terms*. They stated that color categories were not arbitrary and randomly defined by each language. The foci of each basic color category in different languages were all in a close region of the color space. This finding led them to define the set of 11 basic color terms. These terms for English are white, black, red, green, yellow, blue, brown, pink, purple, orange and gray.
2. *Evolutionary Order*. Although languages can have different numbers of basic color terms, they found that the order in which languages encoded color terms in their temporal evolution was not random, but it followed a fixed order that defined seven evolutionary stages:

   Stage I: Terms for only white and black.

   Stage II: A term for red is added.

   Stage III: A term for either green or yellow (but not both) is added.

Stage IV: A term for green or yellow (the one that was not added in the previous stage) is added.

Stage V: A term for blue is added.

Stage VI: A term for brown is added.

Stage VII: Terms for pink, purple, orange and gray are added (in any order).

These sequences can be summarized with the expression:

$$[\text{white, black}] < [\text{red}] < [\text{green, yellow}] < [\text{blue}]$$
$$< [\text{brown}] < [\text{pink, purple, orange, gray}]$$

where symbol '<' indicates temporal precedence, that is, for two categories A and B, $A < B$ means that A is present in the language before B, and order between terms inside '[ ]' depends on each language.

3. *Correlation with Cultural Development.* They noticed a high correlation of color vocabulary of a language with technological and cultural evolution. Languages from developed cultures were all in the last stage of color terms evolution, while languages from isolated and low-developed cultures were at lower stages of color vocabulary evolution.

Figure 17.1 shows the boundaries in the Munsell space of the 11 basic color categories for English that were obtained by Berlin and Kay. This categorization of the Munsell array has been used as a reference in later color-naming studies that confirmed the Berlin and Kay results [331, 332]. Boundaries on the Munsell array obtained by Sturges and Whitfield in their experiment [332] are shown in Figure 17.2.

Although the findings of Berlin and Kay were widely accepted for years, the debate about the existence of universals in color naming was reopened by Roberson et al. [333, 334] in 2000. On their works, Roberson and her colleagues defended that color categories were determined by language and, thus, boundaries between color categories were arbitrarily set by each language.

Despite additional evidences for the universalist theory being presented [335], an intermediate position between universalists and relativists has gained support

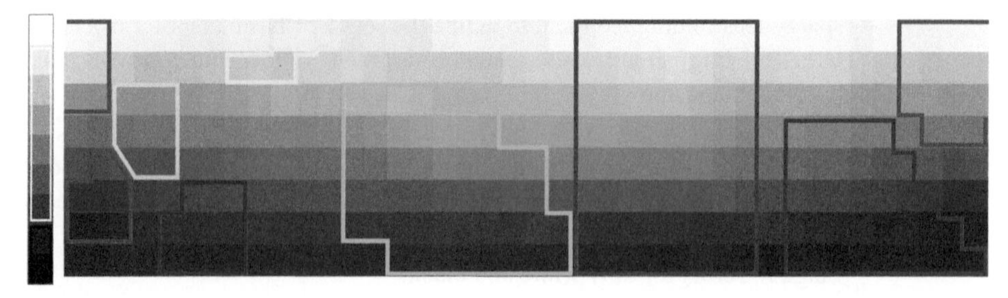

Figure 17.1 Categorization of the Munsell color array obtained by Berlin and Kay in their experiments for English.

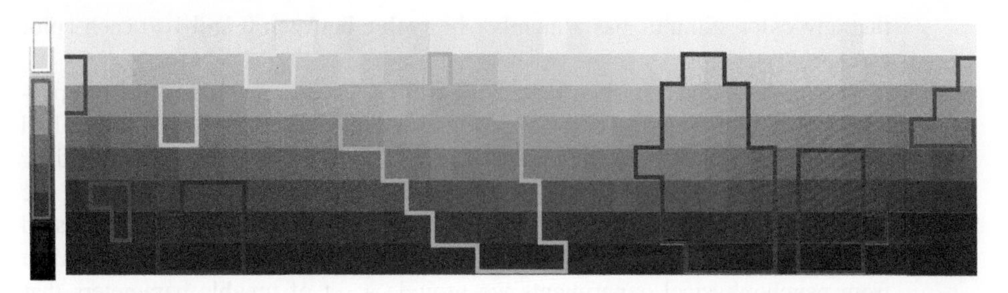

Figure 17.2 Categorization of the Munsell color array obtained in their experiment by Sturges and Whitfield for English.

in the past years [336]. According to this new theory, the organization of the color-naming systems in different languages is ruled by universal tendencies and, therefore, color categories are not arbitrarily located in the color space. On the other hand, the extension and boundaries of the categories are more language dependent and more arbitrarily set by each language [337]. The explanation to this point could be found on the shape of the color space, where some areas seem to be more salient than others. Philipona and O'Regan [338] showed that some properties of the human visual system could support the existence of such relevant parts in the color space. Moreover, they showed that these areas coincide with the location of the focal colors for red, green, blue and yellow, found by Berlin and Kay. Hence, these salient parts of the color space would condition the formation of the color categories to reach a final configuration in each language that tends to be optimal in terms of information representation [339].

In this chapter, we discuss two approaches to computational color naming, Which, starting from a set of basic color terms, will learn models to predict color names in images. The experiments are based on English language color terms. Although the debate about the universality of color categories is outside the scope of this book, this issue may have some influence on how the presented models can be applied in images. If universalists are right, these computational models are equally valid in other languages. However, it is more probably that new models should be relearned for each language, especially to correctly place the boundaries between color names.

17.2 Color Names from Calibrated Data

In this section, we present a fuzzy color-naming model based on the use of parametric membership functions whose advantages are discussed later in the section.

Fuzzy set theory is a useful tool to model human decisions related to perceptual processes and language. A theoretical fuzzy framework was already proposed by Kay and McDaniel [340] in 1978. The basis of such a framework is to consider that any color stimulus has a membership value between 0 and 1 to each color category.

The model defined in this framework will be fitted to data derived from psychophysical experiments. These experiments are usually carried out in controlled environments under stable viewing conditions, with a known illuminant and by using samples with precisely measured reflectances. In such an environment, the naming judgments obtained from the subjects in the experiments are isolated from most other perceptual processes. By fitting a parametric model to data from psychophysical experiments we provide a set of tunable parameters that analytically define the shape of the fuzzy sets representing color categories.

Parametric models have been previously used to model color information [341], and the suitability of such an approach can be summed up in the following points:

Inclusion of Prior Knowledge. Prior knowledge about the structure of the data allows choosing the best model on each case. However, this could turn into a disadvantage if a nonappropriate function for the model is selected.

Compact Categories. Each category is completely defined by a few parameters, and training data do not need to be stored after an initial fitting process. This implies lower memory usage and lower computation when the model is applied.

Meaningful Parameters. Each parameter has a meaning in terms of the characterization of the data, which allows modifying and improving the model by just adjusting the parameters.

Easy Analysis. As a consequence of the previous one, the model can be analyzed and compared by studying the values of its parameters.

We have worked on the CIE $L^*a^*b^*$ color space since it is a quasi-perceptually uniform color space where a good correlation between Euclidean distance between color pairs and perceived color dissimilarity can be observed. It is likely that other spaces could be suitable whenever one of the dimensions correlates with color lightness and the other two with chromaticity components. In this section, we denote any color point in such space as $\mathbf{s} = (I, c_1, c_2)$, where I is the lightness and c_1 and c_2 are the chromaticity components of the color point.

Ideally, color memberships should be modeled by three-dimensional functions, that is, functions defined $\mathbb{R}^3 \rightarrow [0, 1]$, but unfortunately it is not easy to infer precisely the way in which color naming data is distributed in the color space and, hence, finding parametric functions that fit these data is a very complicated task. For this reason, in this proposal the three-dimensional color space has been sliced in a set of N_L levels along the lightness axis (Fig. 17.3), obtaining a set of chromaticity planes over which membership functions have been modeled by two-dimensional functions. Therefore, any specific chromatic category will be defined by a set of functions, each one depending on a lightness component, as it is expressed later in Equation 17.12. Achromatic categories (black, gray and white) will be given as the complementary function of the chromatic ones but weighted

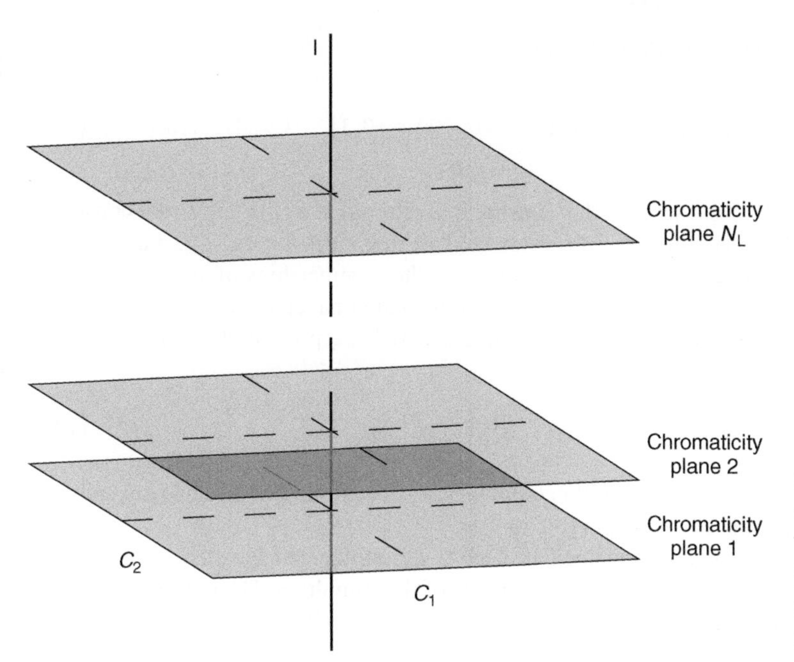

Figure 17.3 Scheme of the model. The color space is divided in N_L levels along the lightness axis.

by the membership function of each one of the three achromatic categories. To go into the details of the approach we will first give the basis of the fuzzy framework and afterwards we will pose the considerations on the function shapes for the chromatic categories. Finally, the complementary achromatic categories will be derived.

17.2.1 Fuzzy Color Naming

A fuzzy set is a set whose elements have a degree of membership [342]. In a more formal way, a fuzzy set A is defined by a crisp set X, called the *universal set*, and a membership function, μ_A, which maps elements of the universal set into the [0,1] interval, that is, $\mu_A : X \to [0, 1]$.

Fuzzy sets are a good tool to represent imprecise concepts expressed in natural language. In color naming, we can consider that any color category, C_k, is a fuzzy set with a membership function, μ_{C_k}, which assigns, to any color sample $\mathbf{s}$ represented in a certain color space, that is, the universal set, a membership value $\mu_{C_k}(\mathbf{s})$ within the [0, 1] interval. This value represents the certainty we have that $\mathbf{s}$ belongs to category C_k, which is associated with the linguistic term t_k.

In the context of color categorization with a fixed number of categories, we need to impose the constraint that, for a given sample $\mathbf{s}$, the sum of its memberships to the n categories must be the unity

$$\sum_{k=1}^{n} \mu_{C_k}(\mathbf{s}) = 1 \qquad \text{with} \qquad \mu_{C_k}(\mathbf{s}) \in [0, 1], \quad k = 1, \ldots, n. \tag{17.1}$$

In the rest of the section, this constraint is referred to as the unity-sum constraint. Although this constraint does not hold in fuzzy set theory, it is interesting in this case because it allows us to interpret the memberships of any sample as the contributions of the considered categories to the final color sensation.

Hence, for any given color sample $\mathbf{s}$ it will be possible to compute a color descriptor, $\mathcal{CD}$, such as

$$\mathcal{CD}(\mathbf{s}) = \left[\mu_{C_1}(\mathbf{s}), \ldots, \mu_{C_n}(\mathbf{s})\right], \tag{17.2}$$

where each component of this n-dimensional vector describes the membership of $\mathbf{s}$ to a specific color category.

The information contained in such a descriptor can be used by a decision function, $\mathcal{N}(\mathbf{s})$, to assign the color name of the stimulus $\mathbf{s}$. The most easy decision rule we can derive is to choose the maximum from $\mathcal{CD}(\mathbf{s})$:

$$\mathcal{N}(\mathbf{s}) = t_{k_{\max}} \quad | \quad k_{\max} = \arg \max_{k=1,\ldots,n} \{\mu_{C_k}(\mathbf{s})\}, \tag{17.3}$$

where t_k is the linguistic term associated with color category C_k.

In this case, the categories considered are the basic categories proposed by Berlin and Kay, that is, $n = 11$ and the set of categories is

$$C_k \in \{\text{red, orange, brown, yellow, green, blue, purple, pink, black, gray, white}\}, \tag{17.4}$$

17.2.2 Chromatic Categories

According to the fuzzy framework defined previously, any function we select to model color categories must map values to the [0, 1] interval, that is, $\mu_{C_k}(\mathbf{s}) \in [0, 1]$. In addition, the observation of the membership values of psychophysical data obtained from a color-naming experiment [343] made us hypothesize about a set of necessary properties that membership functions for the chromatic categories should fulfill:

- *Triangular Basis.* Chromatic categories present a plateau, or an area with no confusion about the color name, with a triangular shape and a principal vertex shared by all the categories.
- *Different Slopes.* For a given chromatic category, the slope of naming certainty toward the neighboring categories can be different on each side of the category (e.g., transition from blue to green can be different from that from blue to purple).

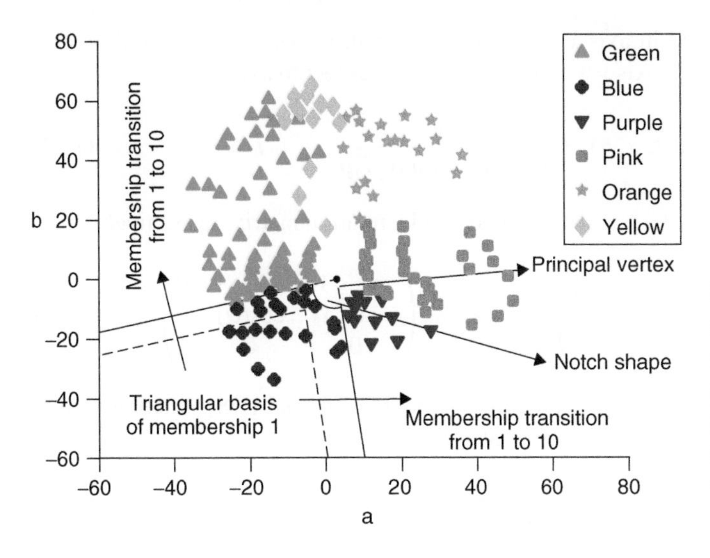

Figure 17.4 Desirable properties of the membership function for chromatic categories plotted on the chromaticity plane (ab-plane). In this case, on the blue category.

- *Central Notch*. The transition from a chromatic category to the central achromatic one has the form of a notch around the principal vertex.

In Figure 17.4, we show a scheme of the preceding conditions on a chromaticity diagram where the samples of a color-naming experiment have been plotted.

Here we will define the membership function, the triple sigmoid with elliptical center (TSE), as a two-dimensional function, $TSE : \mathbb{R}^2 \rightarrow [0, 1]$. Earlier works considered other membership functions [344, 345]. The definition of the TSE starts from the one-dimensional sigmoid function:

$$S^1(x, \beta) = \frac{1}{1 + \exp(-\beta x)}, \tag{17.5}$$

where β controls the slope of the transition from 0 to 1.

This can be extended to a two-dimensional sigmoid function, $S : \mathbb{R}^2 \rightarrow [0, 1]$,

$$S(\mathbf{p}, \beta) = \frac{1}{1 + \exp(-\beta \mathbf{u_i p})}, \qquad i = 1, 2 \tag{17.6}$$

where $\mathbf{p} = (x, y)^T$ is a point in the plane and vectors $\mathbf{u_1} = (1, 0)$ and $\mathbf{u_2} = (0, 1)$ define the axis in which the function is oriented.

By adding a translation, $\mathbf{t} = (t_x, t_y)$, and a rotation, α, to the previous equation, the function can adopt a wide set of shapes. In order to represent the formulation in a compact matrix form, we use homogeneous coordinates [346]. Let us redefine $\mathbf{p}$ to be a point in the plane expressed in homogeneous coordinates as $\mathbf{p} = (x, y, 1)^T$, and let us denote the vectors $\mathbf{u_1} = (1, 0, 0)$ and $\mathbf{u_2} = (0, 1, 0)$. We define S_1 as a function oriented in axis x with rotation α with respect to axis y and S_2 as a function oriented in axis y with rotation α with respect to axis x:

$$S_i(\mathbf{p}, \mathbf{t}, \alpha, \beta) = \frac{1}{1 + \exp(-\beta \mathbf{u_i} R_\alpha T_t \mathbf{p})}, \qquad i = 1, 2 \tag{17.7}$$

where T_t and R_α are a translation matrix and a rotation matrix, respectively:

$$T_t = \begin{pmatrix} 1 & 0 & -t_x \\ 0 & 1 & -t_y \\ 0 & 0 & 1 \end{pmatrix}, \qquad R_\alpha = \begin{pmatrix} \cos(\alpha) & \sin(\alpha) & 0 \\ -\sin(\alpha) & \cos(\alpha) & 0 \\ 0 & 0 & 1 \end{pmatrix} \tag{17.8}$$

By multiplying S_1 and S_2, we define the double-sigmoid (DS) function, which fulfills the first two properties presented before:

$$\mathrm{DS}(\mathbf{p}, \mathbf{t}, \theta_{\mathrm{DS}}) = S_1(\mathbf{p}, \mathbf{t}, \alpha_y, \beta_y) S_2(\mathbf{p}, \mathbf{t}, \alpha_x, \beta_x), \tag{17.9}$$

where $\theta_{\mathrm{DS}} = (\alpha_x, \alpha_y, \beta_x, \beta_y)$ is the set of parameters of the double-sigmoid function. Figure 17.5a shows a plot of a two-dimensional sigmoid oriented in the x-axis direction (S_1 in Eq. 17.7). By multiplying two oriented sigmoid functions, the double-sigmoid function DS is obtained (Fig. 17.5(b)).

To obtain the central notch shape needed to fulfill the third property, let us define the elliptic-sigmoid (ES) function by including the ellipse equation in the sigmoid formula:

$$\mathrm{ES}(\mathbf{p}, \mathbf{t}, \theta_{ES}) = \frac{1}{1 + \exp\left\{ -\beta_e \left[\left(\frac{\mathbf{u_1} R_\phi T_t \mathbf{p}}{e_x}\right)^2 + \left(\frac{\mathbf{u_2} R_\phi T_t \mathbf{p}}{e_y}\right)^2 - 1 \right] \right\}}, \tag{17.10}$$

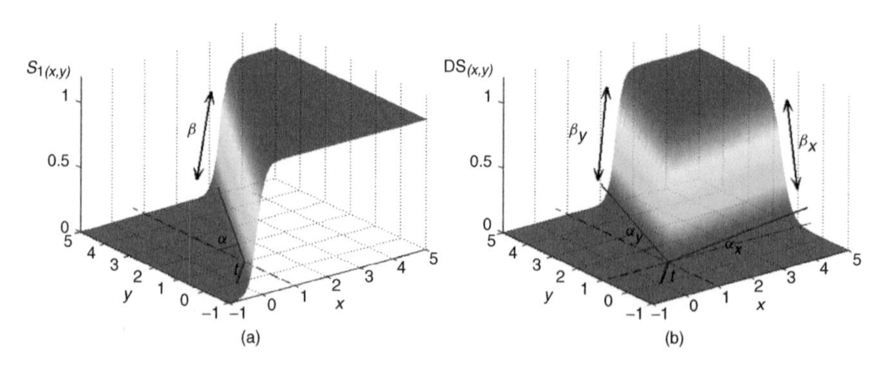

Figure 17.5 Two-dimensional sigmoid functions. (a) S_1: Sigmoid function oriented in axis x direction (b) DS: The product of two differently oriented sigmoid functions generates a plateau with some of the properties needed for the membership function.

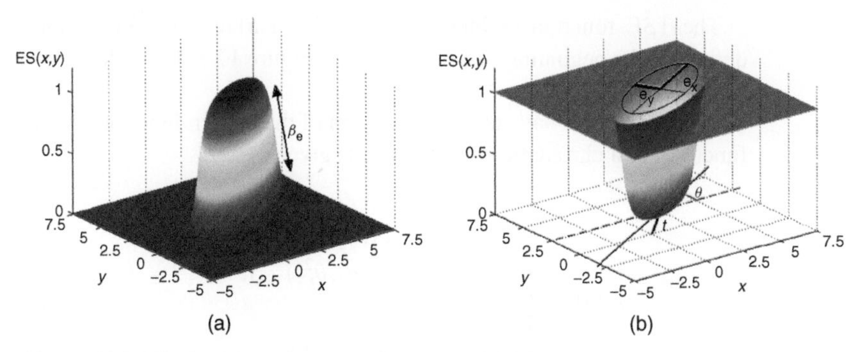

Figure 17.6 Elliptic-sigmoid function $ES(\mathbf{p}, \mathbf{t}, \theta_{\mathrm{ES}})$. (a) ES for $\beta_e < 0$ and (b) ES for $\beta_e > 0$.

where $\theta_{ES} = (e_x, e_y, \phi, \beta_e)$ is the set of parameters of the ES function, e_x and e_y are the semiminor and semimajor axes, respectively, ϕ is the rotation angle of the ellipse, and β_e is the slope of the sigmoid curve that forms the ellipse boundary. The function obtained is an elliptic plateau if β_e is negative and an elliptic valley if β_e is positive. The surfaces obtained can be seen in Figure 17.6.

Finally, by multiplying the double-sigmoid by the ES with a positive β_e, we define the TSE as

$$\mathrm{TSE}(\mathbf{p}, \theta) = \mathrm{DS}(\mathbf{p}, \mathbf{t}, \theta_{\mathrm{DS}})\mathrm{ES}(\mathbf{p}, \mathbf{t}, \theta_{\mathrm{ES}}), \tag{17.11}$$

where $\theta = (\mathbf{t}, \theta_{\mathrm{DS}}, \theta_{\mathrm{ES}})$ is the set of parameters of the TSE.

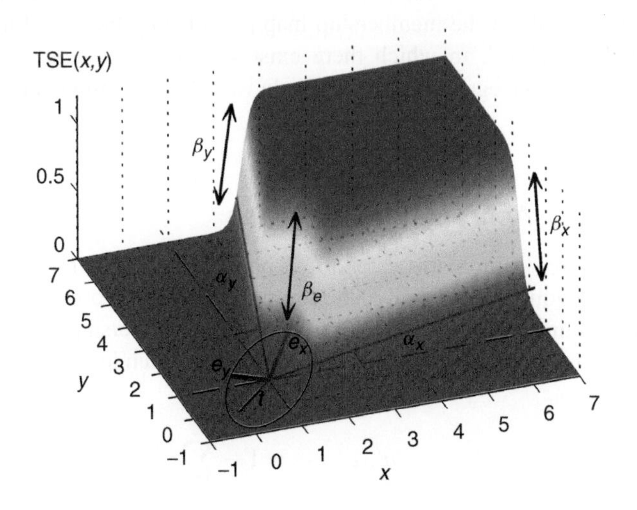

Figure 17.7 Triple sigmoid with elliptical center (TSE).

The TSE function defines a membership surface that fulfills the properties defined at the beginning of Section 17.2.2. Figure 17.7 shows the form of the TSE function.

Hence, once we have the analytic form of the chosen function, the membership function for a chromatic category μ_{C_k} is given by

$$\mu_{C_k}(\mathbf{s}) = \begin{cases} \mu^1_{C_k} = \mathrm{TSE}(c_1, c_2, \theta^1_{C_k}) & \text{if } I \leq I_1, \\ \mu^2_{C_k} = \mathrm{TSE}(c_1, c_2, \theta^2_{C_k}) & \text{if } I_1 < I \leq I_2, \\ \vdots & \vdots \\ \mu^{N_L}_{C_k} = \mathrm{TSE}(c_1, c_2, \theta^{N_L}_{C_k}) & \text{if } I_{N_L-1} < I, \end{cases} \tag{17.12}$$

where $\mathbf{s} = (I, c_1, c_2)$ is a sample on the color space, N_L is the number of chromaticity planes, $\theta^i_{C_k}$ is the set of parameters of the category C_k on the ith chromaticity plane, and I_i are the lightness values that divide the space in the N_L lightness levels.

By fitting the parameters of the functions, it is possible to obtain the variation of the chromatic categories through the lightness levels. By doing this for all the categories, it will be possible to obtain membership maps; that is, for a given lightness level we have a membership value to each category for any color point $\mathbf{s} = (I, c_1, c_2)$ of the level. Notice that since some categories exist only at certain lightness levels (e.g., brown is defined only for low lightness values and yellow only for high values), on each lightness level not all the categories will have memberships different from zero for any point of the level. Figure 17.8 shows an example of the membership map provided by the TSE functions for a given lightness level, in which there exist six chromatic categories. The other two chromatic categories in this example would have zero membership for any point of the level.

17.2.3 Achromatic Categories

The three achromatic categories (black, gray and white) are first considered as a unique category at each chromaticity plane. To ensure that the unity-sum constraint is fulfilled (i.e., the sum of all memberships must be one) a global achromatic membership, μ_A, is computed for each level as

$$\mu^i_A(c_1, c_2) = 1 - \sum_{k=1}^{n_c} \mu^i_{C_k}(c_1, c_2), \tag{17.13}$$

where i is the chromaticity plane that contains the sample $\mathbf{s} = (I, c_1, c_2)$ and n_c is the number of chromatic categories (here, $n_c = 8$). The differentiation among the three achromatic categories must be done in terms of lightness. To model the

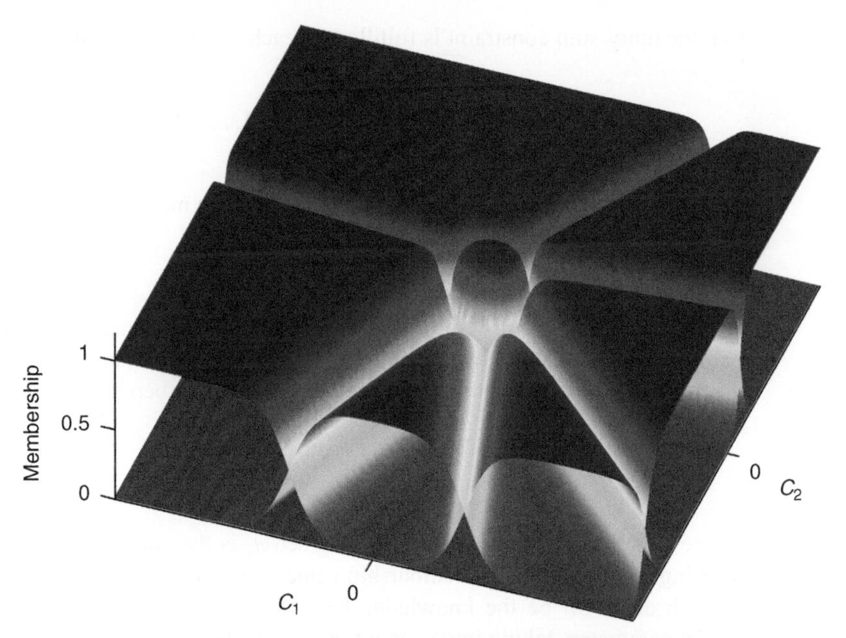

Figure 17.8 TSE function fitted to the chromatic categories defined on a given lightness level. In this case, only six categories have memberships different from zero.

fuzzy boundaries among these three categories we use one-dimensional sigmoid functions along the lightness axis:

$$\mu_{A_{\text{black}}}(I, \theta_{\text{black}}) = \frac{1}{1 + \exp[-\beta_b(I - t_b)]}, \tag{17.14}$$

$$\mu_{A_{\text{gray}}}(I, \theta_{\text{gray}}) = \frac{1}{1 + \exp[\beta_b(I - t_b)]} \frac{1}{1 + \exp[-\beta_w(I - t_w)]}, \tag{17.15}$$

$$\mu_{A_{\text{white}}}(I, \theta_{\text{white}}) = \frac{1}{1 + \exp[\beta_w(I - t_w)]}, \tag{17.16}$$

where $\theta_{\text{black}} = (t_b, \beta_b)$, $\theta_{\text{gray}} = (t_b, \beta_b, t_w, \beta_w)$, and $\theta_{\text{white}} = (t_w, \beta_w)$ are the set of parameters for *black*, *gray*, and *white*, respectively.

Hence, the membership of the three achromatic categories on a given chromaticity plane is computed by weighting the global achromatic membership (Eq. 17.13) with the corresponding membership in the lightness dimension (Eqs. 17.14 and 17.16):

$$\mu_{C_k}(\mathbf{s}, \theta_{C_k}) = \mu_A^i(c_1, c_2)\mu_{A_{C_k}}(I, \theta_{C_k}), \qquad 9 \leq k \leq 11, \quad I_i < I \leq I_{i+1}, \tag{17.17}$$

where i is the chromaticity plane in which the sample is included and the values of k correspond to the achromatic categories (Eq. 17.4). In this way, we can assure that the unity-sum constraint is fulfilled on each specific chromaticity plane,

$$\sum_{k=1}^{11} \mu_{C_k^i}(\mathbf{s}) = 1 \qquad i = 1, \ldots, N_L, \tag{17.18}$$

where N_L is the number of chromaticity planes in the model.

17.2.4 Fuzzy Sets Estimation

After defining the membership functions of the model, the next step is to fit their parameters. To this end, we need a set of psychophysical data, D, composed of a set of samples from the color space and their membership values to the 11 categories,

$$D = \{< \mathbf{s_i}, m_1^i, \ldots, m_{11}^i >\}, \qquad i = 1, \ldots, n_s, \tag{17.19}$$

where $\mathbf{s_i}$ is the ith sample of the learning set, n_s is the number of samples in the learning set, and m_k^i is the membership value of the ith sample to the kth category.

Such data will be the knowledge basis for a fitting process to estimate the model parameters taking into account the unity-sum constraint given in Equation 17.18. In this case, the model will be estimated for the CIE $L^*a^*b^*$ space since it is a standard space with interesting properties. However, any other color space with a lightness dimension and two chromatic dimensions would be suitable for this purpose.

17.2.4.1 Learning Set The data set for the fitting process must be perceptually significant; that is, the judgments should be coherent with results from psychophysical color-naming experiments and the samples should cover all the color space.

To build a wide learning set, we have used the color-naming map proposed by Seaborn et al. in [347]. This color map has been built by making some considerations on the consensus areas of the Munsell color space provided by the psychophysical data from the experiments of Sturges and Whitfield [332]. Using such data and the fuzzy k-means algorithm this method allows us to derive the memberships of any point in the Munsell space to the 11 basic color categories.

In this way, we have obtained the memberships of a wide sample set, and afterwards we have converted this color sampling set to their corresponding CIE $L^*a^*b^*$ representation. The data set was initially composed of the 1269 samples of the Munsell Book of Color [348]. Their reflectances and CIE $L^*a^*b^*$ coordinates, calculated by using the CIE D65 illuminant, are available at the web site of the University of Joensuu in Finland [349]. This data set was extended with selected samples to a total number of 1617 samples (see Reference 350 for more details on how these extra samples were selected).

Hence, with such a data set we accomplish the perceptual significance required for the learning set. First, by using Seaborn's method, we include the results of the psychophysical experiment of Sturges and Whitfield, and, in addition, it covers an area of the color space that suffices for the fitting process.

17.2.4.2 ***Parameter Estimation*** Before starting with the fitting process, the number of chromaticity planes and the values that define the lightness levels (Eq. 17.12) must be set. These values depend on the learning set used and must be chosen while taking into account the distribution of the samples from the learning set. In this case, the number of planes that delivered best results was found to be six, and the values I_i that define the levels were selected by choosing some local minima in the histogram of samples along the lightness axis: $I_1 = 31, I_2 = 41, I_3 = 51, I_4 = 66, I_5 = 76$. However, if a more extensive learning set were available, a higher number of levels would possibly deliver better results.

For each chromaticity plane, the global goal of the fitting process is finding an estimation of the parameters, $\hat{\theta}^j$, that minimizes the mean squared error between the memberships from the learning set and the values provided by the model:

$$\hat{\theta}^j = \arg\min_{\theta^j} \frac{1}{n_{cp}} \sum_{i=1}^{n_{cp}} \sum_{k=1}^{n_c} (\mu^j_{C_k}(\mathbf{s}_i, \theta^j_{C_k}) - m^i_k)^2, \qquad j = 1, \ldots, N_L, \tag{17.20}$$

where $\hat{\theta}^j = (\hat{\theta}^j_{C_1}, \ldots, \hat{\theta}^j_{C_{nc}})$ is the estimation of the parameters of the model for the chromatic categories on the *j*th chromaticity plane, $\theta^j_{C_k}$ is the set of parameters of the category C_k for the *j*th chromaticity plane, n_c is the number of chromatic categories, n_{cp} is the number of samples of the chromaticity plane, $\mu^j_{C_k}$ is the membership function of the color category C_k for the *j*th chromaticity plane, and m^i_k is the membership value of the *i*th sample of the learning set to the *k*th category.

The previous minimization is subject to the unity-sum constraint:

$$\sum_{k=1}^{11} \mu^j_{C_k}(\mathbf{s}, \theta^j_{C_k}) = 1, \qquad \forall \mathbf{s} = (I, c_1, c_2) \mid I_{j-1} < I \le I_j, \tag{17.21}$$

which is imposed to the fitting process through two assumptions. The first one is related to the membership transition from chromatic categories to achromatic categories:

Assumption 17.1. All the chromatic categories in a chromaticity plane share the same ES function, which models the membership transition to the achromatic categories. This means that all the chromatic categories share the set of estimated parameters for ES:

$$\theta^j_{\mathrm{ES}_{C_p}} = \theta^j_{\mathrm{ES}_{C_q}} \quad \text{and} \quad \mathbf{t}^{\mathbf{j}}_{\mathbf{C_p}} = \mathbf{t}^{\mathbf{j}}_{\mathbf{C_q}}, \qquad \forall p, q \in \{1, \ldots, n_c\}, \tag{17.22}$$

where n_c is the number of chromatic categories.

The second assumption refers to the membership transition between adjacent chromatic categories:

Assumption 17.2. Each pair of neighboring categories, C_p and C_q, share the parameters of slope and angle of the double-sigmoid function, which define their boundary:

$$\beta_y^{C_p} = \beta_x^{C_q} \qquad \text{and} \qquad \alpha_y^{C_p} = \alpha_x^{C_q} - \left(\frac{\pi}{2}\right), \tag{17.23}$$

where the superscripts indicate the category to which the parameters correspond.

These assumptions considerably reduce the number of parameters to be estimated. Hence, for each chromaticity plane, we must estimate two parameters for the translation, $\mathbf{t} = (t_x, t_y)$, four for the ES function, $\theta_{ES} = (e_x, e_y, \phi, \beta_e)$, and a maximum of $2 \times n_c$ for the DS functions, since the other two parameters of $\theta_{DS} = (\alpha_x, \alpha_y, \beta_x, \beta_y)$ can be obtained from the neighboring category (Eq. 17.23).

All the minimizations to estimate the parameters are performed by using the simplex search method proposed in Reference 351 (see Reference 350 for more details about the parameters estimation process). After the fitting process, we obtain the parameters that completely define the color-naming model and these are summarized in Table 17.1.

The evaluation of the fitting process is done in terms of two measures. The first one is the mean absolute error (MAE_{fit}) between the learning set memberships and the memberships obtained from the parametric membership functions:

$$\text{MAE}_{fit} = \frac{1}{n_s}\frac{1}{11}\sum_{i=1}^{n_s}\sum_{k=1}^{11} |m_k^i - \mu_{C_k}(\mathbf{s_i})|, \tag{17.24}$$

where n_s is the number of samples in the learning set, m_k^i is the membership of $\mathbf{s_i}$ to the kth category, and $\mu_{C_k}(\mathbf{s_i})$ is the parametric membership of $\mathbf{s_i}$ to the kth category provided by the model.

The value of MAE_{fit} is a measure of the accuracy of the model fitting to the learning data set, and in this case the value obtained was of $\text{MAE}_{fit} = 0.0168$. This measure was also computed for a test data set of 3149 samples. To build the test data set, the Munsell space was sampled at hues 1.25, 3.75, 6.25, and 8.75; values from 2.5 to 9.5 at steps of 1 unit; and chromas from 1 to the maximum available with a step of 2 units. As in the case of the learning set, the memberships of the test set that were considered the ground truth were computed with Seaborn's algorithm. The corresponding CIE $L^*a^*b^*$ values to apply the parametric functions were computed with the Munsell Conversion software. The value of MAE_{fit} obtained was 0.0218, which confirms the accuracy of the fitting that allows the model to provide membership values with very low error even for samples that were not used in the fitting process.

The second measure evaluates the degree of fulfillment of the unity-sum constraint. Considering as error the difference between the unity and the sum of

Table 17.1 Parameters of the triple sigmoid with elliptical center model.[a,b]

Achromatic axis

Black-Gray boundary	$t_b = 28.28$	$\beta_b = -0.71$
Gray-White boundary	$t_w = 79.65$	$\beta_w = -0.31$

Chromaticity Plane 1

$t_a = 0.42$	$e_a = 5.89$	$\beta_e = 9.84$	
$t_b = 0.25$	$e_b = 7.47$	$\phi = 2.32$	

	α_a	α_b	β_a	β_b
Red	−2.24	−56.55	0.90	1.72
Brown	33.45	14.56	1.72	0.84
Green	104.56	134.59	0.84	1.95
Blue	224.59	−147.15	1.95	1.01
Purple	−57.15	−92.24	1.01	0.90

Chromaticity Plane 2

$t_a = 0.23$	$e_a = 6.46$	$\beta_e = 6.03$	
$t_b = 0.66$	$e_b = 7.87$	$\phi = 17.59$	

	α_a	α_b	β_a	β_b
Red	2.21	−48.81	0.52	5.00
Brown	41.19	6.87	5.00	0.69
Green	96.87	120.46	0.69	0.96
Blue	210.46	−148.48	0.96	0.92
Purple	−58.48	−105.72	0.92	1.10
Pink	−15.72	−87.79	1.10	0.52

Chromaticity Plane 3

$t_a = -0.12$	$e_a = 5.38$	$\beta_e = 6.81$	
$t_b = 0.52$	$e_b = 6.98$	$\phi = 19.58$	

	α_a	α_b	β_a	β_b
Red	13.57	−45.55	1.00	0.57
Orange	44.45	−28.76	0.57	0.52
Brown	61.24	6.65	0.52	0.84
Green	96.65	109.38	0.84	0.60
Blue	199.38	−148.24	0.60	0.80
Purple	−58.24	−112.63	0.80	0.62
Pink	−22.63	−76.43	0.62	1.00

Chromaticity Plane 4

$t_a = -0.47$	$e_a = 5.99$	$\beta_e = 7.76$	
$t_b = 1.02$	$e_b = 7.51$	$\phi = 23.92$	

	α_a	α_b	β_a	β_b
Red	26.70	−56.88	0.91	0.76
Orange	33.12	−9.90	0.76	0.48
Yellow	80.10	5.63	0.48	0.73
Green	95.63	108.14	0.73	0.64
Blue	198.14	−148.59	0.64	0.76
Purple	−58.59	−123.68	0.76	5.00
Pink	−33.68	−63.30	5.00	0.91

Chromaticity Plane 5

$t_a = -0.57$	$e_a = 5.37$	$\beta_e = 100.00$	
$t_b = 1.16$	$e_b = 6.90$	$\phi = 24.75$	

	α_a	α_b	β_a	β_b
Orange	25.75	−15.85	2.00	0.84
Yellow	74.15	12.27	0.84	0.86
Green	102.27	98.57	0.86	0.74
Blue	188.57	−150.83	0.74	0.47
Purple	−60.83	−122.55	0.47	1.74
Pink	−32.55	−64.25	1.74	2.00

Chromaticity Plane 6

$t_a = -1.26$	$e_a = 6.04$	$\beta_e = 100.00$	
$t_b = 1.81$	$e_b = 7.39$	$\phi = -1.19$	

	α_a	α_b	β_a	β_b
Orange	25.74	−17.56	1.03	0.79
Yellow	72.44	16.24	0.79	0.96
Green	106.24	100.05	0.96	0.90
Blue	190.05	−149.43	0.90	0.60
Purple	−59.43	−122.37	0.60	1.93
Pink	−32.37	−64.26	1.93	1.03

[a]Angles are expressed in degrees and subscripts x and y are changed to a and b, respectively, in order [b]to make parameter interpretation easier, since parameters have been estimated for the CIE $L^*a^*b^*$ space.

all the memberships at a point, $\mathbf{p_i}$, the measure is

$$\mathrm{MAE}_{\mathrm{unitsum}} = \frac{1}{n_p} \sum_{i=1}^{n_p} |1 - \sum_{k=1}^{11} \mu_{C_k}(\mathbf{p_i}))|, \tag{17.25}$$

where n_p is the number of points considered and μ_{C_k} is the membership function of category C_k.

To compute this measure, we have sampled each one of the six chromaticity planes with values from -80 to 80 at steps of 0.5 units on both a and b axis, which means that $n_p = 153600$. The value obtained of $\mathrm{MAE}_{\mathrm{unitsum}} = 6.41e - 04$ indicates that the model provides a great fulfillment of that constraint, making the model consistent with the presented framework.

Hence, for any point of the CIE $L^*a^*b^*$ space we can compute the membership to all the categories and, at each chromaticity plane, these values can be plotted to generate a membership map. In Figure 17.9, we show the membership maps of the six chromaticity planes considered with the membership surfaces labeled with their corresponding color term.

17.3 Color Names from Uncalibrated Data

In the previous section, we saw an example where the mapping between RGB and color names is inferred from a labeled set of color patches. Other examples of such methods include References 352–356. In such methods, multiple test subjects are asked to label hundreds of color chips within a well-defined experimental setup. From this labeled set of color chips the mapping from RGB values to color names is derived. The main difference of these methods from the one discussed in this section is that they are all based on calibrated data. Color names from calibrated data have been shown to be useful within the linguistic and color science fields. However, when applied to real-world images, these methods were often found to obtain unsatisfactory results. Color naming chips under ideal lighting on a color neutral background greatly differ from the challenge of color-naming in images coming from real-world applications without a neutral reference color and with physical variations such as shading effects and different light sources.

In this section, we discuss a method for color naming in uncalibrated images. More precisely, with *uncalibrated images* we refer to images that are taken under varying illuminants, with interreflections, coming from unknown cameras, colored shadows, compression artifacts, aberrations in acquisition, unknown camera and camera settings, etc. The majority of the image data in computer vision belongs to this category: even in the cases that camera information is available and the images are uncompressed, the physical setting of the acquisition are often difficult to recover, due to unknown illuminant colors, unidentified shadows, view-point changes, and interreflections.

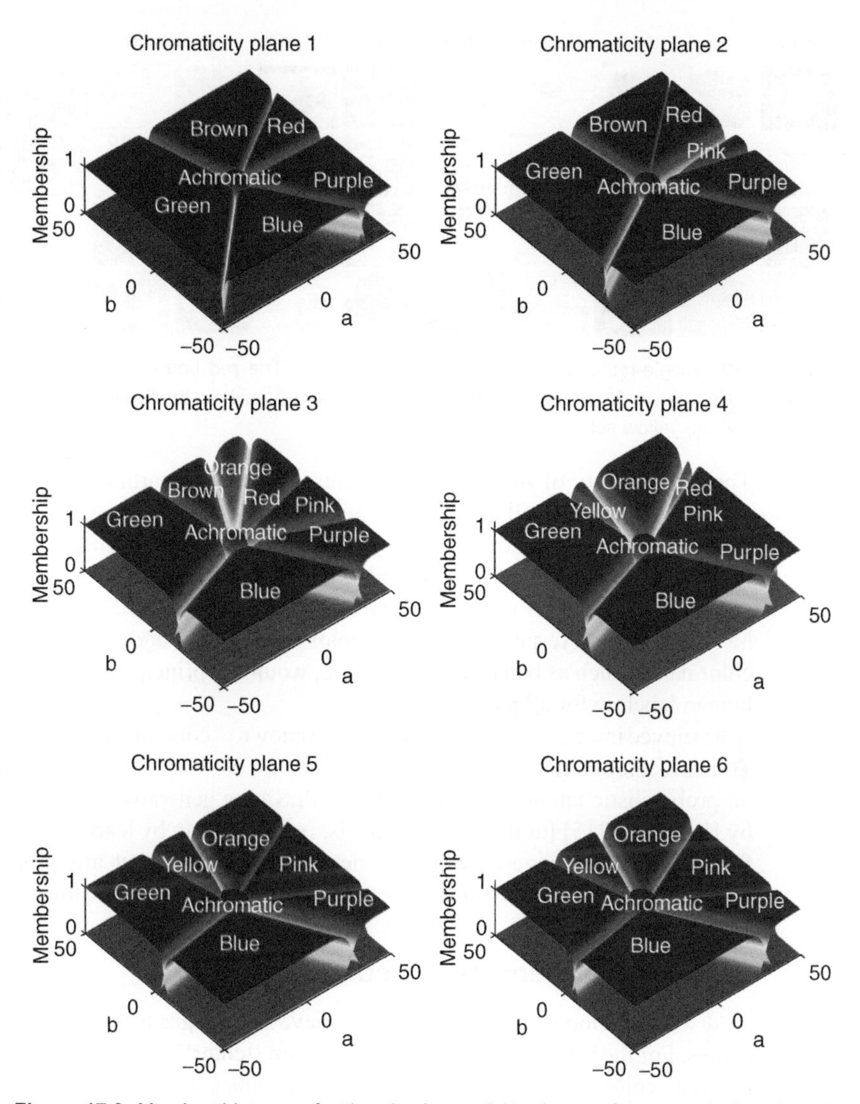

Figure 17.9 Membership maps for the six chromaticity planes of the model plotted on the chromaticity plane (ab-plane).

To infer what RGB values color names take on in real-world images, a large data set of color name labeled images is required. One possible way to obtain such a data set is by means of Google Image search. We retrieve 250 images for each of the 11 basic color terms discussed in Section 17.1 (Figure 17.10). These images contain a large variety of appearances of the queried color name. For example, the query ''red'' will contain images with red objects, taken under varying physical variations, such as different illuminants, shadows, and specularities. The images are taken with different cameras and stored with various compression methods.

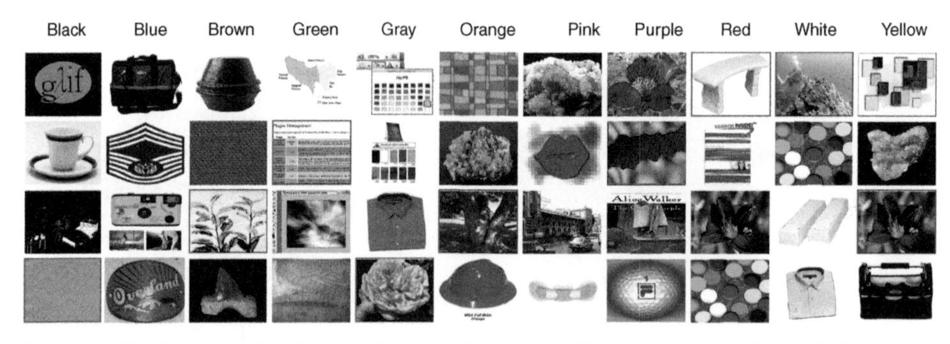

Figure 17.10 Google-retrieved examples for color names. The red bounding boxes indicate false positives. An image can be retrieved with various color names, such as the flower image that appears in the red and the yellow set.

The large variety of this training set suits our goal of learning color names for real-world images well, since we want to apply the color-naming method on uncalibrated images taken under varying physical settings. Furthermore, a system based on Google image has the advantage that it is flexible with respect to variations in the color name set. Methods based on calibrated data are known to be inflexible with respect to the set of color names, since adding for example new color names such as beige, violet, or olive, would, in principle, imply redoing the human labeling for all patches.

Retrieved images from Google search are known to contain many false positives. To learn color names from such a noisy data set, we will discuss a method based on probabilistic latent semantic analysis (PLSA), a generative model introduced by Hofmann [215] for document analysis. In conclusion, by learning color names from real-world images, we aim to derive color names that are applicable on challenging real-world images typical for computer vision applications.

17.3.1 Color Name Data Sets

As discussed, Google Image is used to retrieve 250 images for each of the 11 color names. For the actual search we added the term ‘‘color’’; hence, for red the query is ‘‘red+color.’’ Examples for the 11 color names are given in Figure 17.10. Almost 20 % of the images are false positives, that is, images that do not contain the color of the query. We call such a data set *weakly labeled* since the image labels are global, meaning that no information to which a particular region of the image the label refers is available. Furthermore, in many cases only a small portion, as little as a few percent of the pixels, represents the color label. Our goal is to learn a color-naming system based on the raw results of Google image, that is, we used both true and false positives.

The Google data set contains weakly labeled data, meaning that we only have an image-wide label, indicating that a part of the pixels in the image can be described by the color name of the label. To remove some of the pixels that are not likely indicated by the image label, we remove the background from the

Google images by iteratively removing pixels that have the same color as the border. Furthermore, since the color label often refers to an object in the center of the image, we crop the image to be 70% of its original width and height.

The Google images will be represented by color histograms. We consider the images from the Google data sets to be in *sRGB* format. Before computing the color histograms, these images are gamma corrected with a correction factor of 2.4. Although images might not be correctly white balanced, we do not apply a color constancy algorithm, since color constancy was found to yield unsatisfying results for these images. Furthermore, many Google images lack color calibration information, and regularly break assumptions on which color constancy algorithms are based. The images are converted to the $L^*a^*b^*$ color space, which is a perceptually linear color space, ensuring that similar differences between $L^*a^*b^*$ values are considered equally important color changes to humans. This is a desired property because the uniform binning we apply for histogram construction implicitly assumes a meaningful distance measure. To compute the $L^*a^*b^*$ values we assume a D65 white light source.

17.3.2 Learning Color Names

Here we discuss a method to learn color names, which is based on latent aspect models. Latent aspect models have received considerable interest in the text analysis community as a tool to model documents as a mixture of several semantic—but a priori unknown, and hence "latent"—topics. Latent Dirichlet allocation (LDA) [357] and PLSA [215] are perhaps the most well known among such models.

Here we use the topics to represent the color names of pixels. Latent aspect models are of interest to our problem since they naturally allow for multiple topics in the same image, as is the case in the Google data set where each image contains a number of colors. Pixels are represented by discretizing their $L^*a^*b^*$ values into a finite vocabulary by assigning each value by cubic interpolation to a regular $10 \times 20 \times 20$ grid in the $L^*a^*b^*$-space.[1] An image (document) is then represented by a histogram indicating how many pixels are assigned to each bin (word).

We start by explaining the standard PLSA model, after which an adapted version better suited to the problem of color naming is discussed. Given a set of documents $D = \{d_1, \ldots, d_N\}$ each described in a vocabulary $W = \{w_1, \ldots, w_M\}$, the words are taken to be generated by latent topics $Z = \{z_1, \ldots, z_K\}$. In the PLSA model, the conditional probability of a word w in a document d is given by

$$p(w|d) = \sum_{z \in Z} p(w|z) p(z|d) . \tag{17.26}$$

[1] Because the $L^*a^*b^*$ space is perceptually uniform we discretize it into equal volume bins. Different quantization levels per channel are chosen because of the different ranges: the intensity axis ranges from 0 to 100, and the chromatic axes range from -100 to 100.

Both distributions $p(z|d)$ and $p(w|z)$ are discrete multinomial distributions, and can be estimated with an expectation-maximization (EM) algorithm [215] by maximizing the log-likelihood function

$$L = \sum_{d \in D} \sum_{w \in W} n(d, w) \log p(d, w), \tag{17.27}$$

where $p(d, w) = p(d)p(w|d)$, and $n(d, w)$ is the term frequency, containing the word occurrences for every document.

The method in Equation 17.26 is called a *generative model*, since it provides a model of how the observed data has been generated given hidden parameters (the latent topics). The aim is to find the latent topics that best explain the observed data. In the case of learning color names, we model the color values in an image as being generated by the color names (topics). For example, the color name red generates $L^*a^*b^*$ values according to $p(w|t = red)$. These word-topic distributions $p(w|t)$ are shared between all images. The amount of the various colors we see in an image is given by the mixing coefficients $p(t|d)$, and these are image specific. The aim of the learning process is to find the $p(w|t)$ and $p(t|d)$ that best explain the observations $p(w|d)$. As a consequence, colors that often co-occur are more likely to be found in the same topic. For example, the label red will not only co-occur with highly saturated reds but also with some pinkish-red colors because of specularities on the red object, and dark reds caused by shadows or shading. All the different appearances of the color name red are captured in $p(w|t = red)$.

In Figure 17.11, an overview of applying PLSA to the problem of color naming is provided. The goal of the system is to find the color name distributions $p(w|t)$. First, the weakly labeled Google images are represented by their normalized $L^*a^*b^*$ histograms. These histograms form the columns of the image-specific word distribution $p(w|d)$. Next, the PLSA algorithm aims to find the topics (color names) that best explain the observed data. This process can be understood as a matrix decomposition of $p(w|d)$ into the word-topic distributions $p(w|t)$ and the document-specific mixing proportions $p(t|d)$. The columns of $p(w|t)$ contain the information we are seeking, namely, the distributions of the color names over $L^*a^*b^*$ values. In the remainder of this section, we discuss two adaptations to the standard model.

17.3.2.1 Exploiting Image Labels The standard PLSA model cannot exploit the labels of images. More precisely, the labels have no influence on the maximum likelihood (Eq. 17.27). The topics are hoped to converge to the state where they represent the desired color names. As is pointed out in Reference 358 in the context of discovering object categories using LDA, this is rarely the case. To overcome this shortcoming, we discuss an adapted model that does take into account the label information.

The image labels can be used to define a prior distribution on the frequency of topics (color names) in documents $p(z|d)$. This prior will still allow each color to

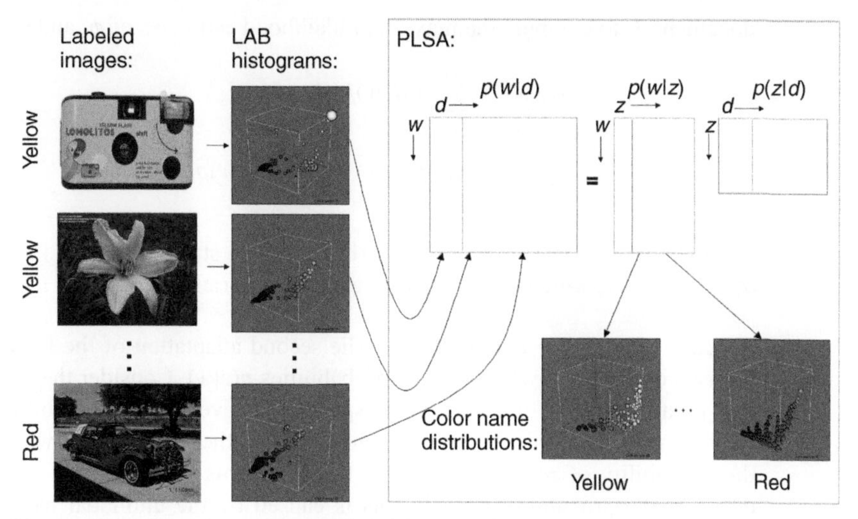

Figure 17.11 Overview of standard PLSA model for learning color names. See text for explanation. *Source*: Reprinted with permission, © 2009 IEEE.

be used in each image, but the topic corresponding to the label of the image—here obtained with Google—is a priori assumed to have a higher frequency than other colors. Below, we use the shorthands $p(w|z) = \phi_z(w)$ and $p(z|d) = \theta_d(z)$.

The multinomial distribution $p(z|d)$ is supposed to have been generated from a Dirichlet distribution of parameter α_{l_d}, where l_d is the label of the document d. The vector α_{l_d} has length K (number of topics), where $\alpha_{l_d}(z) = c \geq 1$ for $z = l_d$, and $\alpha_{l_d}(z) = 1$ otherwise. By varying c we control the influence of the image labels l_d on the distributions $p(z|d)$. The exact setting of c will be learned from the validation data.

For an image d with label l_d, the generative process thus reads:

1. Sample θ_d (distribution over topics) from the Dirichlet prior with parameter α_{l_d}.
2. For each pixel in the image
   (a) Sample z (topic, color name) from a multinomial with parameter θ_d
   (b) Sample w (word, pixel bin) from a multinomial with parameter ϕ_z

The distributions over words ϕ_z associated with the topics, together with the image-specific distributions θ_d, have to be estimated from the training images. This estimation is done using an EM algorithm. In the expectation step we evaluate for each word (color bin) w and document (image) d

$$p(z|w, d) \propto \theta_d(z)\phi_z(w). \tag{17.28}$$

During the maximization step, we use the result of the expectation step together with the normalized word document counts $n(d, w)$ (frequency of word w in document d) to compute the maximum likelihood estimates of ϕ_z and θ_d as

$$\phi_z(w) \propto \sum_d n(d,w)p(z|w,d), \tag{17.29}$$

$$\theta_d(z) \propto (\alpha_{l_d}(z) - 1) + \sum_w n(d,w)p(z|w,d). \tag{17.30}$$

Note that we obtain the EM algorithm for the standard PLSA model when $\alpha_{l_d}(z) = c = 1$, which corresponds to a uniform Dirichlet prior over θ_d.

17.3.2.2 Enforcing Unimodality The second adaptation of the PLSA model is based on prior knowledge of the probabilities $p(z|w)$. Consider the color name red: a particular region of the color space will have a high probability of red, moving away from this region in the direction of other color names will decrease the probability of red. Moving even further in this direction can only further decrease the probability of red. This is caused by the unimodal nature of the $p(z|w)$ distributions. Next, we discuss an adaptation of the PLSA model to enforce unimodality to the estimated $p(z|w)$ distributions.

It is possible to obtain a unimodal version of a function by means of grayscale reconstruction. The grayscale reconstruction of function p is obtained by iterating geodesic grayscale dilations of a marker m under p until stability is reached [359]. Consider the example given in Figure 17.12. In the example, we consider two $1D$ topics $p_1 = p(z_1|w)$ and $p_2 = p(z_2|w)$. By iteratively applying a geodesic dilation from the marker m_1 under the mask function p_1 we obtain the grayscale reconstruction ρ_1. The function ρ_1 is by definition unimodal, since it only has one maximum at the position of the marker m_1. Similarly, we obtain a unimodal version of p_2 by a grayscale reconstruction of p_2 from marker m_2.

Something similar can be done for the color name distributions $p(z|w)$. We can compute a unimodal version $\rho_z^{m_z}(w)$ by performing a grayscale reconstruction of $p(z|w)$ from markers m_z (finding a suitable position for the markers will be explained below). To enforce unimodality, without assuming anything about the shape of the distribution, we add the difference between the distributions

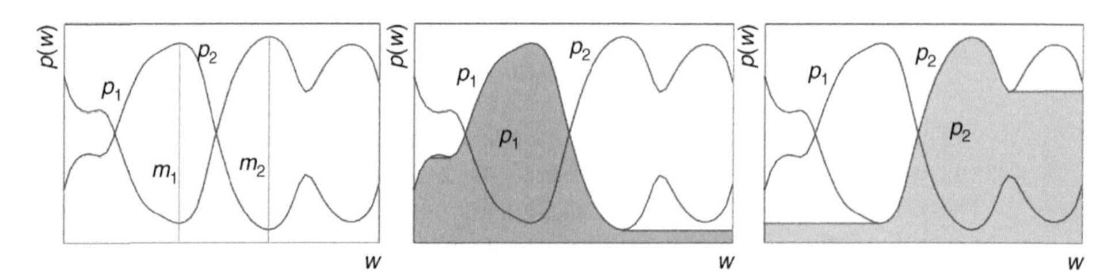

Figure 17.12 Example of grayscale reconstruction. (a) Initial functions $p_1 = p(z_1|w)$, $p_2 = p(z_2|w)$, and markers m_1 and m_2. (b) Grayscale reconstruction ρ_1 of p_1 from m_1. (c) Grayscale reconstruction ρ_2 of p_2 from m_2. Since ρ_1 is by definition a unimodal function, enforcing the difference between p_1 and ρ_1 to be small reduces the secondary modes of p_1. *Source*: Reprinted with permission, © 2009 IEEE.

$p(z|w)$ and their unimodal counterparts $\rho_z^{m_z}(z)$ as a regularization factor to the log-likelihood function:

$$L = \sum_{d \in D} \sum_{w \in W} n(d, w) \log p(d, w) - \gamma \sum_{z \in Z} \sum_{w \in W} \left(p(z|w) - \rho_z^{m_z}(w)\right)^2. \quad (17.31)$$

Adding the regularization factor in Equation 17.27 forces the functions $p(z|w)$ to be closer to $\rho_z^{m_z}(z)$. Since $\rho_z^{m_z}(z)$ is unimodal, this will suppress the secondary modes in $p(z|w)$, that is, the modes that it does not have in common with $\rho_z^{m_z}(z)$.

In the case of the color name distributions $p(z|w)$, the gray reconstruction is performed on the 3D spatial grid in $L^*a^*b^*$ space with a 26 connected structuring element. The markers m_z for each topic are computed by finding the local mode starting from the center of mass of the distribution $p(z|w)$. This was found to be more reliable than using the global mode of the distribution. The regularization functions $\rho_z^{m_z}$, which depend on $p(z|w)$, are updated at every iteration step of the conjugate-gradient-based maximization procedure that is used to compute the maximum likelihood estimates of $\phi_z(w)$. The computation of the maximum likelihood estimate for $\theta_d(z)$ is not directly influenced by the regularization factor and is still computed with Equation 17.30.

In conclusion, two improvements of the standard PLSA model have been discussed. Firstly, the image labels are used to define a prior distribution on the frequency of topics. Secondly, a regularization factor is added to the log likelihood function, which suppresses the secondary modes in the $p(z|w)$ distributions. The two parameters, c and γ, which regularize the strength of the two adaptations, can be learned from validation data.

17.3.3 Assigning Color Names in Test Images

Once we have estimated the distributions over words $p(w|z)$ representing the topics, we can use them to compute the probability of color names corresponding to image pixels in test images. As the test images are not expected to have a single dominant color, we do not use the label-based Dirichlet priors that are used when estimating the topics. The probability of a color name given a pixel is given by

$$p(z|w) \propto p(z)p(w|z), \quad (17.32)$$

where the prior over the color names $p(z)$ is taken to be uniform. The probability over the color names for a region is computed by a simple summation over all pixels in the region of the probabilities $p(z|w)$, computed with Equation 17.32 using a uniform prior.

The impact of the two improvements to standard PLSA discussed in Section 17.3.2 is illustrated in Figure 17.13 (for more detailed analysis, see also [360]). The image shows pixels of constant intensity, with varying hues in the angular direction and varying saturation in the radial direction. On the right side of the image a bar with varying intensity is included. Color names are expected to

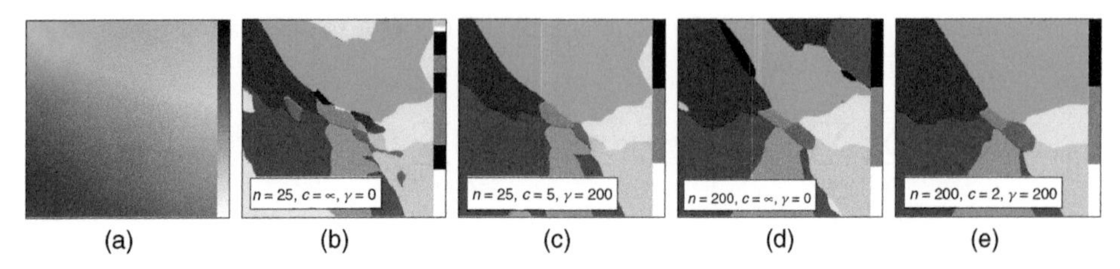

Figure 17.13 (a) A challenging synthetic image: the highly saturated RGB values at the border rarely occur in natural images. (b–e) Results obtain with different settings for c, γ, and n the number of train images per color name. The figure demonstrates that in the PLSA method, images (c) and (e), improve results. *Source*: Reprinted with permission, © 2006 IEEE.

be relatively stable for constant hues, only for low saturation they change to an achromatic color (i.e., in the center of the image). The only exception to this rule is brown, which is low saturated orange. Hence, we expect the color names to form a pielike partitioning with an achromatic color in the center, as in the parametric model that was introduced in Section 17.2. Assigning color names based on the empirical distribution (Fig. 17.13b) leads to many errors, especially in the saturated regions. The extended method trained from only 25 images per color name (Fig. 17.13c) obtains results much closer to what is expected. If we look at the performance as a function of the number of training images from Google Image, we see that the difference between the PLSA method with optimal c-γ settings and the empirical distributions becomes smaller by increasing the number of training images. However, the comparison shows that the extended method obtains significantly better results, especially in saturated regions (Fig. 17.13d,e).

17.3.4 Flexibility Color Name Data Set

An advantage of learning color names from uncalibrated images, collected with Google Image search, is that one can easily vary the set of color names. For the parametric method described in Section 17.2, changing the set of color names would mean that the psychophysical experiment needs to be repeated. Different sets of color names have, for example, been proposed in the work of Mojsilovic [355]. She asked a number of human test subjects to name the colors in a set of images. In addition to the 11 basic color terms beige, violet and olive were also mentioned.

In Figure 17.14, we show prototypes of the 11 basic color terms learned from Google Images. The prototype w_z of a color name is that color which has the highest probability of occurring given the color name $w_z = \text{argmax}_w\, p(w|z)$. In addition, we add a set of 11 extra color names, for which we retrieve 100 images from Google Image each. Again, the images contain many false positives. Then a single extra color name is added to the set of 11 basic color terms, and the color distributions $p(w|z)$ are recomputed, after which the prototype of the newly added color name is derived. This process is repeated for the 11 new color names. The results are depicted in the second row of Figure 17.14 and correspond to the colors we expect to find.

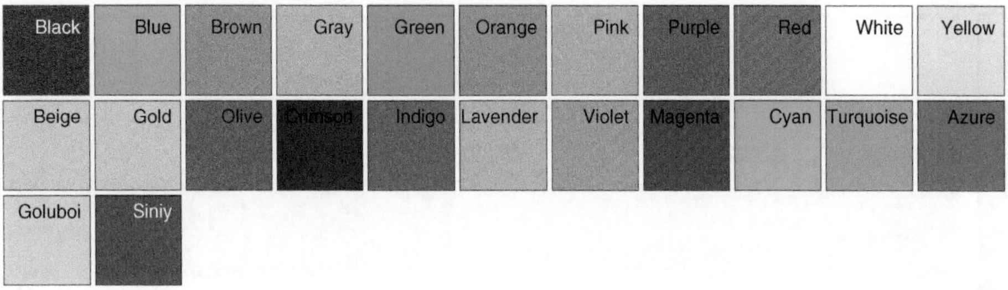

Figure 17.14 First row: prototypes of the 11 basic color terms learned from Google Images based on PLSA. Second row: prototypes of a varied set of color names learned from Google Images. Third row: prototypes of the two Russian blues learned from Google Images. *Source*: Reprinted with permission, © 2009 IEEE.

As a second example of flexibility of data acquisition we look into interlinguistic differences in color naming. The Russian language is one that has 12 basic color terms. The color term blue is split up into two color terms: goluboi (голубой), and siniy (синий). We ran the system on 30 images for both blues, returned by Google Image. Results are given in Figure 17.14, and correspond with the fact that goluboi is a light blue and siniy a dark blue. This example shows the Internet as a potential source of data for the examination of linguistic differences in color naming.

17.4 Experimental Results

In this section, we compare the two computational color naming methods discussed in Sections 17.2 and 17.3. The most relevant difference between the two methods is the training data on which they are based, either calibrated or uncalibrated. The parametric method is based on color name labels given to colored patches, which are taken in a highly controlled environment with known illumination, absence of context, and a gray reference background. The second approach is based on real-world images of objects within a context with unknown camera settings and illumination. We will test the two methods both on calibrated and on uncalibrated data. We will refer to the two methods as the parametric method and PLSA method.

Calibrated Color Naming Data First, we compare both methods on classifying single patches that are presented under white illumination. We have applied both color naming algorithms to the Munsell color array used in the World Color Survey by Berlin and Kay [329]. The results are shown in Figure 17.15. The results based on the parametric method are shown in Figure 17.15a and the results obtained with the PLSA method are shown in Figure 17.15b. The color names are similarly centered, and only on the borders there are some disagreements. The main difference that we can observe is that all chromatic patches are named by chromatic color names by the parametric method, whereas the PLSA method names multiple chromatic patches with achromatic color names.

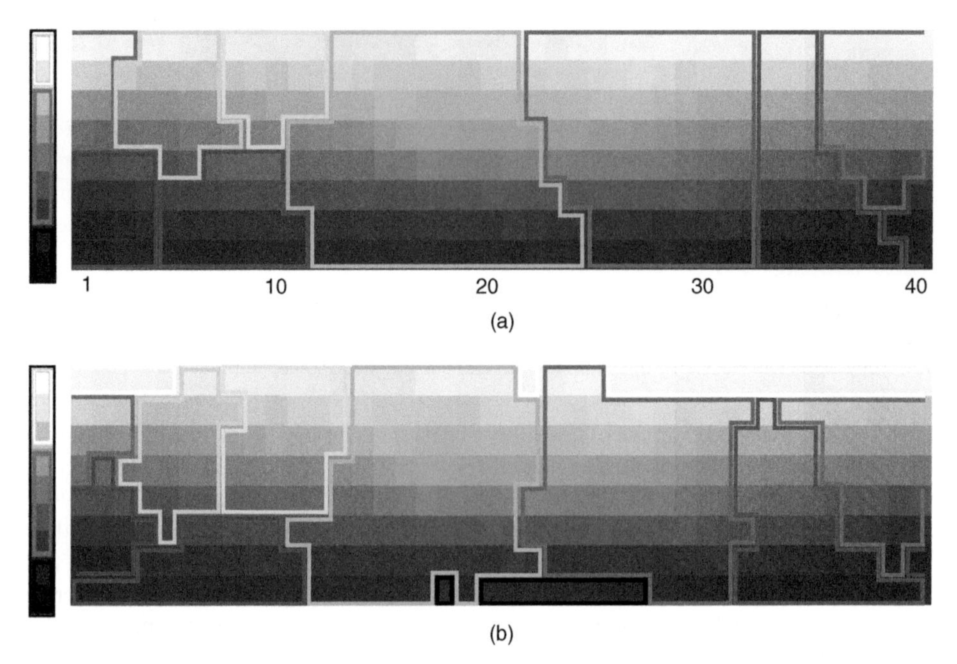

Figure 17.15 (a) Color name categories on the Munsell color array obtained by parametric method. (b) Color names obtained with the PLSA method. Note the differences in chromatic and achromatic assignments. The colored lines indicate the boundaries of the 11 color categories. *Source*: Reprinted with permission, © 2009 IEEE.

To quantitatively compare the two methods on calibrated patches, we compare the outcome of the parametric and the PLSA method against the color name observations from two works of reference: the study of Berlin and Kay [329] and the experiments of Sturges and Whitfield [332] (Figs. 17.1 and 17.2). We count the number of coincidences and dissimilarities between the predictions of the models and their observations. The results are summarized in Table 17.2. We see that the parametric model does significantly better than the PLSA model. This is what we expected since the parametric model is designed to perform color naming in calibrated circumstances. In addition, a comparison was made with the categorization done by an English speaker presented by MacLaury in Reference 361. The results obtained by the English speaker show the variability of the problem, since any individual subject judgments will normally differ from those of a color-naming experiment, which are usually averaged from several subjects. Notice that the performance of the PLSA model is similar to that of an individual human observer, when compared to the averaged results from psychophysical experiments

Uncalibrated Color Naming Data To test the computational color-naming methods on uncalibrated data, a human-labeled set of object images is required. For this purpose, we use a data set of images from the auction website Ebay [362].

Users labeled their objects with a description of the object in text, often including a color name. The data set contains four categories of objects: cars, shoes, dresses, and pottery (Fig. 17.16). For each object category, 121 images where collected, 12 for each color name. The final set is split in a test set of 440 images, and a validation set of 88 images. The images contain several challenges. The reflection properties of the objects differ from matt reflection of dresses to highly specular surfaces of cars and pottery. Furthermore, it comprises both indoor and outdoor scenes. For all images, a hand-segmentation of the object areas that correspond to the color name is provided. The color-naming methods are compared on the task of pixel-wise color name annotation of the Ebay images. All pixels within the segmentation masks are assigned to their most likely color name. We report the *pixel annotation score*, which is the percentage of correctly annotated pixels.

The results are given in Table 17.3. As can be seen, the PLSA method outperforms the parametric method by about 6%. This was to be expected since the PLSA method is learned from real-world Google Images, which look more similar to the Ebay Images. On the other hand, the parametric method based on calibrated data faces difficulties in the real-world where colors are not presented on a color neutral background under a known white light source. Figure 17.17

Table 17.2 Comparison of different munsell categorizations to the results from color-naming experiments of Berlin and Kay [329], and Sturges and Whitfield [332].

	Berlin and Kay Data			Sturges and Whitfield Data		
Model	Coincidences	Errors	% Errors	Coincidences	Errors	% Errors
Parametric	193	17	8.10	111	0	0.00
PLSA	180	30	14.3	106	5	4.50
Human	182	28	13.33	107	4	3.60

Figure 17.16 Examples for the four classes of the Ebay data: green car, pink dress, yellow plate, and orange shoes. For all images, masks with the area corresponding to the color name are hand segmented.

Figure 17.17 Two examples of pixel-wise color name annotation. The color names are represented by their corresponding color. In the center image, the results of the parametric model are given, and in the image on the right the results of the PLSA method.

Table 17.3 Pixel annotation score for the four classes in the ebay data set[a].

Method	Cars	Shoes	Dresses	Pottery	Overall
Parametric	56	72	68	61	64.7
PLSA	56	77	80	70	70.6

[a]The fifth column provides average results over the four classes.

shows results on two real-world images. Both methods obtain similar results, but one can see that especially in the achromatic regions they differ (ground plane below the strawberry and the house). The parametric method assigns more chromatic color names, whereas the PLSA method requires more saturated colors before assigning chromatic color names. Another difference is seen in some parts of the strawberry that are wrongly labeled as brown by the parametric model. Since the psychophysical experiments are carried out in controlled conditions, the parametric model is not able to include the different shades that a color can adopt because of illumination effects. By contrast, the PLSA method labels correctly most parts of the strawberry because learning from real-world images allows the PLSA method to consider the different variations that any color can present.

17.5 Conclusions

In this chapter, we have discussed two approaches to computational color naming. Firstly, we have discussed a method for learning color names from calibrated data. The parametric fuzzy model for color naming is based on the definition of the TSE as a membership function. The main advantage of the parametric model is that it allows to incorporate prior knowledge about the shapes of the color name distributions. Secondly, we have seen a method to learn color names from uncalibrated images collected from Google Image search. We have discussed a PLSA-based learning to cope with the inherently noisy data retrieved from Google Image search (the data contains many false positives). Learning color names from image search engines has the additional advantage that the method can easily vary the set of desired color names, something that is otherwise very costly.

Comparing the results of the two methods, we observed that the parametric method obtains superior results on calibrated data, and that the PLSA method outperforms the parametric model for real-world uncalibrated images. We use the term *uncalibrated* to refer to all kinds of deviations from the perfect setting of a single-color patch on a gray background. Hence, uncalibrated refers not only to unknown camera setting and unknown illuminant but also to the presence of physical events such as shadows and specularities. In the future, when color image understanding has improved, with improved illuminant estimation, and better object segmentations, the fact that the initial image is uncalibrated becomes less relevant. In such a scenario where we will be able to automatically calibrate uncalibrated data, the parametric models will become more important. The robustness that we observe now in the PLSA model will then be a disadvantage because it leads to reduced sensitivity. As a last remark, we would like to point out that we have ignored the interactions between neighboring colors, which can greatly influence the color sensation. With the usage of induction models, the perceived color sensation can be predicted [363, 364]. Therefore, addition of such models in color-naming algorithms is expected to improve results.

18 Segmentation of Multispectral Images

With contributions by Harro M. G. Stokman

Spectral information has become an important quality factor in many imaging processes because of its high accuracy. Spectral imaging is used, for example, in remote sensing, computer vision, and industrial applications. Spectral images can be obtained, for example, by a CCD camera with narrow-band interference filters [365]. Photometric invariance can be derived from multispectral images. In fact, the techniques presented in the previous chapters can be used to detect regions in multispectral images. To obtain robustness against noise, noise propagation can be adopted as discussed in Chapter 4. More information can be found in Reference 366.

In this chapter, methods are discussed to obtain photometric invariant region detection. In Section 18.3, the effect of sensor noise is discussed. Region detection is described in Section 18.4. In Section 18.5, the theoretical estimated uncertainty in polar angular representation is compared empirically to the real uncertainty. Experiments are carried out to evaluate the segmentation method, which are discussed in Section 18.6.

Color in Computer Vision: Fundamentals and Applications, First Edition.
Theo Gevers, Arjan Gijsenij, Joost van de Weijer, and Jan-Mark Geusebroek.

18.1 Reflection and Camera Models

In this section, we discuss the camera and image formation model. On the basis of the models, we examine cluster shapes drawn by uniformly colored objects in multispectral color space.

18.1.1 Multispectral Imaging

We use the Imspector V7 spectrograph from Spectral Imaging Ltd. The spectrograph transforms the monochrome CCD camera to a line scanner: one axis displays the spatial information, whereas along the other axis the visible wavelength range is recorded generating an image $h(x, \lambda)$ for each position (x, λ). The Jain CV-M300 camera is used with 576 pixels along the optical axis. We use the Imspector V7 spectrograph with shortest observable wavelength of 410 nm and longest wavelength of 700 nm. The wavelength interval corresponds to 5 nm.

18.1.2 Camera and Image Formation Models

We use a linear camera model to describe the relation between input signal h_i and the output signal c^i for the ith color channel at position $\vec{x}$ as

$$c^i(\vec{x}, \lambda) = \gamma_i f^i(\vec{x}, \lambda) + \mathrm{d}(\vec{x}), \tag{18.1}$$

where $d(\vec{x})$ denotes the dark current independent of the wavelength and γ_i denotes the camera gain for the ith color channel. For the moment, we ignore the dark current for notational simplicity. For the same reason, the notation for the position is left out.

The camera gain may further be refined as consisting of two terms

$$\gamma_i = \gamma_e \cdot \gamma_{w,i}, \tag{18.2}$$

where γ_e denotes the electronic gain and $\gamma_{w,i}$ denotes the white-balancing gain.

For inhomogeneous, dielectric materials, the measured input signal h_i of Equation 18.1 is described by the dichromatic reflection model [26] (Chapter 3). According to Equation 3.6, we obtain

$$f^i(\lambda) = m^b(\vec{n}, \vec{s}) \int_\lambda \rho^c(\lambda) e(\lambda) c^b(\lambda) \mathrm{d}\lambda + m^i(\vec{n}, \vec{s}, \vec{v}) \int_\lambda \rho^c(\lambda) e(\lambda) c^s(\lambda) \mathrm{d}\lambda, \tag{18.3}$$

denoting the camera output (without the camera gain) for filter ρ^c with central wavelength c. Further, $c^b(\lambda)$ and $c^s(\lambda)$ are the surface albedo and Fresnel reflectance respectively, $\vec{n}$ is the surface patch normal, $\vec{s}$ is the direction of the illumination source, and $\vec{v}$ is the direction of the viewer. Geometric terms m^b and m^i denote the geometric dependencies on $\vec{n}, \vec{s}$ and $\vec{v}$. Finally, $e(\vec{x}, \lambda)$ is the spectral power distribution of the incident (ambient) light at the object surface at $\vec{x}$.

18.1.3 White Balancing

According to Equation 18.3, a matte, white reference standard with constant spectral response can be described by $c^b(\lambda) = 1$ and $m^b(\vec{n}, \vec{s}) = 1$. Furthermore, assume that the camera is not white balanced so $\gamma_{w,i} = 1$, say, for all color channels i. The measured sensor values are obtained substituting the body reflection of Equation 18.3 in Equation 18.1 as

$$w_i(\lambda) = \gamma_e \int_\lambda \rho^c(\lambda) e(\lambda) \mathrm{d}\lambda, \tag{18.4}$$

denoting the sensor response for the white matte reference standard. The gain parameter $\gamma_{w,i}$ of Equation 18.2 is adjusted, either by the white-balancing procedure of the CCD camera or else manually, as

$$\gamma_{w,i} = \frac{1}{w_i(\lambda)}. \tag{18.5}$$

Then the output of a white-balanced camera system is as follows:

$$c^i(\lambda) = \frac{\gamma_e m^b(\vec{n}, \vec{s}) \int_\lambda \rho^c(\lambda) e(\lambda) c^b(\lambda) \mathrm{d}\lambda}{\gamma_e \int_\lambda \rho^c(\lambda) e(\lambda) \mathrm{d}\lambda} + \frac{\gamma_e m^i(\vec{n}, \vec{s}, \vec{v}) \int_\lambda \rho^c(\lambda) e(\lambda) c^s(\lambda) \mathrm{d}\lambda}{\gamma_e \int_\lambda \rho^c(\lambda) e(\lambda) \mathrm{d}\lambda}. \tag{18.6}$$

Considering the neutral interface reflection (NIR) model [26] (assuming that $c^s(\lambda)$ has a nearly constant value independent of the wavelength), we obtain $c^s(\lambda) = c^s$. Then the specular term of Equation 18.6 rewrites to

$$s_i(\lambda) = \frac{m^i(\vec{n}, \vec{s}, \vec{v}) c^s \int_\lambda \rho^c(\lambda) e(\lambda) \mathrm{d}\lambda}{\int_\lambda \rho^c(\lambda) e(\lambda) \mathrm{d}\lambda} = m^i(\vec{n}, \vec{s}, \vec{v}) c^s, \tag{18.7}$$

making the surface reflection term of Equation 18.3 independent of the spectral distribution of the light source. Owing to the white-balancing operation and the NIR assumption, the color channels $c^i(\lambda)$ produce equal output when an achromatic object is imaged.

Further, in case of the Imspector V7 spectrograph, we have narrow band filters $\rho(\lambda_i)$, which can be modeled as a unit impulse that is shifted over i wavelengths: the transmission at $\lambda_i = \delta$ and zero elsewhere. Note the subtle difference between $\rho^c(\lambda)$ and $\rho(\lambda^c)$. $\rho^c(\lambda)$ denotes a broad-band color filter (integrating over various wavelengths) with central wavelength c. $\rho(\lambda^c)$ denotes a narrow-band filter of unit impulse at wavelength c. Then Equation 18.7 rewrites to

$$s(\lambda_i) = \frac{m^i(\vec{n}, \vec{s}, \vec{v}) e(\lambda_i) c^s}{e(\lambda_i)} = m^i(\vec{n}, \vec{s}, \vec{v}) c^s, \tag{18.8}$$

again independent of λ and consequently invariant to the spectral distribution of the light source. Further, assuming narrow-band filters, Equation 18.6 rewrites to

$$c(\lambda_i) = \frac{m^b(\vec{n},\vec{s})e(\lambda_i)c^b(\lambda_i)}{e(\lambda_i)} + \frac{m^i(\vec{n},\vec{s},\vec{v})e(\lambda_i)c^s}{e(\lambda_i)} = m^b(\vec{n},\vec{s})c^b(\lambda_i) + m^i(\vec{n},\vec{s},\vec{v})c^s, \tag{18.9}$$

corresponding to the camera output at wavelength λ_i, making the whole dichromatic reflection model of Equation 18.3 independent of the spectral distribution of the light source (i.e. color constancy). In vector notation, a spectrum is denoted as

$$\vec{c} = m^b(\vec{n},\vec{s})\vec{c}^b + m^i(\vec{n},\vec{s},\vec{v})\vec{c}^s. \tag{18.10}$$

The vectors $\vec{n}, \vec{s}$, and $\vec{v}$ are three-dimensional. The vectors $\vec{c}, \vec{c}^b$, and $\vec{c}^s$ are N-dimensional, with N the number of samples taken in the wavelength range.

18.2 Photometric Invariant Distance Measures

From Chapter 4, we know that uniformly colored matte objects draw unit-constrained vectors (half rays) in (multispectral) color space due to changes in the surface orientation, illumination intensity, and shading. In addition, because of specularities, shiny objects draw half planes in multispectral space. Hence, for photometric invariant region detection, the shape of the clusters can be modeled as either a half ray or a half plane. In the next section, the angular representation of spectra is discussed.

18.2.1 Distance between Chromaticity Polar Angles

Spectra can be transformed into polar coordinates. To define polar coordinate descriptors, the origin O and a positive horizontal axis are fixed. Then each N-dimensional point $\vec{P}$ can be located by assigning to it polar coordinates $(\rho, \vec{\theta})$ where the one-dimensional term ρ gives the distance from O to $\vec{P}$ and the $(N-1)$-dimensional term $\vec{\theta}$ gives the angles from the initial axis to $\vec{P}$.

A spectrum defined by Equation 18.9 is transformed to its polar coordinate representation as

$$\rho_t = |\vec{c}|, \tag{18.11}$$

$$\theta_c(\lambda_i) = \arctan\left(\frac{c(\lambda_i)}{c(\lambda_N)}\right), 1 \leq i \leq N-1 \tag{18.12}$$

where ρ_t encodes the intensity of the spectrum and $\theta_c(\lambda_i)$ the chromaticity of the spectrum. $\theta_c(\lambda_i)$ takes on values in the range $0 \leq \theta_c \leq \frac{\pi}{2}$.

For the analysis of photometric invariance indexMultispectral images!angular representation of the chromaticity angular representation of spectra, substitution of the body reflection term of Equation 18.9 in 18.12 gives

$$\theta_c(\lambda_i) = \arctan\left(\frac{m^b(\vec{n},\vec{s})c^b(\lambda_i)}{m^b(\vec{n},\vec{s})c^b(\lambda_N)}\right) = \arctan\left(\frac{c^b(\lambda_i)}{c^b(\lambda_N)}\right), \tag{18.13}$$

independent of geometry term $m^b(\vec{n},\vec{s})$.

The quadratic distance, e, between any two M-dimensional vectors of angles $\vec{\theta}_1$ and $\vec{\theta}_2$ is defined as follows:

$$e^2(\vec{\theta}_1,\vec{\theta}_2) = \sum_{i=1}^{M} \left(\Delta(\theta_{1i},\theta_{2i})\right)^2, \qquad 0 \leq \theta_{1i},\theta_{2i} < 2\pi. \tag{18.14}$$

Here, θ_{1i} denotes the ith of M angles for the first vector. The distance $\Delta(\theta_i,\theta_j)$ takes values in the interval [0, 1] and is defined as follows:

$$\Delta(\theta_i,\theta_j) = [(\cos(\theta_i) - \cos(\theta_j))^2 + (\sin(\theta_i) - \sin(\theta_j))^2]^{1/2}. \tag{18.15}$$

The angular difference Δ is indeed a distance because it satisfies the following metric criteria:

- $\Delta(\theta_i,\theta_j) \geq 0$ for all θ_i and θ_j,
- $\Delta(\theta_i,\theta_j) = 0$ if and only if $\theta_i = \theta_j$,
- $\Delta(\theta_i,\theta_j) = \Delta(\theta_j,\theta_i)$ for all θ_i and θ_j,
- $\Delta(\theta_i,\theta_j) + \Delta(\theta_j,\theta_k) \geq \Delta(\theta_i,\theta_k)$ for all θ_i,θ_j, and θ_k.

The proof of the first three conditions is trivial. To see the triangular inequality, consider two angles θ_i,θ_j. Define

$$\vec{\theta}_i = [\ \cos(\theta_i) \quad \sin(\theta_i)\ ]^T, \tag{18.16}$$

and define $\vec{\theta}_j$ in a similar manner. Since $\Delta(\theta_i,\theta_j) = d(\vec{\theta}_i,\vec{\theta}_j)$ where d denotes the well-known Euclidean distance, the triangular inequality is proved.

Since chromaticity polar angles are independent of the geometry of the object, as was shown in Equation 18.13, the distance between two chromaticity angles is photometric invariant as well.

18.2.2 Distance between Hue Polar Angles

Consider an N-dimensional spectrum $\vec{c}$ defined by Equation 18.9 transformed to a different polar coordinate representation as follows:

$$\rho_s = 1 - \min\{c(\lambda_1),\ldots,c(\lambda_N)\}, \tag{18.17}$$

$$\theta_h = \alpha[\ c(\lambda_i) - [1 - \rho_s],\quad \phi(i, N)\ ], \tag{18.18}$$

where θ_h takes on values in the range $0 \le \theta_h < 2\pi$ and where

$$\phi(i, N) = \frac{i-1}{N-1} \cdot \frac{4}{3}\pi, \tag{18.19}$$

and

$$\alpha(w_i, \theta_i) = \arctan\left(\frac{\sum_{i=1}^{N} w_i \sin(\theta_i)}{\sum_{i=1}^{N} w_i \cos(\theta_i)}\right). \tag{18.20}$$

The function ϕ takes on values in the range $0 \le \phi(i, N) \le \frac{4}{3}\pi$. The function α takes on values in the range $0 \le \alpha < 2\pi$. The function denotes the weighted average of a series of N angles θ_i with corresponding weight w_i. The average is computed by decomposing the angular value into a horizontal and vertical component. The saturation of the spectrum is encoded by ρ_s. The angle θ_h can be thought of as the hue obtained directly from multispectral data. The function $\phi(i, N)$ assigns a hue angle to the ith of N spectral samples. The range from $0 \ldots 4/3\pi$ is reserved for the colors ranging from red through green through blue, so that the range from $4/3\pi \ldots 2\pi$ represents the purplish colors. The choice of $4/3\pi$ is somewhat arbitrary but can be defended by taking the hue computation into account based on conventional red-green-blue colors where a similar division is employed. For example, reconsider the definition of hue according to Equation 3.62:

$$\theta = \arctan\left(\frac{\sqrt{3}(G - B)}{(R - G) + (R - B)}\right). \tag{18.21}$$

Equation 18.19 assigns hue-angle $\theta_h = 0$ to the red channel, hue-angle $\theta_h = 2/3\pi$ to the green channel, and $\theta_h = 4/3\pi$ to the blue channel. Let $\rho_s = 1 - \min\{R, G, B\}$, then the weights of Equation 18.20 are defined as $w_1 = R - \rho_s$ for the red channel, $w_2 = G - \rho_s$ for the green channel, and $w_3 = B - \rho_s$ for the blue channel. Substitution of these results into Equation 18.18 gives

$$\begin{aligned}\theta &= \arctan\left(\frac{(R - \rho_s)\sin(0) + (G - \rho_s)\sin(2/3\pi) + (B - \rho_s)\sin(4/3\pi)}{(R - \rho_s)\cos(0) + (G - \rho_s)\cos(2/3\pi) + (B - \rho_s)\cos(4/3\pi)}\right), \\ &= \arctan\left(\frac{\frac{1}{2}\sqrt{3}G - \frac{1}{2}\sqrt{3}B}{R - \frac{1}{2}G - \frac{1}{2}B}\right) = \arctan\left(\frac{\sqrt{3}(G - B)}{(R - G) + (R - B)}\right),\end{aligned} \tag{18.22}$$

identical to Equation 18.21.

The polar coordinates are illustrated in Figure 18.1. The hue polar angle θ_h is invariant to the geometry and specularities: For a multispectral camera

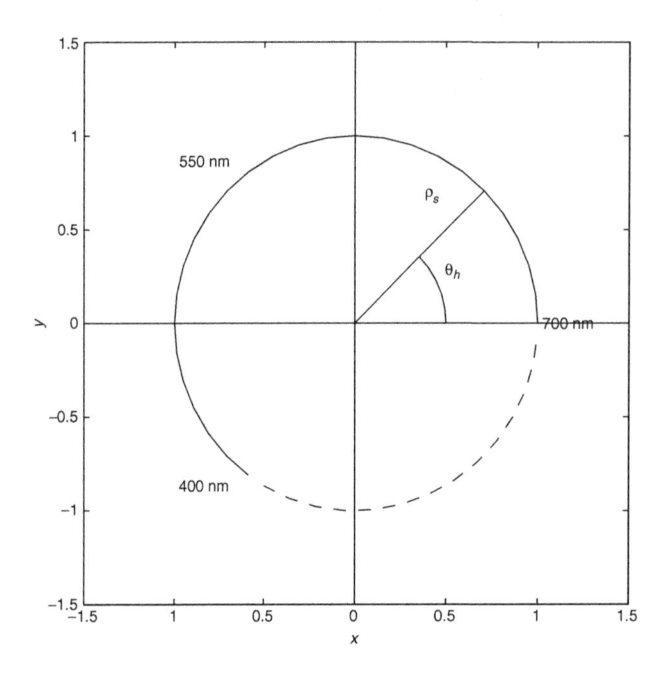

Figure 18.1 Polar coordinate representation of a spectrum depicted as a Euclidean map. ρ_s encodes the saturation of the spectrum and θ_h encodes the hue. The hue polar angle forms a unit half plane emanating from the origin in multispectral space. The hue range from $0\ldots4/3\pi$ is reserved for the colors ranging from red (700 nm) through green (550 nm) to blue (400 nm). The range from $4/3\pi\ldots2\pi$ (dashed part of the hue circle) represents purplish colors.

with narrow-band filters, consider substitution of Equation 18.9 in the term $c(\lambda_i) - [1 - \rho_s]$ of 18.18 as

$$c(\lambda_i) - [1 - \rho_s] = m^b(\vec{n}, \vec{s})[c^b(\lambda_i)c^b(\lambda_\rho)], \tag{18.23}$$

where $c^b(\lambda_\rho) = \min\{c(\lambda_1), \ldots, c(\lambda_N)\}$. This term is clearly independent of the specularity term $m^i(\vec{n}, \vec{s}, \vec{v})$. Moreover, the hue polar angle is independent of shadows (i.e., assuming that the light in the shadow has the same spectral characteristics as the light in the nonshadow area) and geometry as the substitution of Equation 18.23 in Equation 18.18 gives

$$\theta_h = \arctan\left(\frac{\sum_{i=1}^{N}[c^b(\lambda_i) - c^b(\lambda_\rho)]\sin[\phi(i, N)]}{\sum_{i=1}^{N}[c^b(\lambda_i) - c^b(\lambda_\rho)]\cos[\phi(i, N)]}\right), \tag{18.24}$$

independent of the geometric term $m^b(\vec{n}, \vec{s})$. Similar arguments hold for the white-balanced spectral sharpened *RGB* camera.

The distance between two hue polar angles $\theta_{h,i}, \theta_{h,j}$ is computed as $\Delta(\theta_{h,i}, \theta_{h,j})$ where Δ is defined in Equation 18.15. Because the hue polar angle is independent of the geometry of the object and independent of shadows and specularities, the distance between two hue polar angles is therefore photometric invariant as well.

18.2.3 Discussion

The distance between chromaticity angles, $\Delta(\theta_{c,i}, \theta_{c,j})$, is a photometric invariant for shadows (i.e., assuming that a shadow area has the same spectral characteristics as the light source) and the geometry of objects. Similarly, the distance between hue-angles, $\Delta(\theta_{h,i}, \theta_{h,j})$, is a photometric invariant to shadows, the geometry, and highlights (i.e., again assuming spectrally uniform illumination). These novel invariant measures come at the expense of requiring white balancing.

18.3 Error Propagation

For noise propagation, we reconsider Section 4.5.1 where the result of the measurement of a quantity u is properly stated as

$$\hat{u} = u_{\mathrm{e}} \pm \sigma_u, \tag{18.25}$$

where u_{e} is the best estimate for the quantity u and σ_u is the uncertainty or error in the measurement of u. Suppose that $u, \ldots, w$ are measured with corresponding uncertainties $\sigma_u, \ldots, \sigma_w$, and the measured values are used to compute the function $q(u, \ldots, w)$. If the uncertainties in $u, \ldots, w$ are independent, random and small, then the estimated uncertainty in $\hat{q}$ [42] is

$$\sigma_q = \sqrt{\left(\frac{\partial q}{\partial u}\sigma_u\right)^2 + \cdots + \left(\frac{\partial q}{\partial w}\sigma_w\right)^2}. \tag{18.26}$$

The uncertainty in q is never larger than the ordinary sum

$$\sigma_q \leq \left|\frac{\partial q}{\partial u}\right| \sigma_u + \cdots \left|\frac{\partial q}{\partial w}\right| \sigma_w. \tag{18.27}$$

In fact, this equation is really the *upper limit* on the uncertainty as proved by Taylor [42]. Therefore, whether or not the errors in $u, \ldots, w$ are dependent (or normally distributed), the uncertainty in q will never exceed the right side of Equation 18.27. Therefore, Equation 18.27 can be used for independent *and* dependent errors and will be used in the following sections to propagate noise through the polar angle representations of spectra.

18.3.1 Propagation of Uncertainties due to Photon Noise

Modern CCD cameras are sensitive enough to be able to count individual photons. Photon noise arises from the fundamentally stochastical nature of photon production. The probability distribution for counting ρ photons during t seconds is known to follow the Poisson distribution. The number of photons measured at pixel x is given by its average as

$$\hat{h}(x) = \rho t \pm \sqrt{\rho t}. \tag{18.28}$$

Let σ_d denote the dark current uncertainty. Incorporating σ_d and the uncertainty of Equation 18.28 in Equation 18.1 gives

$$c(x) \pm \sigma_{c(x)} = \gamma[\rho t \pm \sqrt{\rho t}] + [d(x) \pm \sigma_d]. \tag{18.29}$$

Our interest is in computing $\sigma_{c(x)}$. Let the dark current variance be denoted as $\mathrm{var}(d) = \sigma_d^2$. Let the average image intensity measured over a homogeneously colored patch be $\hat{I} = \gamma \rho t$; then the associated variance $\mathrm{var}(\hat{I}) = \gamma^2 \rho t$. We have the linear relation between $\hat{I}$ and $\mathrm{var}(\hat{I})$ based on [367] as

$$\mathrm{var}(\hat{I}) + \mathrm{var}(\hat{d}) = \gamma \hat{I} + \mathrm{var}(\hat{d}). \tag{18.30}$$

Linear regression among some intensity-variance pairs gives a robust estimation of the gain γ. It follows that the uncertainty in the number of photons measured at an arbitrary pixel $c(x)$ is given by

$$\sigma_c^2(x) = [\gamma \cdot c(x)]^2 + \sigma_d^2. \tag{18.31}$$

18.3.2 Propagation of Uncertainty

As stated before, the Jain CV-M300 camera is used with 576 pixels along the optical axis. We use the Imspector V7 spectrograph, from 410 nm to 700 nm, with a wavelength interval of 5 nm. Then the number of spectral samples obtained is 59. In fact, these 59 samples are recorded (uniformly spaced) over the 576 pixels along the optical axis. Therefore, the pixels at position (x, λ) of image h can be averaged in spectral direction by a uniform filter depending on the number of pixels. Let $K' = \mathrm{round}(576/59)$. If K' is odd, then the size of the filter $K = K'$, else $K = K' - 1$. The averaged spectral image h' is

$$h'(x, \lambda) = \frac{1}{K} \sum_{i=y_\lambda - \lfloor K/2 \rfloor}^{y_\lambda + \lfloor K/2 \rfloor} h(x, \lambda_i). \tag{18.32}$$

The uncertainty in a pixel value is propagated to the uncertainty in polar angles as follows. First, it is assumed that the pixel values in the spectral image are dependent. Therefore, using Equation 18.27 instead of Equation 18.26, the uncertainty due to the smoothing operation of Equation 18.32 reduces to

$$\sigma_{h'}^2(x, \lambda) = \frac{1}{K} \sum_{i=y_\lambda - \lfloor K/2 \rfloor}^{y_\lambda + \lfloor K/2 \rfloor} \sigma_h^2(x, \lambda_i). \tag{18.33}$$

From Equation 18.9, it follows that the uncertainty in the white-balanced camera output is

$$\sigma_{c'}^2(x,\lambda_i) = \frac{c^2(x,\lambda_i)\sigma_w^2(x,\lambda_i) + w^2(x,\lambda_i)\sigma_c^2(x,\lambda_i)}{w^4(x,\lambda_i)}, \tag{18.34}$$

where c' denotes the white-balanced camera output and c denotes the observed camera output. w denotes the output for the white matte reference standard.

For the general function $q(u, v) = \arctan(u/v)$, where the parameters u, v are dependent and have associated uncertainties σ_u, σ_v, the uncertainty in output σ_q is obtained using Equation 18.27 as

$$\sigma_q \leq \left|\frac{v\sigma_u}{u^2+v^2}\right| + \left|\frac{u\sigma_v}{u^2+v^2}\right|. \tag{18.35}$$

The function is shown in Figure 18.2. Large uncertainties occur if u and v both approach the value zero. The polar angles of Equation 18.12 are interdependent as each angle is obtained by division through the same value $\theta(\lambda_N)$. The uncertainty in chromaticity polar angle of Equation 18.12 therefore follows straightforward from Equation 18.35 by substituting $u = c(\lambda_i)$, $v = c(\lambda_N)$, and where both σ_u and σ_v are obtained from Equation 18.34.

To obtain an estimate of the uncertainty of the hue polar angle, consider the term $c(\lambda_i) - [1 - \rho_s]$ of Equation 18.18. The parameters $c(\lambda_i)$ and ρ_s are assumed independent because the reflectance factor $c(\lambda_i)$ is assumed to be obtained independent from the reflectance factor $\rho_s = 1 - \min\{c(\lambda_1), \ldots, c(\lambda_N)\}$. Thus, the resulting uncertainty is obtained using Equation 18.26 as

$$\sigma_{c-[1-\rho]}^2(\lambda_i) = \sigma_c^2(\lambda_i) + \sigma_\rho^2. \tag{18.36}$$

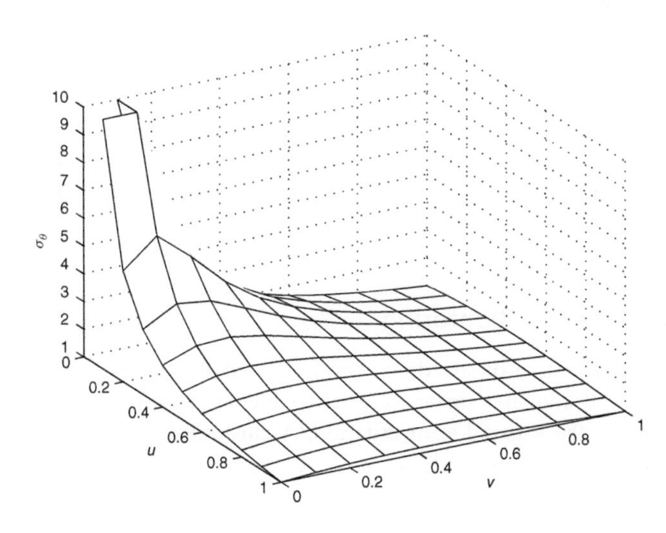

Figure 18.2 Uncertainty in the function arctan(u/v) as the function of u and v. The uncertainties σ_u and σ_v are set equal to one. Large uncertainties occur if both u and v approach the value zero, indicating the instability of the function around the origin.

The uncertainty for the hue polar angle of Equation 18.18 follows from Equation 18.20. The exact number generated by Equation 18.19 has no associated uncertainty, and therefore $\sin[\phi(i,N)]$ has no associated uncertainty. However, the weights $w_i = c(\lambda_i)$ do have uncertainty $\sigma_c(\lambda_i)$, again specified by Equation 18.34. The individual terms $w_i \sin(\theta_i)$ are considered independent because the reflectance factor $w_i = c(\lambda_i)$ is assumed to be obtained independent from the reflectance factor at wavelength $w_j = c(\lambda_j)$. Therefore, the uncertainty of the enumerator term $u = \sum_i w_i \sin(\theta_i)$ is

$$\sigma_u^2 = \sum_i (\sigma_c(\lambda_i) \cdot \sin[\phi(i,N)])^2. \tag{18.37}$$

A similar argument holds for the denominator term $u = \sum_i w_i \cos(\theta_i)$ yielding σ_v. The uncertainty for Equation 18.20 is then obtained by straightforward substitution of u, σ_u, and v, σ_v in Equation 18.35.

In summary, the uncertainty in the reflectance factors of a spectrum is determined in theory by converting a pixel color value into the number of photons counted at that pixel. Under the assumption that counting photons follows a Poisson distribution, the uncertainty associated with a pixel value is determined. The obtained uncertainty is propagated to the uncertainty in the two polar angle representations of the spectrum.

18.4 Photometric Invariant Region Detection by Clustering

In Chapter 4, it was shown that uniformly colored objects of matte material draw half rays in *RGB* and multispectral color space due to changes in the surface orientation, illumination intensity, and shading. In Section 18.2, it was derived that the distance from a spectrum to such half rays is a photometric invariant. Furthermore, in Section 18.3, we derived for the *i*th spectrum $\vec{c}_i$, $i = 1, \cdots, n$ that the uncertainty $\vec{\sigma}_i$ can be obtained using Equation 18.34. In this section, the uncertainty of the polar angle representations will be incorporated into the image segmentation scheme.

18.4.1 Robust *K*-Means Clustering

Let a multispectral image consist of spectra $\vec{c}_i$, $i = 1, \ldots, n$, with corresponding uncertainties $\vec{\sigma}_i$. The well-known *K*-means clustering method [368] segments the image by minimizing the squared error criterion. A clustering is a partition $[\vec{v}_1, \ldots, \vec{v}_K]$ that assigns each spectrum to a single partition $\vec{v}_j$, $1 \leq j \leq K$. The spectra assigned to $\vec{v}_j$ form the *j*th cluster. We assume that the number K is given.

We compute the cluster center as the weighted average [42]. If M spectra $\vec{c}_i$ with corresponding uncertainties $\vec{\sigma}_i$, $i = 1, \ldots, M$, are assigned to a cluster, then the weighted average is computed as

$$\vec{v} = \frac{\sum_{i=1}^{M} \vec{w}_i \cdot \vec{c}_i}{\sum_{i=1}^{M} \vec{w}_i}, \tag{18.38}$$

where the weights are the inverse squares of the uncertainties

$$\vec{w}_i = \frac{1}{\vec{\sigma}_i \cdot \vec{\sigma}_i}. \tag{18.39}$$

Since the weight attached to each measurement involves the square of the corresponding uncertainty σ_i, any measurement that is much less precise than the others contributes very much less to the final answer (Eq. 18.38). With $\vec{c}_i$ the series of M spectra assigned to the j-th cluster, and with $\vec{v}_j$ the weighted average of the spectra, the squared error for the j-th cluster is

$$e_j^2 = \sum_{i=1}^{M} (\vec{c}_i - \vec{v}_j) \cdot (\vec{c}_i - \vec{v}_j), \tag{18.40}$$

and the squared error for the clustering is

$$E^2 = \sum_{i=1}^{K} e_i^2. \tag{18.41}$$

The objective of the K-means clustering method is to define, for given K, a clustering that minimizes E^2 by moving spectra from one cluster to another.

18.4.2 Photometric Invariant Segmentation

To obtain photometric invariant region detection, we cluster on K straight lines from the origin. Assume that the N-dimensional spectrum $\vec{c}$ is described by Equation 18.10 with associated uncertainty $\vec{\sigma}_c$ obtained from Equation 18.34. The spectrum is transformed to chromaticity polar angles $\vec{\theta}$ by Equation 18.12 with associated uncertainty $\vec{\sigma}_\theta$ obtained from Equation 18.35.

To cluster in polar angle space, the angular distance of Equation 18.14 replaces Equation 18.40 and the weighted angular average of Equation 18.20 replaces Equation 18.38. Given K clusters v, the spectrum is then assigned to the closest cluster. In the next step of the clustering algorithm, new partitions are obtained by moving spectra from one cluster to another. Kender [96] pointed out that color space transforms are unstable for sensor input values near the singularity. As is clear from Equation 18.35, the instability of the polar angle transformation is the drawback of the polar angle representation of the spectrum. The instability is dealt with by updating the cluster by the weighted sum as defined by Equation 18.20, where the weights w_j are derived by Equation 18.39. In other words, transformed polar angles with higher uncertainty contribute much less to the final estimate of the cluster than polar angles with small uncertainty. It was shown that the chromaticity polar angle representation is invariant to changes in the geometry of a uniformly colored object. Therefore, clustering in chromaticity polar angle space yields regions invariant to the geometry. In conclusion, using the uncertainties, we obtain segmentation results invariant to photometric effects and robust against noise.

To find homogeneously colored surfaces from glossy materials, we cluster in the hue polar angle representation.

For shiny surfaces, the spectrum is transformed to the hue polar angle θ_h by Equation 18.18 with associated uncertainty σ_{θ_h} obtained from Equation 18.35. Given K clusters, the distance from the cluster υ_j to the spectrum is derived by Equation 18.14. The spectrum is then assigned to the closest cluster υ_i. The instability of the polar angle transformation is again dealt with by updating the cluster using the weighted sum (Eq. 18.20). In other words, transformed polar angles with higher uncertainty contribute much less to the final estimate of the cluster than polar angles with small uncertainty. In conclusion, it was shown that the hue polar angle representation is invariant to changes in the geometry and specularities. Therefore, clustering in hue polar angle space yields regions invariant to the geometry and specularities. Using the weighted sum for the updating of cluster centroids achieves robustness against noise.

18.5 Experiments

All multispectral images are grabbed using a Jain CV-M300 monochrome CCD camera, Matrox Corona Frame-grabber, Navitar 7000 zoom lens, and Imspector V7 spectrograph under 500 Watt halogen illumination. The *RGB* images are grabbed using a Sony 3CCD color camera XC-003P and four Osram 18 Watt ''Lumilux deLuxe daylight'' fluorescent light sources.

To estimate the values of the electronic gain parameter γ_e (Eq. 18.2) and the value of the dark current variance of Equation 18.30 for the monochrome camera, 19 images are taken of a white reference while varying the lens aperture such that each image has a different intensity as shown in Figure 18.3. A line is fitted through the intensity-variance data yielding an electronic gain of $\gamma = 0.0069$, and dark current variance of $\sigma_d^2 = 0.87$.

The *RGB* camera has a white-balancing option. The goal is therefore to establish the overall value of the camera gain γ_i, where $i \in \{R, G, B\}$. To that end, 26 images are taken of a white reference while repeating the procedure to obtain different intensity images. The data are shown in Figure 18.4. Fitting three lines through a common origin yields a camera gain of $\gamma_R = 0.040$, of $\gamma_G = 0.014$, of $\gamma_B = 0.021$ and dark current variance $\sigma_d^2 = 2.7$.

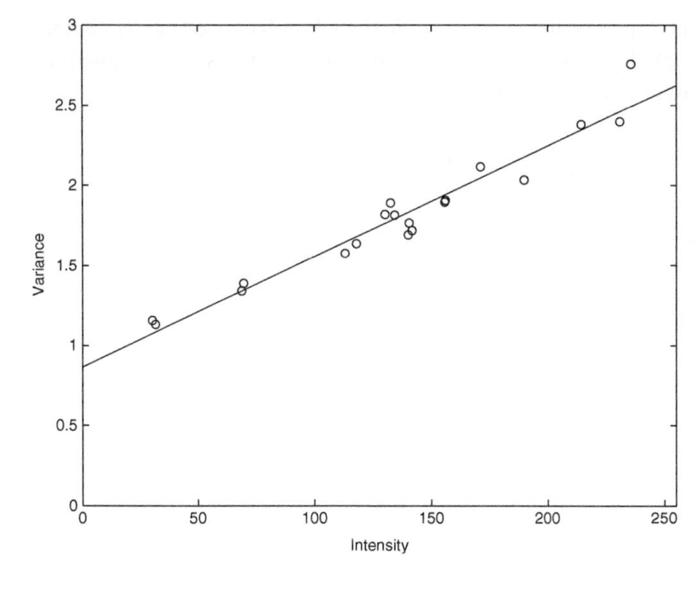

Figure 18.3 Visualization of the fitted line $\mathrm{var}(I) = \gamma I + \mathrm{var}(d)$ for the Jain monochrome camera.

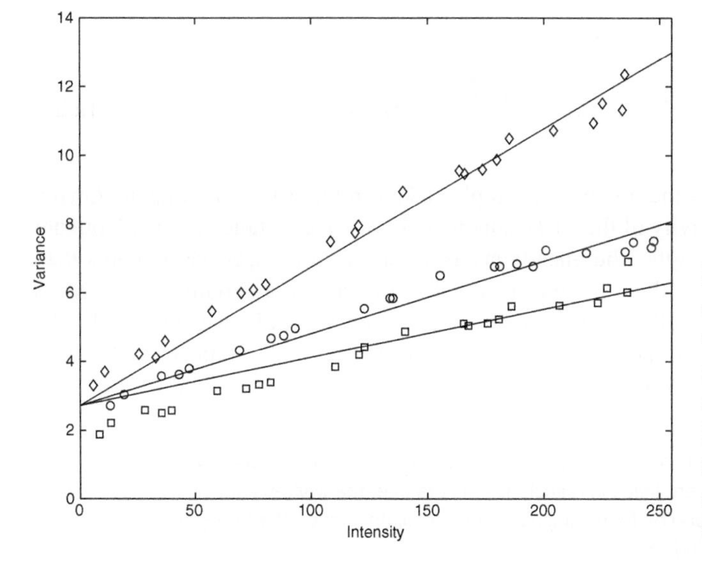

Figure 18.4 Visualization of the fitted lines $\mathrm{var}(I_i) = \gamma_i I_i + \mathrm{var}(d)$, where $i \in \{R, G, B\}$ for the Sony color camera. Diamonds correspond to the red color channel, squares to green, and circles to blue.

18.5.1 Propagation of Uncertainties in Transformed Spectra

Models were presented in (Eqs. 18.31, 18.33, and 18.34) to estimate uncertainties resulting from sensor noise in a spectrum for white-balanced camera systems. The goal of the experiment is to verify empirically the validity of the equations. Therefore, five multispectral images are taken from uniformly colored sheets of paper such that the entire spectral image exhibits one single color. The colors are red, yellow, green, cyan, and blue.

Using the gain parameters, the uncertainty in the white-balanced camera output $\hat{\sigma}_c(\lambda)$ can be estimated (Eq. 18.31). The estimated uncertainties are averaged for each wavelength over the spatial range as

$$\hat{\sigma}_c(\lambda) = \frac{1}{M} \sum_{i=1}^{M} \sigma_c(x_i, \lambda), \tag{18.42}$$

for M pixels along the one-dimensional spatial axis of the multispectral image. The real uncertainty is derived from the standard deviation of reflectance factors $c(\lambda)$ over the spatial range as

$$\sigma^2(\lambda) = \frac{1}{M-1} \sum_{i=1}^{M} \left(c(x_i, \lambda) - \overline{c}(\lambda)\right)^2, \tag{18.43}$$

where $\overline{c}(\lambda)$ denotes the average reflectance factor. The absolute difference $\delta(\hat{\sigma}(\lambda), \sigma(\lambda))$ between the real and estimated error is obtained as

$$\delta(\hat{\sigma}_c(\lambda), \sigma_c(\lambda)) = |\hat{\sigma}_c(\lambda) - \sigma_c(\lambda)|, \tag{18.44}$$

and then averaged over the wavelength range as

$$\delta(\hat{\sigma}, \sigma) = \frac{1}{N} \sum_{i=1}^{N} \delta(\hat{\sigma}_c(\lambda_i), \sigma_c(\lambda_i)), \tag{18.45}$$

where N denotes the number of samples taken in the wavelength range. Owing to the low sensitivity of the CCD camera and low transmittance of the illuminant at lower wavelengths, the uncertainty is greater at the lower wavelengths than at higher wavelengths. The reflectance of a spectrum at a certain wavelength is expressed as the reflectance factor $c(\lambda)$ taking on values between 0 and 1. The difference between the estimated and real uncertainty in the reflectance factor is given in Table 18.1 and is approximately 0.01, corresponding to 1%. Therefore,

Table 18.1 Results differentiated for the estimated and real uncertainties in reflectance factors after the white-balancing operation for multispectral images of uniformly colored paper as indicated.

Color	Multispectral $\delta(\hat{\sigma}(\lambda), \sigma(\lambda))$ **(Eq. 18.44), (Eq. 18.45)**
Red	0.011 ± 0.011
Yellow	0.011 ± 0.011
Green	0.009 ± 0.011
Cyan	0.008 ± 0.011
Blue	0.006 ± 0.011

the table shows a very reasonable correspondence between the measured and real uncertainty. This conclusion can be confirmed visually by examination of Figure 18.5.

The estimation of the uncertainty in the chromaticity and hue polar angles by Equation 18.35 is verified empirically in a similar way. The average of a series of M angular values θ_i, $i = 1, \ldots, M$, with equal weights w_i is computed using Equation 18.20 and is denoted $\overline{\theta}$. The standard deviation is computed as

$$\sigma_\theta = \frac{1}{N-1} \sum_{i=1}^{N} \left[\Delta(\overline{\theta}, \theta_i)\right]^2, \qquad 0 \leq \overline{\theta}, \theta_i < 2\pi \tag{18.46}$$

where Δ is defined by Equation 18.15. Similarly, the difference $\Delta(\hat{\sigma}_\theta(\lambda), \sigma_\theta(\lambda))$ between the real and estimated error between chromaticity angles at a certain wavelength is obtained using Equation 18.15. The results are averaged over the wavelength range as

$$\delta(\hat{\sigma}_\theta, \sigma_\theta) = \frac{1}{N} \sum_{i=1}^{N} \Delta(\hat{\sigma}_\theta(\lambda_i), \sigma_\theta(\lambda_i)). \tag{18.47}$$

The results of the experiment are given in Table 18.2. The dimension of chromaticity polar angles is the number of spectral samples minus one. The second column of the table specifies the results for the spectrograph. The results are averaged over 58 chromaticity angles; therefore, the standard deviation is given as well. The third column specifies the results for the RGB camera averaged over two chromaticity angles. The chromaticity angles are in the range of 0–90°, the difference between

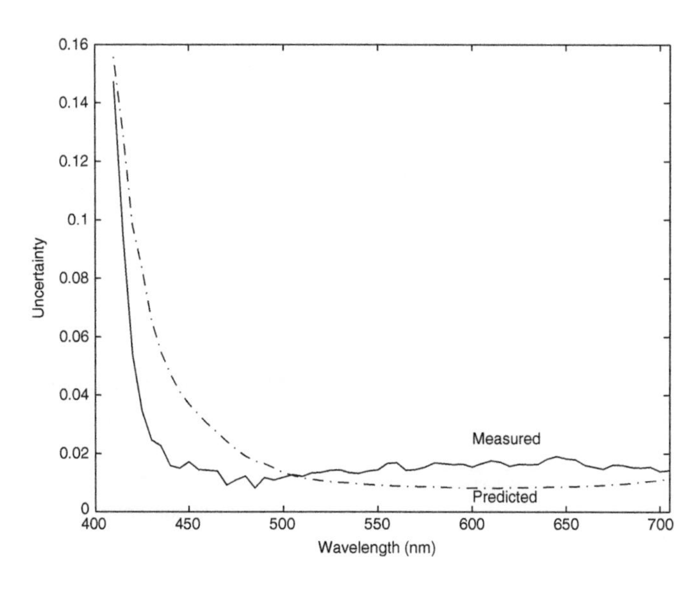

Figure 18.5 Yellow paper. Experiment: Comparison of the estimated uncertainty (dashed line) versus the real uncertainty (solid line) of the reflection factors for a yellow paper.

Table 18.2 Results differentiated for the estimated and measured uncertainties in chromaticity polar angles using equation 18.47.

Color	**Multispectral** $\delta(\hat{\sigma}_\theta, \sigma_\theta)$	***RGB*** $\delta(\hat{\sigma}_\theta, \sigma_\theta)$
Red	0.6 ± 0.8	1.26
Yellow	0.5 ± 0.7	0.01
Green	1.6 ± 2.4	0.36
Cyan	0.7 ± 0.7	0.11
Blue	0.9 ± 1.1	0.07

the estimated and real uncertainty is less than 1%. Consequently, there is a very reasonable correspondence between the measured and real uncertainty. A more detailed example is given in Figure 18.6 for the results for the yellow color.

Similarly, for the hue polar angle, the results are given in Table 18.3. The hue angles are in the range of 0–360°, the difference between the estimated and real uncertainty is less than 1%. Consequently, there is a very reasonable correspondence between the measured and real uncertainty.

18.5.2 Photometric Invariant Clustering

18.5.2.1 Multispectral Images Figure 18.7a shows a multispectral image of a textile sample. The spectral information is on the vertical axis. The top of the picture corresponds to 410 nm, the bottom to 700 nm. The left-hand side of the

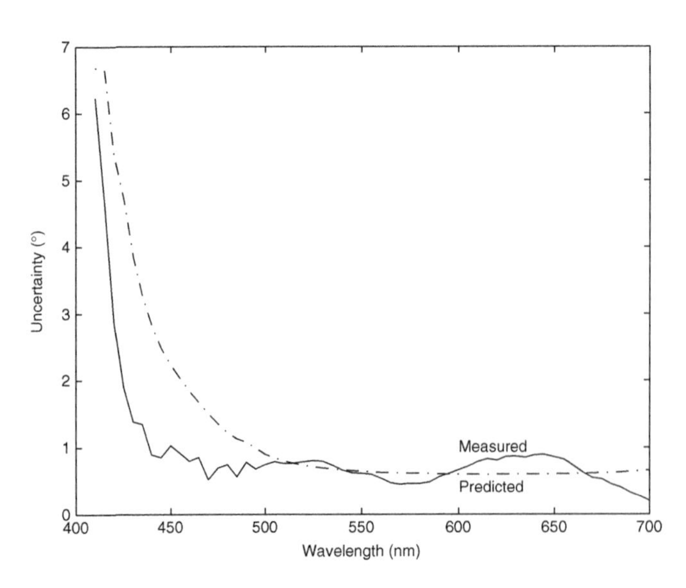

Figure 18.6 Yellow paper, uncertainty in chromaticity angle. The absolute difference averaged over the wavelength range is $0.5 \pm 0.7°$.

(a) (b)

Figure 18.7 (a) A multispectral image of a textile sample. The spatial information is on the horizontal axis and the spectral information is on the vertical axis. The top corresponds to 410 nm wavelength, the bottom to 700 nm wavelength. The left-hand side of the image is colored homogeneously red and the right-hand side is colored green. The structure of the textile is visible through the intensity fluctuations occurring in the further homogeneous spectra. (b) The spectra of two plastic objects. The left-hand side object is colored orange and the right-hand side object is green. The objects are smooth and structureless, but reflect specularities showing up as the vertical bright streaks in the spectral image. Furthermore, the intensity of the spectra gradually reduces toward the right-hand side of the image because of a change in the surface orientation of the objects.

Table 18.3 Results differentiated for the estimated and measured uncertainties in hue polar angles using equation 18.15.

Color	**Multispectral** $\delta(\hat{\sigma}_\theta, \sigma_\theta)$	***RGB*** $\delta(\hat{\sigma}_\theta, \sigma_\theta)$
Red	0.7	1.1
Yellow	0.4	0.5
Green	2.1	1.8
Cyan	0.5	0.7
Blue	1.7	0.5

image is from homogeneously red colored textile, the right-hand side is colored green. The structure of the textile is visible in intensity fluctuations occurring in otherwise homogeneous spectra. The result of clustering in the chromaticity polar angle space is shown in Figure 18.8. The figure shows how the spectra form half rays because of the geometry changes of the structure of the textile. Fitting of half rays through the chromaticity angle representation results in invariance for shadows and surface orientation changes.

Figure 18.7b shows the spectra of two plastic objects. The left-hand side object is colored orange, the right-hand side object is green. The objects are smooth and structureless, but reflect specularities showing up as the vertical bright streaks in

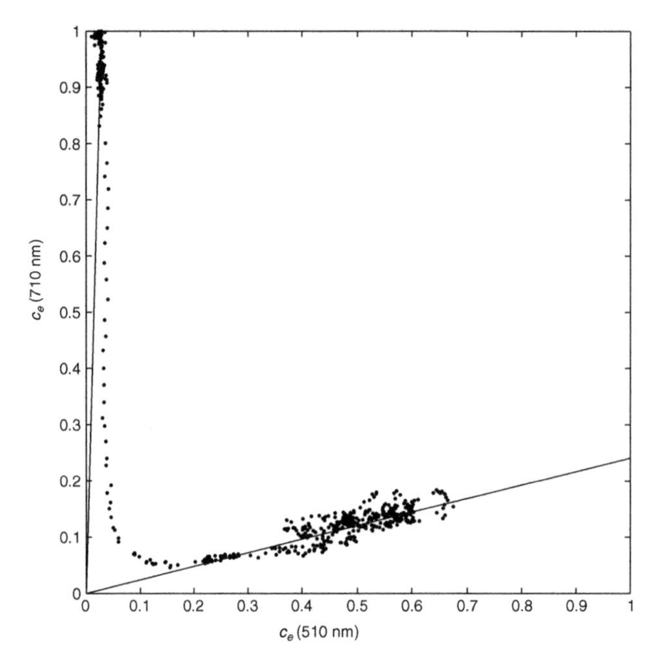

Figure 18.8 Result of clustering in the chromaticity polar angle space for the multispectral image shown in Figure 18.7a. The result is shown for the angle between the 510 and 710 nm wavelength. The spectra form half rays because of the geometry changes of the structure of the textile. Fitting of half rays through the chromaticity angle representation results in invariance for shadows and surface orientation changes.

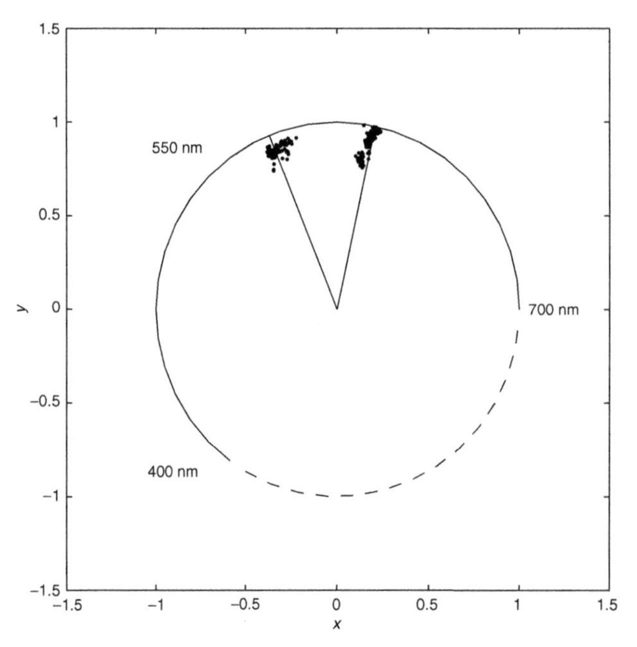

Figure 18.9 Result of clustering in hue polar angle space for the image shown in Figure 18.7b. Clustering in hue polar angle representation results in independence to the highlights and surface orientation changes.

the spectral image. Furthermore, the intensity of the spectra gradually reduces toward the right-hand side of the image because of a change in the surface orientation of the objects.

The result of clustering in hue polar angle space is shown in Figure 18.9. Clustering in hue polar angle representation results in independence to the highlights and surface orientation changes..

18.5.2.2* RGB *images Figure 18.10 shows an *RGB* image of several toys against a background consisting of four squares. The upper left quadrant of the

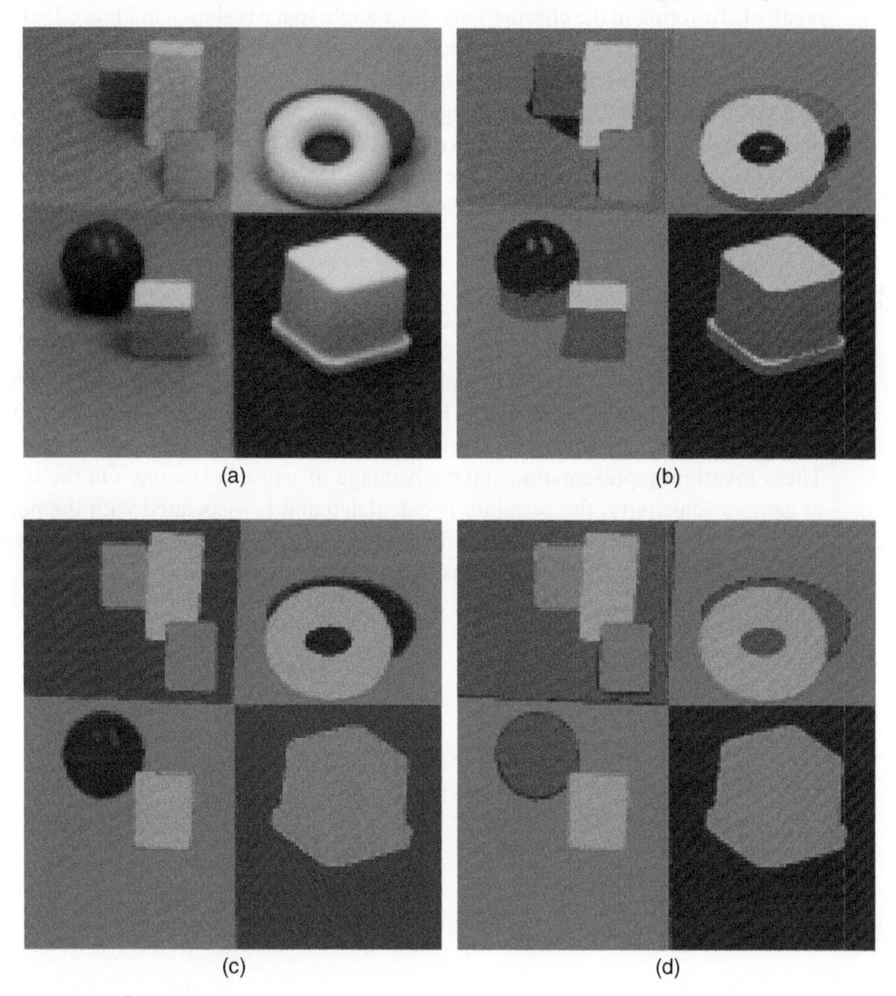

Figure 18.10 Segmentation results for the *K*-means clustering method. (a) *RGB* image. (b) Cluster model is a point, region detection is sensitive to intensity changes, shadows, geometry, highlights, and color transitions. (c) Cluster model is a half ray, region detection is sensitive to highlights and color transitions. (d) Cluster model is a triangularly shaped plane, region detection is sensitive only to color transitions.

image consists of three uniformly painted matte cubes of wood. The upper right quadrant contains two specular plastic donuts on top of each other. In the bottom left quadrant, a red highlighted ball and a matte cube are shown, while the last quadrant contains two matte cubes. Each individual object is painted uniformly with a distinct color. The image is contaminated by noise, shadows, shading, and specularities.

In Figure 18.10b, the segmentation result is shown obtained by the K-means clustering method in RGB data. False regions are detected because of abrupt surface orientations, shadows, inter-reflections, and highlights. In contrast, the result of clustering in the chromaticity polar angle space is shown in Figure 18.10c. Regions are detected insensitive to shadows and surface orientation changes but are affected by highlights. The result of clustering in the hue polar angle space is shown in Figure 18.10d. Here, computed region edges correspond to material boundaries discounting the disturbing influences of surface orientation, illumination, shadows, and highlights. The difference between Figure 18.10c,18.10d is the invariance of the latter to the specularities reflected at the red ball.

18.6 Summary

We have discussed the detection of photometric invariant regions in multispectral images robust against sensor noise. Therefore, different polar angle representations of a spectrum are examined for invariance using the dichromatic reflection model. These invariant representations take advantage of white balancing. On the basis of camera sensitivity, the certainty is calculated and is associated with the polar angular representations under the influence of noise. The expression is employed by the segmentation technique to ensure robustness against sensor noise.

Citation Guidelines

Throughout our research career, we have collaborated with many people, without whose help the writing of this book would have not been possible. Here, we provide the citation information for those chapters that are written with additional authors.

Chapter 2

Marcel P. Lucassen. (2012) Color Vision. In Theo Gevers, Arjan Gijsenij, Joost van de Weijer, and Jan-Mark Geusebroek, Color in Computer Vision, John Wiley & Sons, Inc., Hoboken, NJ, USA.

Chapter 5

Joost van de Weijer, Theo Gevers, and Cordelia Schmid. (2012) Color Ratios. In Theo Gevers, Arjan Gijsenij, Joost van de Weijer, and Jan-Mark Geusebroek, Color in Computer Vision, John Wiley & Sons, Inc., Hoboken, NJ, USA.

Chapter 6

Jan-Mark Geusebroek, Joost van de Weijer, Rein van den Boomgaard, Theo Gevers, and Arnold W. M. Smeulders. (2012) Derivative-based Photometric Invariance. In Theo Gevers, Arjan Gijsenij, Joost van de Weijer, and Jan-Mark Geusebroek, Color in Computer Vision, John Wiley & Sons, Inc., Hoboken, NJ, USA.

Chapter 7

José M. Àlvarez, Theo Gevers, and Antonio M. López. (2012) Invariance by Learning. In Theo Gevers, Arjan Gijsenij, Joost van de Weijer, and Jan-Mark Geusebroek, Color in Computer Vision, John Wiley & Sons, Inc., Hoboken, NJ, USA.

Color in Computer Vision: Fundamentals and Applications, First Edition.
Theo Gevers, Arjan Gijsenij, Joost van de Weijer, and Jan-Mark Geusebroek.

Chapter 13

Joost van de Weijer, Theo Gevers, Arnold W. M. Smeulders, and Andrew D. Bagdavnov (2012) Color Feature Detection. In Theo Gevers, Arjan Gijsenij, Joost van de Weijer, and Jan-Mark Geusebroek, Color in Computer Vision, John Wiley & Sons, Inc., Hoboken, NJ, USA.

Chapter 14

Gertjan J. Burghouts and Jan-Mark Geusebroek. (2012) Color Descriptors. In Theo Gevers, Arjan Gijsenij, Joost van de Weijer, and Jan-Mark Geusebroek, Color in Computer Vision, John Wiley & Sons, Inc., Hoboken, NJ, USA.

Chapter 15

Jan-Mark Geusebroek, Theo Gevers, and Gertjan J. Burghouts. (2012) Color Image Segmentation. In Theo Gevers, Arjan Gijsenij, Joost van de Weijer, and Jan-Mark Geusebroek, Color in Computer Vision, John Wiley & Sons, Inc., Hoboken, NJ, USA.

Chapter 16

Koen E. A. van de Sande, Theo Gevers, and Cees G. M. Snoek. (2012) Retrieval and Recognition of Digital Content. In Theo Gevers, Arjan Gijsenij, Joost van de Weijer, and Jan-Mark Geusebroek, Color in Computer Vision, John Wiley & Sons, Inc., Hoboken, NJ, USA.

Chapter 17

Robert Benavente, Joost van de Weijer, Maria Vanrell, Cordelia Schmid, Ramon Baldrich, Jakob Verbeek, and Diane Larlus. (2012) Color Names. In Theo Gevers, Arjan Gijsenij, Joost van de Weijer, and Jan-Mark Geusebroek, Color in Computer Vision, John Wiley & Sons, Inc., Hoboken, NJ, USA.

Chapter 18

Theo Gevers and Harro M. G. Stokman (2012) Segmentation of Multispectral Images. In Theo Gevers, Arjan Gijsenij, Joost van de Weijer, and Jan-Mark Geusebroek, Color in Computer Vision, John Wiley & Sons, Inc., Hoboken, NJ, USA.

References

1. L. T. Sharpe, A. Stockman, H. Jägle, and J. Nathans. Opsin genes, cone photopigments, color vision and colorblindness. In K. Gegenfurtner and L. T. Sharpe, editors, *Color vision: From Genes to Perception*, pages 3–50. Cambridge University Press, Cambridge, 1999.
2. A. Stockman and L. T. Sharpe. Cone spectral sensitivities and color matching. In K. Gegenfurtner and L. T. Sharpe, editors, *Color vision: From Genes to Perception*, pages 53–87. Cambridge University Press, Cambridge, 1999.
3. R. Marois and J. Ivanoff. Capacity limits of information processing in the brain. *Trends in Cognitive Sciences*, 9(6): 296–305, 2005.
4. T. N. Wiesel and D. H. Hubel. Spatial and chromatic interactions in the lateral geniculate body of the rhesus monkey. *Journal of Neurophysiology*, 29(6): 1115–1156, 1966.
5. D. M. Dacey and B. B. Lee. The 'blue-on' opponent pathway in primate retina originates from a distinct bistratified ganglion cell type. *Nature*, 367(6465): 1115–1156, 1994.
6. A. R. Hill. How we see colour. In R. McDonald, editor, *Colour physics for industry*, pages 211–281. H. Charlesworth & Co Ltd, Huddersfield, 1987.
7. R. Shapley and M. J. Hawken. Color in the cortex: single- and double-opponent cells. *Vision Research*, 51(7): 701–717, 2011.
8. B. R. Conway. *Neural Mechanisms of Color Vision*. Kluwer Academic Publishers, Boston (MA), 2002.
9. D. Jameson, L. M. Hurvich, and F. D. Varner. Receptoral and postreceptoral visual processes in recovery from chromatic adaptation. *Proceedings of the National Academy of Sciences of the United States of America*, 76(6): 3034–3038, 1979.
10. O. Rinner and K. R. Gegenfurtner. Time course of chromatic adaptation for color appearance and discrmination. *Vision Research*, 40(14): 1813–1826, 2000.

Color in Computer Vision: Fundamentals and Applications, First Edition.
Theo Gevers, Arjan Gijsenij, Joost van de Weijer, and Jan-Mark Geusebroek.

11. H. E. Smithson. Review. sensory, computational and cognitive components of human color constancy. *Philosophical Transactions of the Royal Society*, 360(1458): 1329–1346, 2005.
12. D. H. Foster. Color constancy. *Vision Research*, 51(7): 674–700, 2011.
13. E. H. Land and J. J. McCann. Lightness and retinex theory. *Journal of the Optical Society of America A*, 61: 1–11, 1971.
14. E. H. Land. The retinex theory of color vision. *Scientific American*, 237(6): 108–128, 1977.
15. A. Hurlbert. Formal connections between lightness algorithms. *Journal of the Optical Society of America A*, 3(10): 1684–1693, 1986.
16. J. von Kries. Die gesichtsempfindungen. In W. Nagel, editor, *Handbuch der Physiologie des Menschen, Physiologie der Sinne*, Volume 3, Vieweg und Sohn, Braunschweig, 1905.
17. H. Helson. Fundamental problems in color vision. i. the principle governing changes in hue saturation and lightness of non-selective samples in chromatic illumination. *Journal of Experimental Psychology*, 23(5): 439–476, 1938.
18. S. K. Shevell and F. A. A. Kingdom. Color in complex scenes. *Annual Review of Psychology*, 59: 143–166, 2008.
19. K. T. Mullen. The contrast sensitivity of human colour vision to red-green and blue-yellow chromatic gratings. *Neurotoxicology and Teratology*, 359(1): 381–400, 1985.
20. D. L. MacAdam. Sensitivities to color differences in daylight. *Journal of the Optical Society of America A*, 32(5): 247–273, 1942.
21. R. G. Kuehni. *Color Space and Its Divisions: Color Order from Antiquity to the present*. Wiley, New York, 2003.
22. J. S. Werner, D. H. Peterzell, and A. J. Scheetz. Light, vision and aging. *Optometry and Vision Science*, 67(3): 214–229, 1990.
23. D. Mergler, L. Blain, J. Lemaire, and F. Lalande. Colour vision impairment and alcohol consumption. *Neurotoxicology and Teratology*, 10(3): 255–260, 1988.
24. H. D. Abraham. A chronic impairment of colour vision in users of lsd. *Neurotoxicology and Teratology*, 140(5): 518–520, 1982.
25. G. Wyszecki and W. S. Stiles. *Color Science: Concepts and Methods, Quantitative Data and Formulae*. John Wiley & Sons, New York, 1982.
26. S. A. Shafer. Using color to separate reflection components. *Color Research and Application*, 10(4): 210–218, 1985.
27. B. A. Maxwell, R. M. Friedhoff, and C. A. Smith. A bi-illuminant dichromatic reflection model for understanding images. In *IEEE Computer Society Conference on Computer Vision and Pattern Recognition*, pages 1–8, June 2008.
28. G. J. Klinker and S. A. Shafer. A physical approach to color image understanding. *International Journal of Computer Vision*, 4(1): 7–38, 1990.
29. B. A. Maxwell and S. A. Shafer. Physics-based segmentation of complex objects using multiple hypothesis of image formation. *Computer Vision and Image Understanding*, 65(2): 265–295, 1997.
30. D. B. Judd and G. Wyszecki. *Color in Business, Science, and Industry*. John Wiley & sons, New York, 1975.

31. P. Kubelka and F. Munk. Ein beitrag zur optik der farbanstriche. *Zeitung fur Technische Physik*, 12: 593, 1999.
32. P. Kubelka. New contribution to the optics of intensely light- scattering materials, part i. *Journal of the Optical Society of America A*, 38(5): 448–457, 1948.
33. J. von Kries. Influence of adaptation on the effects produced by luminous stimuli. In D. L. MacAdam, editor, *Sources of Color Vision*, pages 109–119. MIT Press, Cambridge (MA), 1970.
34. G. D. Finlayson, S. D. Hordley, and R. Xu. Convex programming colour constancy with a diagonal-offset model. In *IEEE International Conference on Image Processing*, pages 948–951, 2005.
35. H. Y. Chong, S. J. Gortler, and T. Zickler. The von kries hypothesis and a basis for color constancy. In *IEEE International Conference on Computer Vision*, pages 1–8, 2007.
36. G. D. Finlayson, M. S. Drew, and B. V. Funt. Spectral sharpening: sensor transformations for improved color constancy. *Journal of the Optical Society of America A*, 11(5): 1553–1563, 1994.
37. B. A. Wandell. *Foundations of Vision*. Sinauer Associates, Inc., Sunderland (MA), 1995.
38. Th. Gevers and A. W. M. Smeulders. Pictoseek: combining color and shape invariant features for image retrieval. *IEEE Transactions on Image Processing*, 9(1): 102–119, 2000.
39. P. D. Burns and R. S. Berns. Error propagation analysis in color measurement and imaging. *Color Research and Application*, 22(4): 280–289, 1997.
40. L. Shafarenko, M. Petrou, and J. Kittler. Histogram-based segmentation in a perceptually uniform color space. *IEEE Transactions on Image Processing*, 7(9): 1354–1358, 1998.
41. Th. Gevers and H. M. G. Stokman. Robust histogram construction from color invariants for object recognition. *IEEE Transactions on Pattern Analysis and Machine Intelligence*, 26(1): 113–118, 2004.
42. J. R. Taylor. *An Introduction to Error Analysis*. University Science Books, Mill Valley (CA), 1982.
43. M. Swain and D. Ballard. Color indexing. *International Journal of Computer Vision*, 7(1): 11–32, 1991.
44. K. Barnard, L. Martin, B. Funt, and A. Coath. A data set for color research. *Color Research and Application*, 27(3): 147–151, 2002.
45. S. K. Nayar and R. M. Bolle. Reflectance based object recognition. *International Journal of Computer Vision*, 17(3): 219–240, 1996.
46. B. V. Funt and G. D. Finlayson. Color constant color indexing. *IEEE Transactions on Pattern Analysis and Machine Intelligence*, 17(5): 522–529, 1995.
47. Th. Gevers and A. W. M. Smeulders. Color-based object recognition. *Pattern Recognition*, 32: 453–464, 1999.
48. J. van de Weijer and C. Schmid. Blur robust and color constancy image description. In *IEEE International Conference on Image Processing*, pages 993–996, 2006.
49. J. van de Weijer and C. Schmid. Coloring local feature extraction. In *European Conference on Computer Vision*, pages 334–348, 2006.

50. G. D. Finlayson, M. S. Drew, and B. Funt. Color constancy: generalized diagonal transforms suffice. *Journal of the Optical Society of America A*, 11(11): 3011–3019, 1994.

51. R. L. Lagendijk and J. Biemond. Basic methods for image restoration and identification. In A. Bovik, editor, *The Image and Video Processing Handbook*, pages 125–139. Academic Press, Burlington (MA), 1999.

52. J. Canny. A computational approach to edge detection. *IEEE Transactions on Pattern Analysis and Machine Intelligence*, 8(6): 679–698, 1986.

53. C. Harris and M. Stephans. A combined corner and edge detector. In *Proceedings of the Alvey Vision Conference*, pages 189–192, 1988.

54. T. Lindeberg. Feature detection with automatic scale selection. *International Journal of Computer Vision*, 30(2): 117–154, 1998.

55. D. G. Lowe. Distinctive image features from scale-invariant keypoints. *International Journal of Computer Vision*, 60(2): 91–110, 2004.

56. S. Belongie, J. Malik, and J. Puzicha. Shape matching and object recognition using shape contexts. *IEEE Transactions on Pattern Analysis and Machine Intelligence*, 24(4): 509–522, 2002.

57. K. Mikolajczyk and C. Schmid. A performance evaluation of local descriptors. *IEEE Transactions on Pattern Analysis and Machine Intelligence*, 27(10): 1615–1630, 2005.

58. J. van de Weijer, Th. Gevers, and J. M. Geusebroek. Edge and corner detection by photometric quasi-invariants. *IEEE Transactions on Pattern Analysis and Machine Intelligence*, 27(4): 625–630, 2005.

59. J. J. Koenderink. *Color for the Sciences*. The MIT Press, Cambridge (MA), 2010.

60. J. J. Koenderink. The structure of images. *Biological Cybernetics*, 50(5): 363–3710, 1984.

61. J. M. Geusebroek, R. van den Boomgaard, A. W. M. Smeulders, and A. Dev. Color and scale: the spatial structure of color images. In *European Conference on Computer Vision*, pages 331–341, 2000.

62. L. M. J. Florack, B. ter Haar Romeny, J. J. Koenderink, and M. A. Viergever. Scale and the differential structure of images. *Image and Vision Computing*, 10(6): 1992, 376–388.

63. E. Hering. *Outlines of a Theory of the Light Sense*. Harvard University Press, Cambridge (MA), 1964.

64. Basic Parameter Values for the (HDTV) Standard for the Studio and for International Programme Exchange. Technical Report ITU-R Rec. BT. 709, International Telecommunications Union, Switzerland, 1990.

65. Th. Gevers and H. M. G. Stokman. Reflectance based edge classification. In *Proceedings of Vision Interface*, pages 25–32, 1999.

66. L. M. J. Florack, B. ter Haar Romeny, J. J. Koenderink, and M. A. Viergever. Cartesian differential invariants in scale-space. Journal *of Mathematical Imaging and Vision*, 3(4): 1993, 327–348.

67. J. M. Geusebroek, R. van den Boomgaard, A. W. M. Smeulders, and H. Geerts. Color invariance. *IEEE Transactions on Pattern Analysis and Machine Intelligence*, 23(12): 1338–1350, 2001.

68. T. Zickler, S. P. Mallick, D. J. Kriegman, and P. N. Belhumeur. Color subspaces as photometric invariants. *International Journal of Computer Vision*, 79(1): 13–30, 2008.

69. L. T. Maloney and B. A. Wandell. Color constancy: a method for recovering surface spectral reflectance. *Journal of the Optical Society of America A*, 3(1): 29–33, 1986.

70. The PANTONE Color Formula Guide, editor. 1992–1993, Group Basf, Paris, France, Pantone is a trademark of Patone Inc.

71. T. Gevers, J. M. Alvarez, and A. M. Lopez. Learning photometric invariance for object detection. *International Journal of Computer Vision*, 90(1): 45–61, 2010.

72. G. Brown, J. Wyatt, R. Harris, and X. Yao. Diversity creation methods: a survey and categorisation. *Journal of Information Fusion* , 6(1): 5–20, 2005.

73. J. V. Kittler, M. Hatef, R. P. W. Duin, and J. Matas. On combining classifiers. *IEEE Transactions on Pattern Analysis and Machine Intelligence (PAMI)*, 20(3): 226–239, 1998.

74. L. I. Kuncheva. *Combining Pattern Classifiers: Methods and Algorithms*. Wiley-Interscience, New York, 2004.

75. P. Melville and R. Mooney. Creating diversity in ensembles using artificial data. *Information Fusion*, 6(3): 1553–1563, 2005.

76. R. A. Jacobs. Methods for combining experts' probability assessments. *Neural Computation*, 7(5): 867–888, 1995.

77. H. M. G. Stokman and T. Gevers. Selection and fusion of color models for image feature detection. *IEEE Transactions on Pattern Analysis and Machine Intelligence (PAMI)*, 29(3): 371–381, 2007.

78. H. M. Markowitz. *Portfolio Selection: Efficient Diversification of Investments*. Wiley, New York, 1959.

79. B. Scherer. *Portfolio Construction and risk Budgeting*. Rosk Books, London, 2002, Chapter 4.

80. D. M. J. Tax and R. P. W. Duin. Uniform object generation for optimizing one-class classifiers. *Journal of Machine Learning Research*, 2: 155–173, 2002.

81. S. Boyd and L. Vandenberghe. *Convex Optimization*. Cambridge University Press, Cambridge, MA, 2004.

82. R. Michaud. Estimation error and portfolio optimization: a resampling solution. *Journal of Investment Management*, 6(1): 8–28, 2008.

83. R. O. Michaud. *Efficient Asset Management: A Practical Guide to Stock Portfolio Optimization and Asset Allocation*. Oxford University Press, USA, 1998.

84. N. Usmen and H. Markowitz. Resampled frontiers versus diffuse bayes: an experiment. *Journal of Investment Management*, 1(4): 1–17, 2003.

85. K. Dowd. *Beyond Value at Risk: The New Science of Risk Management*. Wiley, New York, 1998.

86. P. Best. *Implementing Value at Risk*. John Wiley & sons, Chichester, UK, 1998.

87. Y. K. Tse. Stock returns volatility in the tokyo stock exchange. *Japan and the World Economy*, 3(3): 285–298, 1991.

88. T. W. Ridler and S. Calvard. Picture thresholding using an iterative selection method. *IEEE Transactions on Systems, Man, and Cybernetics*, 8(8): 630–632, 1978.

89. G. D. Finlayson, S. D. Hordley, C. Lu, and M. S. Drew. On the removal of shadows from images. *IEEE Transactions on Pattern Analysis and Machine Intelligence (PAMI)*, 28(1): 59–68, 2006.

90. G. D. Finlayson, M. S. Drew, and C. Lu. Intrinsic images by entropy minimization. In *Proceedings of the European Conference on Computer Vision (ECCV) (3)*, pages 582–595, 2004.

91. J. M. Álvarez, A. M. L ópez, and R. Baldrich. Illuminant-invarariant model-based road segmentation. In *Proceedings of the 2008 IEEE Intelligent Vehicles Symposium (IV'08)*, 2008.

92. K. E. A. van de Sande, T. Gevers, and C. G. M. Snoek. Evaluation of color descriptors for object and scene recognition. In *Proceedings IEEE Conference on Computer Vision and Pattern Recognition (CVPR)*, pages 453–464, 2008.

93. I. T. Jolliffe. *Principal Component Analysis, Springer Series in Statistics, 2nd Ed.* Springer, New York, 2002.

94. J. A. Hartigan and P. M. Hartigan. The dip test of unimodality. *Annals of Statistics*, 13(1): 70–84, 1985.

95. The Caltech Frontal Face Dataset Computational Vision: Archive, by M. Weber. California Institute of Technology, USA, 2001. Available online at http://www.vision.caltech.edu/html-files/archive.html. Accessed 2008.

96. J. R. Kender. Saturation, hue and normalized color: calculation, digitation effects, and use. Technical Report CMU-RI-TR-05-40, Robotics Institute, Carnegie Mellon University, Pittsburgh, PA, September 2005.

97. F. Wilcoxon. Individual comparisons by ranking methods. *Biometrics Bulletin*, 1(6): 80–83, 1945.

98. M. M. Fleck, D. A. Forsyth, and C. Bregler. Finding naked people. In *Proceedings of the European Conferece on Computer Vision (ECCV) (2)*, Volume 1065, pages 593–602. Springer, 1996.

99. D. Chai and K. N. Ngan. Face segmentation using skin-color map in videophone applications. *IEEE Transactions on Circuits and Systems for Video Technology*, 9(4): 551–564, 1999.

100. K. Sobottka and I. Pitas. A novel method for automatic face segmentation, facial feature extraction and tracking. *Signal Processing-Image Communication*, 12(3): 263–281, 1998.

101. J. Kovac, P. Peer, and F. Solina. Human skin color clustering for face detection. In *International Conference on Computer as a Tool (EUROCON)*, 2003.

102. M. J. Jones and J. M. Rehg. Statistical color models with application to skin detection. *International Journal of Computer Vision (IJCV)*, 46(1): 81–96, 2002.

103. M. A. Sotelo, F. Rodriguez, L. M. Magdalena, L. Bergasa, and L Boquete. A color vision-based lane tracking system for autonomous driving in unmarked roads. *Autonomous Robots*, 16(1), 2004.

104. C. Rotaru, T. Graf, and J. Zhang. Color image segmentation in hsi space for automotive applications. *Journal of Real-Time Image Processing*, 3(4): 1164–1173, 2008.

105. N. Ikonomakis, K. N. Plataniotis, and A. N. Venetsanopoulos. Color image segmentation for multimedia applications. *Journal of Intelligent & Robotic Systems*, 28(1–2): 5–20, 2000.

106. L. Sigal, S. Sclaroff, and V. Athitsos. Skin color-based video segmentation under time-varying illumination. *IEEE Trans. on Pattern Analysis and Machine Intelligence (PAMI)*, 26(7): 862–877, 2004.

107. C. Tan, T. Hong, T. Chang, and M. Shneier. Color model-based real-time learning for road following. In *Proceedings of IEEE Intelligent Transportation Systems*, pages 939–944, 2006.

108. L. E. Arend, A. Reeves, J. Schirillo, and R. Goldstein. Simultaneous color constancy: papers with diverse munsell values. *Journal of the Optical Society of America A*, 8(4): 661–672, 1991.

109. D. H. Brainard, J. M. Kraft, and P. Longere. Color constancy: developing empirical tests of computational models. In R. Mausfeld and D. Heyer, editors, *Colour Perception: From Light To Object*, pages 307–334. Oxford University Press, Oxford, UK, 2003.

110. P. B. Delahunt and D. H. Brainard. Does human color constancy incorporate the statistical regularity of natural daylight? *Journal of Vision*, 4(2): 57–81, 2004.

111. D. H. Foster, K. Amano, and S. M. C. Nascimento. Color constancy in natural scenes explained by global image statistics. *Visual Neuroscience*, 23(3–4): 341–349, 2006.

112. E. H. Land. Recent advances in retinex theory. *Vision Research*, 26: 7–21, 1986.

113. D. J. Jobson, Z. Rahman, and G. A. Woodell. Properties and performance of a center/surround retinex. *IEEE Transactions on Image Processing*, 6(93): 451–462, 1997.

114. D. J. Jobson, Z. Rahman, and G. A. Woodell. A multiscale retinex for bridging the gap between color images and the human observation of scenes. *IEEE Transactions on Image Processing*, 6(7): 965–976, 1997.

115. E. Provenzi, C. Gatta, M. Fierro, and A. Rizzi. A spatially variant white-patch and gray-world method for color image enhancement driven by local contrast. *IEEE Transactions on Pattern Analysis and Machine Intelligence*, 30(10): 1757–1770, 2008.

116. J. Kraft and D. Brainard. Mechanisms of color constancy under nearly natural viewing. *Proceedings of the National Academy of Sciences of the United States of America*, 96(1): 307–312, 1999.

117. J. Golz and D. I. A. MacLeod. Influence of scene statistics on colour constancy. *Nature*, 415: 637–640, 2002.

118. J. Golz. The role of chromatic scene statistics in color constancy: Spatial integration. *Journal of Vision*, 8(13): 1–16, 2008.

119. F. Ciurea and B. V. Funt. Failure of luminance-redness correlation for illuminant estimation. In *IS&T/SID's Color Imaging Conference*, pages 42–46. IS&T-The Society for Imaging Science and Technology, 2004.

120. A. Gijsenij and Th. Gevers. Color constancy using natural image statistics. In *IEEE Computer Society Conference on Computer Vision and Pattern Recognition*, pages 1–8, 2007.

121. A. Gijsenij and Th. Gevers. Color constancy using natural image statistics and scene semantics. *IEEE Transactions on Pattern Analysis and Machine Intelligence*, 33(4): 687–698, 2011.

122. P. Bradley. Constancy, categories and bayes: a new approach to representational theories of color constancy. *Philosophical Psychology*, 21(5): 601–627, 2008.

123. V. M. N. de Almeida and S. M. C. Nascimento. Perception of illuminant colour changes across real scene. *Perception*, 38(8): 1109–1117, 2009.

124. T. W. Lin and C. W. Sun. Representation or context as a cognitive strategy in colour constancy? *Perception*, 37(9): 1353–1367, 2008.

125. M. Hedrich, M. Bloj, and A. I. Ruppertsberg. Color constancy improves for real 3d objects. *Journal of Vision*, 9(4): 2009, 1–16.

126. D. H. Foster, S. M. C. Nascimento, and K. Amano. Information limits on neural identification of colored surfaces in natural scenes. *Visual Neuroscience*, 21: 331–336, 2004.

127. G. D. Finlayson, B. V. Funt, and K. Barnard. Color constancy under varying illumination. In *IEEE International Conference on Computer Vision*, pages 720–725, IEEE Computer Society, Washington, DC, USA, 1995.

128. K. Barnard, G. D. Finlayson, and B. V. Funt. Color constancy for scenes with varying illumination. *Computer Vision and Image Understanding*, 65(2): 311–321, 1997.

129. W. Xiong and B. Funt. Stereo retinex. *Image and Vision Computing*, 27(1–2): 178–188, 2009.

130. M. Ebner. Color constancy based on local space average color. *Machine Vision and Applications*, 20(5): 283–301, 2009.

131. E. Hsu, T. Mertens, S. Paris, S. Avidan, and F. Durand. Light mixture estimation for spatially varying white balance. In *ACM SIGGRAPH*, pages 1–7, 2008.

132. M. D. Fairchild. *Color Appearance Models, Wiley-IS&T Series in Imaging Science and Technology, 2nd Ed.* John Wiley & sons, Chichester, UK, 2005.

133. G. West and M. H. Brill. Necessary and sufficient conditions for von kries chromatic adaptation to give color constancy. *Journal of Mathematical Biology*, 15(2): 249–258, 1982.

134. B. V. Funt and B. C. Lewis. Diagonal versus affine transformations for color correction. *Journal of the Optical Society of America A*, 17(11): 2108–2112, 2000.

135. K. M. Lam. Metamerism and colour constancy. PhD thesis, University of Bradford, 1985.

136. C. Li, M. R. Luo, B. Rigg, and R. W. G. Hunt. Cmc 2000 chromatic adaptation transform: Cmccat2000. *Color Research and Application*, 26: 49–58, 2002.

137. G. Buchsbaum. A spatial processor model for object colour perception. *Journal of the Franklin Institute*, 310(1): 1–26, 1980.

138. G. D. Finlayson and E. Trezzi. Shades of gray and colour constancy. In *IS & T/SID's Color Imaging Conference*, pages 37–41. IS & T-The Society for Imaging Science and Technology, 2004.

139. J. van de Weijer, Th. Gevers, and A. Gijsenij. Edge-based color constancy. *IEEE Transactions on Image Processing*, 16(9), 2007.

140. R. Gershon, A. D. Jepson, and J. K. Tsotsos. From [r, g, b] to surface reflectance: computing color constant descriptors in images. In *Proceedings of the International Joint Conference on Artificial Intelligence*, pages 755–758, Milan, Italy, 1987.
141. K. Barnard, L. Martin, A. Coath, and B. V Funt. A comparison of computational color constancy algorithms; part ii: Experiments with image data. *IEEE Transactions on Image Processing*, 11(9): 985–996, 2002.
142. W. Xiong, B. V. Funt, L. Shi, S. S. Kim, B. H. Kang, S. D. Lee, and C. Y. Kim. Automatic white balancing via gray surface identification. In *IS &T/SID's Color Imaging Conference*, pages 143–146. IS&T-The Society for Imaging Science and Technology, 2007.
143. W. Xiong and B. V. Funt. Cluster based color constancy. In *IS&T/SID's Color Imaging Conference*, pages 210–214. IS&T-The Society for Imaging Science and Technology, 2008.
144. B. Li, D. Xu, W. Xiong, and S. Feng. Color constancy using achromatic surface. *Color Research and Application*, 35(4): 304–312, 2010.
145. J. van de Weijer, C. Schmid, and J. J. Verbeek. Using high-level visual information for color constancy. In *IEEE International Conference on Computer Vision*, pages 1–8, 2007.
146. A. Gijsenij and Th. Gevers. Color constancy by local averaging. In *2007 Computational Color Imaging Workshop (CCIW'07), in conjunction with ICIAP'07*, pages 1–4, 2007.
147. B. V. Funt and L. Shi. The rehabilitation of maxrgb. In *IS&T/SID's Color Imaging Conference*. IS&T-The Society for Imaging Science and Technology, 2010.
148. B. V. Funt and L. Shi. The effect of exposure on maxrgb color constancy. In *Proceedings SPIE, Volume 7527 Human Vision and Electronic Imaging XV*, 2010.
149. J. van de Weijer, Th. Gevers, and A. Bagdanov. Boosting color saliency in image feature detection. *IEEE Transactions on Pattern Analysis and Machine Intelligence*, 28(1): 150–156, 2006.
150. E. H. Land. An alternative technique for the computation of the designator in the retinex theory of color vision. *Proceedings of the National Academy of Sciences of the United States of America*, 83(10): 3078–3080, 1986.
151. H. H. Chen, C. H. Shen, and P. S. Tsai. Edge-based automatic white balancing linear illuminant constraint. In *Visual Communications and Image Processing*, 2007.
152. A. Chakrabarti, K. Hirakawa, and T. Zickler. Color constancy beyond bags of pixels. In *IEEE Computer Society Conference on Computer Vision and Pattern Recognition*, pages 1–8, 2008.
153. A. Gijsenij, Th. Gevers, and J. van de Weijer. Physics-based edge evaluation for improved color constancy. In *IEEE Computer Society Conference on Computer Vision and Pattern Recognition*, 2009.
154. H. C. Lee. Method for computing the scene-illuminant chromaticity from specular highlights. *Journal of the Optical Society of America A*, 3(10): 1694–1699, 1986.
155. S. Tominaga and B. A. Wandell. Standard surface-reflectance model and illuminant estimation. *Journal of the Optical Society of America A*, 6(4): 576–584, 1989.
156. G. Healey. Estimating spectral reflectance using highlights. *Image and Vision Computing*, 9(5): 333–337, 1991.

157. R. T. Tan, K. Nishino, and Ka. Ikeuchi. Color constancy through inverse-intensity chromaticity space. *Journal of the Optical Society of America A*, 21(3): 321–334, 2004.

158. G. D. Finlayson and G. Schaefer. Solving for colour constancy using a constrained dichromatic reflection model. *International Journal of Computer Vision*, 42(3): 127–144, 2001.

159. J. Toro and B. V. Funt. A multilinear constraint on dichromatic planes for illumination estimation. *IEEE Transactions on Image Processing*, 16(1): 92–97, January 2007.

160. J. Toro. Dichromatic illumination estimation without pre-segmentation. *Pattern Recognition Letters*, 29(7): 871–877, 2008.

161. L. Shi and B. V. Funt. Dichromatic illumination estimation via hough transforms in 3d. In *IS&T's European Conference on Color in Graphics, Imaging and Vision*, 2008.

162. D. A. Forsyth. A novel algorithm for color constancy. *International Journal of Computer Vision*, 5(1): 5–36, 1990.

163. K. Barnard. Improvements to gamut mapping colour constancy algorithms. In *European Conference on Computer Vision*, pages 390–403, 2000.

164. G. D. Finlayson. Color in perspective. *IEEE Transactions on Pattern Analysis and Machine Intelligence*, 18(10): 1034–1038, 1996.

165. G. D. Finlayson and S. D. Hordley. Improving gamut mapping color constancy. *IEEE Transactions on Image Processing*, 9(10): 1774–1783, 2000.

166. G. D. Finalyson and S. D. Hordley. Selection for gamut mapping colour constancy. *Image and Vision Computing*, 17(8): 597–604, 1999.

167. G. D. Finlayson and R. Xu. Convex programming color constancy. In *IEEE Workshop on Color and Photometric Methods in Computer Vision, in conjunction with ICCV'03*, pages 1–8, 2003.

168. M. Mosny and B. V. Funt. Cubical gamut mapping colour constancy. In *IS&T's European Conference on Color in Graphics, Imaging and Vision*, 2010.

169. K. Barnard, V. C. Cardei, and B. V. Funt. A comparison of computational color constancy algorithms; part i: Methodology and experiments with synthesized data. *IEEE Transactions on Image Processing*, 11(9): 972–984, 2002.

170. K. Barnard and B. Funt. Color constancy with specular and non-specular surfaces. In *IS&T/SID's Color Imaging Conference*, pages 114–119, 1999.

171. S. Tominaga, S. Ebisui, and B. A. Wandell. Scene illuminant classification: brighter is better. *Journal of the Optical Society of America A*, 18(1): 55–64, 2001.

172. G. D. Finlayson, S. D. Hordley, and I. Tastl. Gamut constrained illuminant estimation. *International Journal of Computer Vision*, 67(1): 93–109, 2006.

173. A. Gijsenij, Th. Gevers, and J. van de Weijer. Generalized gamut mapping using image derivative structures for color constancy. *International Journal of Computer Vision*, 86(2–3): 127–139, 2010.

174. J. J. Koenderink and A. J. van Doom. Representation of local geometry in the visual system. *Biological Cybernetics*, 55(6): 367–375, 1987.

175. M. Kass and A. Witkin. Analyzing oriented patterns. *Computer Vision Graphics and Image Processing*, 37(3): 362–385, 1987.

176. S. D. Hordley. Scene illuminant estimation: past, present, and future. *Color Research and Application*, 31(4): 303–314, 2006.

177. S. Bianco, F. Gasparini, and R. Schettini. Consensus-based framework for illuminant chromaticity estimation. *Journal of Electronic Imaging*, 17(2): 023013–1–9, 2008.

178. G. Schaefer, S. Hordley, and G. Finlayson. A combined physical and statistical approach to colour constancy. In *IEEE Computer Society Conference on Computer Vision and Pattern Recognition*, pages 148–153, IEEE Computer Society, Washington (DC) USA, 2005.

179. V. C. Cardei, B. V. Funt, and K. Barnard. Estimating the scene illumination chromaticity using a neural network. *Journal of the Optical Society of America A*, 19(12): 2374–2386, 2002.

180. B. V. Funt and W. Xiong. Estimating illumination chromaticity via support vector regression. In *IS&T/SID's Color Imaging Conference*, pages 47–52. IS&T-The Society for Imaging Science and Technology, 2004.

181. W. Xiong and B. V. Funt. Estimating illumination chromaticity via support vector regression. *Journal of Imaging Science and Technology*, 50(4): 341–348, 2006.

182. N. Wang, D. Xu, and B. Li. Edge-based color constancy via support vector regression. *IEICE Transactions on Information and Systems*, E92-D(11): 2279–2282, 2009.

183. V. Agarwal, A. V. Gribok, A. Koschan, and M. A. Abidi. Estimating illumination chromaticity via kernel regression. In *IEEE International Conference on Image Processing*, pages 981–984, 2006.

184. V. Agarwal, A. V. Gribok, and M. A. Abidi. Machine learning approach to color constancy. *Neural Networks*, 20(5): 559–563, 2007.

185. V. Agarwal, A. V. Gribok, A. Koschan, B. Abidi, and M. A. Abidi. Illumination chromaticity estimation using linear learning methods. *Journal of Pattern Recognition Research*, 4(1): 92–109, 2009.

186. W. Xiong, L. Shi, B. V. Funt, S. S Kim, B. Kan, and S. D. Lee. Illumination estimation via thin-plate spline interpolation. In *IS&T/SID's Color Imaging Conference*, 2007.

187. G. D. Finlayson, S. D. Hordley, and P. M. Hubel. Color by correlation: a simple, unifying framework for color constancy. *IEEE Transactions on Pattern Analysis and Machine Intelligence*, 23(11): 1209–1221, 2001.

188. C. Rosenberg, M. Hebert, and S. Thrun. Color constancy using kl-divergence. In *IEEE International Conference on Computer Vision*, pages 239–246, 2001.

189. M. D'Zmura, G. Iverson, and B. Singer. Probabilistic color constancy. In *Geometric Representations of Perceptual Phenomena*, pages 187–202. Lawrence Erlbaum Associates, 1995.

190. D. H. Brainard and W. T. Freeman. Bayesian color constancy. *Journal of the Optical Society of America A*, 14: 1393–1411, 1997.

191. G. Sapiro. Color and illuminant voting. *IEEE Transactions on Pattern Analysis and Machine Intelligence*, 21(11): 1210–1215, 1999.

192. Y. Tsin, R. T. Collins, V. Ramesh, and T. Kanade. Bayesian color constancy for outdoor object recognition. In *IEEE Computer Society Conference on Computer Vision and Pattern Recognition*, pages 1132–1139, 2001.

193. C. Rosenberg, T. Minka, and A. Ladsariya. Bayesian color constancy with non-gaussian models. In *Advances in Neural Information Processing Systems*, 2003.

194. P. V. Gehler, C. Rother, A. Blake, T. P. Minka, and T. Sharp. Bayesian color constancy revisited. In *IEEE Computer Society Conference on Computer Vision and Pattern Recognition*, pages 1–8, 2008.

195. V. C. Cardei and B. V. Funt. Committee-based color constancy. In *IS&T/SID's Color Imaging Conference*, pages 311–313. IS&T-The Society for Imaging Science and Technology, 1999.

196. S. D. Hordley and G. D. Finlayson. Reevaluation of color constancy algorithm performance. *Journal of the Optical Society of America A*, 23(5): 1008–1020, 2006.

197. A. Torralba and A. Oliva. Statistics of natural image categories. *Network-Computation in Neural Systems*, 14(3): 391–412, 2003.

198. J-M. Geusebroek, G. J. Burghouts, and A. W. M. Smeulders. The amsterdam library of object images. *International Journal Computer Vision (IJCV)*, 61(1): 103–112, 2005.

199. A. Torralba. Contextual priming for object detection. *International Journal of Computer Vision*, 53(2): 169–191, 2003.

200. J. C. van Gemert, J. M. Geusebroek, C. J. Veenman, C. G. M. Snoek, and A. W. M. Smeulders. Robust scene categorization by learning image statistics in context. In *CVPR Workshop on Semantic Learning Applications in Multimedia (SLAM)*, New York, June 2006.

201. F. Ciurea and B. V. Funt. A large image database for color constancy research. In *IS&T/SID's Color Imaging Conference*, pages 160–164. IS&T-The Society for Imaging Science and Technology, 2003.

202. C. M. Bishop. *Neural Networks for Pattern Recognition*. Oxford University Press, Oxford, UK, 1996.

203. D. L. Ruderman, T. W. Cronin, and C. C. Chiao. Statistics of cone responses to natural images: implications for visual coding. *Journal of the Optical Society of America A*, 15(8): 2036–2045, 1998.

204. B. Li, D. Xu, and C. Lang. Colour constancy based on texture similarity for natural images. *Coloration Technology*, 125(6): 328–333, 2009.

205. S. Bianco, F. Gasparini, and R. Schettini. Region-based illuminant estimation for effective color correction. In *Proceedings of the International Conference on Image Analysis and Processing*, pages 43–52, 2009.

206. S. Bianco, G. Ciocca, C. Cusano, and R. Schettini. Automatic color constancy algorithm selection and combination. *Pattern Recognition*, 43(3): 695–705, 2010.

207. M. Wu, J. Sun, J. Zhou, and G. Xue. Color constancy based on texture pyramid matching and regularized local regression. *Journal of the Optical Society of America A*, 27(10): 2097–2105, 2010.

208. A. Gijsenij and Th. Gevers. Color constancy using image regions. In *IEEE International Conference on Image Processing*, San Antonio, Tx, USA, September 2007.

209. A. Oliva and A. Torralba. Modeling the shape of the scene: a holistic representation of the spatial envelope. *International Journal of Computer Vision*, 42(3): 145–175, 2001.

210. J. M. Geusebroek. Compact object descriptors from local colour invariant histograms. In *British Machine Vision Conference*, pages 1029–1038, 2006.

211. S. Bianco, G. Ciocca, C. Cusano, and R. Schettini. Improving color constancy using indoor-outdoor image classification. *IEEE Transactions on Image Processing*, 17(12): 2381–2392, 2008.

212. R. Lu, A. Gijsenij, Th. Gevers, K. E. A. van de Sande, J. M. Geusebroek, and D. Xu. Color constancy using stage classification. In *IEEE International Conference on Image Processing*, 2009.

213. R. Lu, A. Gijsenij, Th. Gevers, D. Xu, V. Nedovic, and J. M. Geusebroek. Color constancy using 3d stage geometry. In *IEEE International Conference on Computer Vision*, 2009.

214. V. Nedovic, A. W. M. Smeulders, A. Redert, and J. M. Geusebroek. Stages as models of scene geometry. *IEEE Transactions on Pattern Analysis and Machine Intelligence*, 32(9): 1673–1687, 2010.

215. T. Hofmann. Probabilistic latent semantic indexing. In *Proceedings ACM SIGIR Conference on Research and Development in Information Retrieval*, pages 50–57, 1999.

216. J. Verbeek and B. Triggs. Region classification with markov field aspect models. In *IEEE Computer Society Conference on Computer Vision and Pattern Recognition*, pages 1–8, 2007.

217. R. Manduchi. Learning outdoor color classification. *IEEE Transactions on Pattern Analysis and Machine Intelligence*, 28(11): 1713–1723, 2006s.

218. E. Rahtu, J. Nikkanen, J. Kannala, L. Lepist ö, and J. Heikkil ä. Applying visual object categorization and memory colors for automatic color constancy. In *Proceedings of the International Conference on Image Analysis and Processing*, pages 873–882, 2009.

219. B. V. Funt, K. Barnard, and L. Martin. Is machine colour constancy good enough? In *European Conference on Computer Vision*, pages 445–459, 1998.

220. S. M. C. Nascimento, F. P. Ferreira, and D. H. Foster. Statistics of spatial cone-excitation ratios in natural scenes. *Journal of the Optical Society of America A*, 19(8): 1484–1490, 2002.

221. C. A. P árraga, G. Brelstaff, T. Troscianko, and I. R. Moorehead. Color and luminance information in natural scenes. *Journal of the Optical Society of America A*, 15(3): 563–569, 1998.

222. J. Vazquez-Corral, C. A. P ŕrage, M. Vanrell, and R. Baldrich. Color constancy algorithms: psychophysical rvaluation on a new dataset. *Journal of Imaging Science and Technology*, 53(3): 1–9, 2009.

223. C. A. P árraga, J. Vazquez-Corral, and M. Vanrell. A new cone activation-based natural images dataset. *Perception*, 36(ECVP Abstract Supplement): 180, 2009.

224. A. Gijsenij, Th. Gevers, and M. P. Lucassen. A perceptual analysis of distance measures for color constancy. *Journal of the Optical Society of America A*, 26(10): 2243–2256, 2009.

225. Commission Internationale de L'Eclairage (CIE). *Colorimetry, 2nd Ed.* CIE Publication No. 15.2, Central Bureau of the CIE, Vienna, Austria, 1986.

226. Commission Internationale de L'Eclairage (CIE). *Improvement to Industrial Colour-difference Evaluation.* CIE Publication No. 142–2001, Central Bureau of the CIE, Vienna, Austria, 2001.

227. E. Brunswik. Zur entwicklung der albedowahrnehmung. *Zeitschrift fur Psychologie*, 109: 40–115, 1928.

228. R. V. Hogg and E. A. Tanis. *Probability and Statistical Inference.* Prentice Hall, Upper Saddle River (NJ), 2001.

229. J. W. Tukey. *Exploratory data analysis.* Addison-Wesley, Reading (MA), 1977.

230. H. F. Weisberg. *Central Tendency and Variability.* Sage Publications, Inc., Newbury Park (CA), 1992.

231. A. Gijsenij, Th. Gevers, and J. van de Weijer. Computational color constancy: Survey and experiments. *IEEE Transactions on Image Processing*, 20(9): 2475–2489, 2011.

232. L. Shi and B. V. Funt. Re-processed version of the Gehler color constancy database of 568 images. Available at http://www.cs.sfu.ca/colour/data/. Accessed 2010 Nov 1.

233. C.-C. Chang and C-J. Lin. LIBSVM: a library for support vector machines, 2001. Software Available at http://www.csie.ntu.edu.tw/cjlin/libsvm. Accessed 2010.

234. S. Di Zenzo. A note on the gradient of a multi-image. *Computer Vision Graphics and Image Processing*, 33(1): 116–125, 1986.

235. J. van de Weijer, Th. Gevers, and A. W. M. Smeulders. Robust photometric invariant features from the color tensor. *IEEE Transactions on Image Processing*, 15(1): 118–127, 2006.

236. J. Bigun, G. Granlund,, and J. Wiklund. Multidimensional orientation estimation with applications to texture analysis and optical flow. *IEEE Transactions on Pattern Analysis and Machine Intelligence*, 13(8): 775–790, 1991.

237. J. Bigun. Pattern recognition in images by symmetry and coordinate transformations. *Computer Vision and Image Understanding*, 68(3): 290–307, 1997.

238. O. Hansen and J. Bigun. Local symmetry modeling in multi-dimensional images. *Pattern Recognition Letters*, 13(4): 253–262, 1992.

239. M. S. Lee and G. Medioni. Grouping into regions, curves, and junctions. *Computer Vision and Image Understanding*, 76(1): 54–69, 1999.

240. J. van de Weijer, L. J. van Vliet, P. W. Verbeek, and M. van Ginkel. Curvature estimation in oriented patterns using curvilinear models applied to gradient vector fields. *IEEE Transactions on Pattern Analysis and Machine Intelligence*, 23(9): 1035–1042, 2001.

241. G. Sapiro and D. L. Ringach. Anisotropic diffusion of multivalued images with applications to color filtering. *IEEE Transactions on Image Processing*, 5(11): 1582–1586, 1996.

242. J. Shi and C. Tomasi. Good features to track. In *IEEE Computer Society Conference on Computer Vision and Pattern Recognition*, 1994.

243. D. H. Ballard. Generalizing the hough transform to detect arbitrary shapes. *Pattern Recognition*, 12(2): 111–122, 1981.

244. E. P. Simoncelli, E. H. Adelson, and D. J. Heeger. Probability distributions of optical flow. In *IEEE Computer Society Conference on Computer Vision and Pattern Recognition*, 1991.

245. J. Barron and R. Klette. Quantitative color optical flow. In *International Conference on Pattern Recognition*, pages 251–255, 2002.

246. P. Golland and A. M. Bruckstein. Motion from color. *Computer Vision and Image Understanding*, 68(3): 346–362, 1997.

247. B. K. P. Horn and B. G. Schunk. Determing optical flow. *Artificial Intelligence*, 17(1–3): 185–203, 1981.

248. B. Lucas and T. Kanade. An iterative image registration technique with an application to stereo vision. In *Proceedings of the International Joint Conference on Artificial Intelligence*, pages 674–679, 1981.

249. D. Koubaroulis, J. Matas, and J. Kittler. Evaluating colour-based object recognition algorithms using the soil-47 database. In *Asian Conference on Computer Vision*, 2002.

250. L. Itti, C. Koch, and E. Niebur. A model of saliency-based visual attention for rapid scene analysis. *IEEE Transactions on Pattern Analysis and Machine Intelligence*, 20(11): 1254–1259, 1998.

251. L. Itti and C. Koch. Computational modelling of visual attention. *Nature Reviews Neuroscience*, 2(3): 194–203, 2001.

252. K. Mikolajczyk and C. Schmid. Scale and affine invariant interest point detectors. *International Journal of Computer Vision*, 60(1): 63–86, 2004.

253. C. Schmid and R. Mohr. Local grayvalue invariants for image retrieval. *IEEE Transactions on Pattern Analysis and Machine Intelligence*, 19(5): 530–535, 1997.

254. G. Heidemann. Focus-of-attention from local color symmetries. *IEEE Transactions on Pattern Analysis and Machine Intelligence*, 26(7): 817–830, 2004.

255. Corel Gallery. Available at http://www.corel.com. Accessed 2004.

256. C. Schmid, R. Mohr, and C. Backhage. Evaluation of interest point detectors. *International Journal of Computer Vision*, 37(2): 151–172, 2000.

257. I. Jermyn and H. Ishikawa. Globally optimal regions and boundaries as minimum ratio weight cycles. *IEEE Transactions on Pattern Analysis and Machine Intelligence*, 23(10): 1075–1088, 2001.

258. G. Burghouts and J-M. Geusebroek. Performance evaluation of local colour invariants. *Computer Vision and Image Understanding*, 113(1): 48–62, 2009.

259. K. E. A. van de Sande, Th. Gevers, and C. G. M. Snoek. Evaluation of color descriptors for object and scene recognition. *IEEE Transactions on Pattern Analysis and Machine Intelligence*, 32(9): 1582–1596, 2010.

260. S. Odbrzalek and J. Matas. Object recognition using local affine frames on distinguished regions. In *British Machine Vision Conference*, 2002.

261. F. Rothganger, S. Lazebnik, C. Schmid, and J. Ponce. 3d object modeling and recognition using local affine-invariant image descriptors and multi-view spatial constraints. *International Journal of Computer Vision*, 66(3): 231–259, 2006.

262. J. Sivic and A. Zisserman. Video google: a text retrieval approach to object matching in videos. In *Proceedings of the International Conference on Computer Vision*, pages 1470–1477, 2003.

263. S. Lazebnik, C. Schmid, and J. Ponce. A sparse texture representation using local affine regions. *IEEE Transactions on Pattern Analysis and Machine Intelligence*, 27(8): 1265–1278, 2005.

264. L. M. J. Florack, B. ter Haar Romeny, M. Viergever, and J. J. Koenderink. The gaussian scale-space paradigm and the multiscale local jet. *International Journal of Computer Vision*, 18(1): 1996, 61–75.

265. W. T. Freeman and E. H. Adelson. The design and use of steerable filters. *IEEE Transactions on Pattern Analysis and Machine Intelligence*, 13(9): 891–906, 1991.

266. B. Schiele and J. L. Crowley. Recognition without correspondence using multidimensional receptive field histograms. *International Journal of Computer Vision*, 36(1): 31–50, 2000.

267. A. Ferencz, E. Learned-Miller, and J. Malik. Building a classification cascade for visual identification from one example. In *IEEE International Conference on Computer Vision*, pages 286–293, 2003.

268. P. Montesinos, V. Gouet, R. Deriche, and D. Pele. Matching color uncalibrated images using differential invariants. *Image and Vision Computing*, 18(9): 659–671, 2000.

269. A. W. M. Smeulders, M. Worring, S. Santini, A. Gupta, and R. Jain. Content based image retrieval at the end of the early years. *IEEE Transactions on Pattern Analysis and Machine Intelligence*, 22(12): 1349–1380, 2000.

270. C. Carson, S. Belongie, H. Greenspan, and J. Malik. Blobworld: image segmentation using expectation-maximization and its application to image querying. *IEEE Transactions on Pattern Analysis and Machine Intelligence*, 24(8): 1026–1038, 2002.

271. Y. Ke and R. Sukthankar. Pca-sift: A more distinctive representation for local image descriptors. In *IEEE Computer Society Conference on Computer Vision and Pattern Recognition*, pages 506–513, 2004.

272. M. Grabner, H. Grabner, and H. Bischof. Fast approximated sift. In *Asian Conference on Computer Vision*, pages 918–927, 2006.

273. H. Bay, T. Tuytelaars, and L. Van Gool. Surf: speeded up robust features. In *European Conference on Computer Vision*, 2006.

274. F. Jurie and B. Triggs. Creating efficient codebooks for visual recognition. In *Proceedings of International Conference on Computer Vision*, 2005.

275. J. Matas, O. Chum, M. Urban, and T. Pajdla. Robust wide baseline stereo from maximally stable extremal regions. In *British Machine Vision Conference*, pages 384–393, 2002.

276. T. Lindeberg and J. Garding. Shape-adapted smoothing in estimation of 3d shape cues from affine deformations of local 2d brightness structure. *Image and Vision Computing*, 15(6): 415–434, 1997.

277. T. Kadir and M. Brady. Scale, saliency and image description. *International Journal of Computer Vision*, 45(2): 83–106, 2001.

278. T. Tuytelaars and L. Van Gool. Matching widely separated views based on affine invariant regions. *International Journal of Computer Vision*, 59(1): 61–85, 2004.

279. K. Mikolajczyk, T. Tuytelaars, C. Schmid, A. Zisserman, J. Matas, F. Schaffalitzky, T. Kadir, and L. Van Gool. A comparison of affine region detectors. *International Journal of Computer Vision*, 65(1–2): 43–72, 2005.

280. A. E. Abdel-Hakim and A. A. Farag. Csift: a sift descriptor with color invariant characteristics. In *IEEE Computer Society Conference on Computer Vision and Pattern Recognition*, 2006.

281. A. Bosch, A. Zisserman, and X. Munoz. Scene classification via plsa. In *European Conference on Computer Vision*, 2006.

282. K. J. Dana, B. van Ginneken, S. K. Nayar, and J. J. Koenderink. Reflectance and texture of real world surfaces. *ACM Transactions on Graphics*, 18(1): 1–34, 1999.

283. A. C Bovik, M. Clark, and W. S. Geisler. Multichannel texture analysis using localized spatial filters. *IEEE Transactions On Pattern Analysis And Machine Intelligence*, 12(1): 55–73, 1990.

284. J. G årding and T. Lindeberg. Direct computation of shape cues using scale-adapted spatial derivative operators. *International Journal of Computer Vision*, 17(2): 163–191, 1996.

285. M. Varma and A. Zisserman. A statistical approach to texture classification from single images. *International Journal of Computer Vision*, 62(1–2): 61–81, 2005.

286. B. Julesz. Textons, the elements of texture perception, and their interactions. *Nature*, 290: 91–97, 1981.

287. J. Malik, S. Belongie, T. Leung, and J. Shi. Contour and texture analysis for image segmentation. *International Journal of Computer Vision*, 43(1): 7–27, 2001.

288. M. Mirmehdi and M. Petrou. Segmentation of color textures. *IEEE Transactions on Pattern Analysis and Machine Intelligence PAMI*, 22(2): 142–159, 2000.

289. X. Zhang and B. A. Wandell. A spatial extension of cielab for digital color image reproduction. In *Proceedings of the Society of Information Display Symposium*, 1996.

290. B. Thai and G. Healey. Modeling and classifying symmetries using a multi-scale opponent color representation. *IEEE Transactions on Pattern Analysis and Machine Intelligence*, 20(11): 1224–1235, 1998.

291. M. A. Hoang, J. M. Geusebroek, and A. W. M. Smeulders. Color texture measurement and segmentation. *Signal Processing*, 85(2): 265–275, 2005.

292. A. K. Jain and F. Farrokhnia. Unsupervised texture segmentation using gabor filters. *Pattern Recognition*, 24(12): 1167–1186, 1991.

293. T. Weldon, W. E. Higgins, and D. F. Dunn. Efficient gabor-filter design for texture segmentation. *Pattern Recognition*, 29(12): 2005–2016, 1996.

294. B. S. Manjunath and W. Y. Ma. Texture features for browsing and retrieval of image data. *IEEE Transactions on Pattern Analysis and Machine Intelligence*, 18(8): 837–842, 1996.

295. H. T. Nguyen, M. Worring, and A. Dev. Detection of moving objects in video using a robust motion similarity measure. *IEEE Transactions on Image Processing*, 9(1): 137–141, 2000.

296. J. J. Koenderink, A. J. van Doorn, K. J. Dana, and S. Nayar. Bidirectional reflection distribution function of thoroughly pitted surfaces. *International Journal of Computer Vision*, 31: 129–144, 1999.

297. G. J. Burghouts and J-M. Geusebroek. Material-specific adaptation of color invariant features. *Pattern Recognition Letters*, 30: 306–313, 2009.

298. T. Leung and J. Malik. Representing and recognizing the visual appearance of materials using three-dimensional textons. *International Journal of Computer Vision*, 43(1): 29–44, 2001.

299. E. Hayman, B. Caputo, M. Fritz, and J. O. Eklundh. On the significance of real-world conditions for material classification. In *Proceedings of the European Conference Computer Vision*, pages 253–266. Springer-Verlag, 2004.

300. J. Zhang, M. Marsza, S. Lazebnik, and C. Schmid. Local features and kernels for classification of texture and object categories: a comprehensive study. *International Journal of Computer Vision*, 73(2): 213–238, 2007.

301. E. Nowak, F. Jurie, and B. Triggs. Sampling strategies for bag-of-features image classification. In *Proceedings of the European Conference on Computer Vision*, pages 490–503. Springer Verlag, 2006.

302. J. Winn, A. Criminisi, and T. Minka. Object categorization by learned universal visual dictionary. In *Proceedings of the International Conference Computer Vision*, pages 1800–1807. IEEE Computer Society, 2005.

303. J. Shotton, J. Winn, C. Rother, and A. Criminisi. Textonboost: Joint appearance, shape and context modeling for multi-class object recognition and segmentation. In *Proceedings of the European Conference on Computer Vision*, Springer-Verlag, 2006.

304. P. Suen and G. Healey. The analysis and recognition of real-world textures in three dimensions. *IEEE Transactions on Pattern Analysis and Machine Intelligence*, 22(5): 491–503, 2000.

305. G. Csurka, C. Dance, L. Fan, J. Willamowski, and C. Bray. Visual categorization with bags of keypoints. In *Proceedings of the European Conference on Computer Vision*, 2004.

306. R. Datta, D. Joshi, J. Li, and J. Z. Wang. Image retrieval: ideas, influences, and trends of the new age. *ACM Computing Surveys*, 40(2): 1–60, 2008.

307. R. Fergus, L. Fei-Fei, P. Perona, and A. Zisserman. Learning object categories from Google's image search. In *IEEE International Conference on Computer Vision*, Beijing, China, 2005.

308. S. Lazebnik, C. Schmid, and J. Ponce. Beyond bags of features: Spatial pyramid matching for recognizing natural scene categories. In *IEEE Computer Society Conference on Computer Vision and Pattern Recognition*, pages 2169–2178, 2006.

309. S.-F. Chang, D. Ellis, W. Jiang, K. Lee, A. Yanagawa, A. C. Loui, and J. Luo. Large-scale multimodal semantic concept detection for consumer video. In *ACM International Workshop on Multimedia Information Retrieval*, pages 255–264, 2007.

310. A. F. Smeaton, P. Over, and W. Kraaij. Evaluation campaigns and trecvid. In *ACM International Workshop on Multimedia Information Retrieval*, pages 321–330, 2006.

311. M. Everingham, L. Van Gool, C. K. I. Williams, J. Winn, and A. Zisserman. The pascal visual object classes challenge 2007 (voc2007) results, 2007 In [Online]. Available at http://www.pascal-network.org/challenges/VOC/voc2007/. Accessed 2008.

312. C. G. M. Snoek, M. Worring, J. C. van Gemert, J.-M. Geusebroek, and A. W. M. Smeulders. The challenge problem for automated detection of 101 semantic concepts in multimedia. In *ACM International Workshop on Multimedia Information Retrieval*, pages 421–430, 2006.

313. K. E. A. van de Sande, T. Gevers, and C. G. M. Snoek. Evaluation of color descriptors for object and scene recognition. In *IEEE Computer Society Conference on Computer Vision and Pattern Recognition*, 2008.

314. A. Bosch, A. Zisserman, and X. Munoz. Scene classification using a hybrid generative/discriminative approach. *IEEE Transactions on Pattern Analysis and Machine Intelligence*, 30(4): 712–727, 2008.

315. J. Matas, O. Chum, M. Urban, and T. Pajdla. Robust wide-baseline stereo from maximally stable extremal regions. *Image and Vision Computing*, 22(10): 761–767, 2004.

316. P.-E. Forssen. Maximally stable colour regions for recognition and matching. In *IEEE Computer Society Conference on Computer Vision and Pattern Recognition*, June 2007.

317. R. Fergus, P. Perona, and A. Zisserman. Object class recognition by unsupervised scale-invariant learning. In *IEEE Computer Society Conference on Computer Vision and Pattern Recognition*, pages 264–271, 2003.

318. B. Leibe and B. Schiele. Interleaved object categorization and segmentation. In *British Machine Vision Conference*, pages 759–768, 2003.

319. M. Naphade, J. R. Smith, J. Tesic, S.-F. Chang, W. Hsu, L. Kennedy, A. Hauptmann, and J. Curtis. Large-scale concept ontology for multimedia. *IEEE Multimedia*, 13(3): 86–91, 2006.

320. C. M. Bishop. *Pattern Recognition and Machine Learning*. Springer, New York, 2006.

321. B. Efron. Bootstrap methods: another look at the jackknife. *Annals of Statistics*, 7: 1–26, 1979.

322. M. Marszalek, C. Schmid, H. Harzallah, and J. van de Weijer. Learning object representations for visual object class recognition. In *IEEE International Conference on Computer Vision*, pages 239–246, 2007.

323. M. Everingham, L. Van Gool, C. K. I. Williams, J. Winn, and A. Zisserman. The pascal visual object classes challenge 2008 (voc2008) results, 2008 In [Online]. Available at http://www.pascal-network.org/challenges/VOC/voc2008/. Accessed 2008.

324. M. A. Tahir, K. E. A. van de Sande, J. R. R. Uijlings, et al. University of amsterdam and university of surrey at pascal voc 2008. In *IEEE Computer Society Conference on Computer Vision and Pattern Recognition*, 2008.

325. J. C. van Gemert, C. J. Veenman, A. W. M. Smeulders, and J.-M. Geusebroek. Visual word ambiguity. *IEEE Transactions on Pattern Analysis and Machine Intelligence*, 32(7): 1271–1283, 2010.

326. C. G. M. Snoek, K. E. A. van de Sande, O. de Rooij, B. Huurnink, J. C. van Gemert, J. R. R. Uijlings, et al. The mediamill trecvid 2008 semantic video search engine. In *Proceedings of the 6th TRECVID Workshop*, 2008.

327. C. L. Hardin and L. Maffi, editors. *Color Categories in Thought and Language*. Cambridge University Press, Cambridge, England, 1997.

328. L. Steels and T. Belpaeme. Coordinating perceptually grounded categories through language: a case study for colour. *Behavioral and Brain Sciences*, 28: 469–529, 2005.

329. B. Berlin and P. Kay. *Basic Color Terms: Their Universality and Evolution*. University of California, Berkeley, 1969.

330. A. Maerz and M. R. Paul. *A Dictionary of Color, 1st Ed*. McGraw-Hill, New York, 1930.

331. R. M. Boynton and C. X. Olson. Locating basic colors in the OSA space. *Color Research and Application*, 12(2): 94–105, 1987.

332. J. Sturges and T. W. A. Whitfield. Locating basic colors in the Munsell space. *Color Research and Application*, 20(6): 364–376, 1995.

333. D. Roberson, I. Davies, and J. Davidoff. Color categories are not universal: replications and new evidence from a stone-age culture. *Journal of Experimental Psychology-General*, 129(3): 369–398, 2000.

334. D. Roberson, J. Davidoff, I. R. L. Davies, and L. R. Shapiro. Color categories: Evidence for the cultural relativity hypothesis. *Cognitive Psychology*, 50(4): 378–411, 2005.

335. T. Regier, P. Kay, and R. S. Cook. Focal colors are universal after all. *Proceedings of the National Academy of Sciences of the United States of America*, 102(23): 8386–8391, 2005.

336. T. Regier and P. Kay. Language, thought, and color: whorf was half right. *Trends in Cognitive Sciences*, 13(10): 439–446, 2009.

337. T. Regier, P. Kay, and R. S. Cook. Universal foci and varying boundaries in linguistic color categories. In B. G. Gara, L. Barsalou, and M. Bucciarelli, editors, *Proceedings of the 27th Meeting of the Cognitive Science Society*, Cognitive Science Society, Hillsdale (NJ), pages 1827–1832, 2005.

338. D. L. Philipona and J. K. O'Regan. Color naming, unique hues, and hue cancellation predicted from singularities in reflection properties. *Visual Neuroscience*, 23(3–4): 331–339, 2006.

339. T. Regier, P. Kay, and N. Khetarpal. Color naming reflects optimal partitions of color space. *Proceedings of the National Academy of Sciences of the United States of America*, 104(4): 1436–1441, 2007.

340. P. Kay and C. K. McDaniel. The linguistic significance of the meanings of basic color terms. *Language*, 3(54): 610–646, 1978.

341. D. Alexander. Statistical Modelling of Colour Data and Model Selection for Region Tracking. PhD thesis, Department of Computer Science, University College London, 1997.

342. L. A. Zadeh. Fuzz sets. *Information and Control*, 8(3): 338–353, 1965.

343. R. Benavente, M. Vanrell, and R. Bladrich. A data set for fuzzy colour naming. *Color Research and Application*, 31(1): 48–56, 2006.

344. R. Benavente, M. Vanrell, and R. Baldrich. Estimation of fuzzy sets for computational colour categorization. *Color Research and Application*, 29(5): 342–353, 2004.

345. R. Benavente and M. Vanrell. Fuzzy colour naming based on sigmoid membership functions. In *Proceedings of the 2nd European Conference on Colour in Graphics, Imaging, and Vision (CGIV'2004)*, pages 135–139, Aachen (Germany), 2004.

346. W. C. Graustein. Homogeneous Cartesian Coordinates. Linear Dependence of Points and Lines. *Introduction to Higher Geometry*, pages 29–49. Macmillan, New York, 1930, Chapter 3.

347. N. Seaborn, L. Hepplewhite, and J. Stonham. Fuzzy colour category map for the measurement of colour similarity and dissimilarity. *Pattern Recognition*, 38(1): 165–177, 2005.

348. Munsell Color Company, Inc. *Munsell Book of Color-Matte Finish Collection.* Munsell Color Company, Baltimore (MD), 1976.

349. Spectral database, university of joensuu color group. Available at http://spectral.joensuu.fi. Accessed 2011 September 20.

350. R. Benavente, M. Vanrell, and R. Baldrich. Parametric fuzzy sets for automatic color naming. *Journal of the Optical Society of America A*, 25(10): 2582–2593, 2008.

351. J. C. Lagarias, J. A. Reeds, M. H. Wright, and P. E. Wright. Convergence properties of the Nelder-Mead simplex method in low dimensions. *SIAM Journal of Optimization*, 9(1): 112–147, 1998.

352. D. M. Conway. An experimental comparison of three natural language colour naming models. In *Proceedings East-West International Conference on Human-computer Interaction*, pages 328–339, 1992.

353. L. D. Griffin. Optimality of the basic colour categories for classification. *Journal of the Royal Society Interface*, 3(6): 71–85, 2006.

354. J. M. Lammens. A computational model of color perception and color naming. PhD thesis, University of Buffalo, 1994.

355. A. Mojsilovic. A computational model for color naming and describing color composition of images. *IEEE Transactions on Image Processing*, 14(5): 690–699, 2005.

356. G. Menegaz, A. Le Troter, J. Sequeira, and J. M. Boi. A discrete model for color naming. *EURASIP Journal on Advances in Signal Processing*, 2007:Article ID 29125, 2007.

357. D. Blei, A. Ng, and M. Jordan. Latent Dirichlet allocation. *Journal of Machine Learning Research*, 3: 993–1022, 2003.

358. D. Larlus and F. Jurie. Latent mixture vocabularies for object categorization. In *British Machine Vision Conference*, 2006.

359. L. Vincent. Morphological grayscale reconstruction in image analysis: applications and efficient algorithms. *IEEE Transactions on Image Processing*, 2(2): 176–201, 1993.

360. J. van de Weijer, C. Schmid, J. Verbeek, and D. Larlus. Learning color names for real-world applications. *IEEE Transactions on Image Processing*, 18(7): 1512–1524, 2009.

361. R. E. MacLaury. From brightness to hue: An explanatory model of color-category evolution. *Current Anthropology*, 33: 137–186, 1992.

362. J. van de Weijer, C. Schmid, and J. Verbeek. Learning color names from real-world images. In *IEEE Computer Society Conference on Computer Vision and Pattern Recognition*, Minneapolis (MN) USA, 2007.

363. X. Otazu and M. Vanrell. Building perceived colour images. In *Proceedings of the 2nd European Conference on Colour in Graphics, Imaging, and Vision (CGIV'2004)*, pages 140–145, April 2004.

364. X Otazu, C. A P árraga, and M Vanrell. Toward a unified chromatic induction model. *Journal of Vision*, 10(12)(6), 2010.

365. S. Kawata, K. Sasaki, and S. Minami. Component analysis of spatial and spectral patterns in multispectral images. *Journal of the Optical Society of America A*, 4: 2101–2106, 1987.

366. H. M. G. Stokman and Th. Gevers. Robust photometric invariant segmentation of multispectral images. *International Journal of Computer Vision*, 53(2): 135–151, 2003.

367. J. C. Mullikin, L. J. van Vliet, H. Netten, F. R. Boddeke, G. van der Feltz, and I. T. Young. Methods for CCD camera characterization. In H. C. Titus and A. Waks, editors, *Image Acquisition and Scientific Imaging Systems*, San Jose (CA), volume 2173, pages 73–84. SPIE, 1994.

368. R. Dubes and A. K. Jain. Clustering techniques: the user's dilemma. *Pattern Recognition*, 8: 247–260, 1976.

Index

Applications
- color constancy, 143, 152, 161
- feature detection, 189
- feature extraction, 189
- image retrieval, 221, 233, 282
  - color names, 313
- image segmentation, 189, 244
- object recognition, 221, 282
- road detection, 129
- skin detection, 132

Chromatic adaptation
- Bradford transform, 141
- CMCCAT2000 transform, 141

Color boosting, 212
- photometric robustness, 217

Color constancy, 137
- chromatic adaptation, 137–8, 141
- color ratios, 69
- computational, 138
- diagonal offset model, 140
- human, 137
  - chromatic adaptation, 20
  - ratios, 19
- human color constancy, 18
- illuminant estimation, 137–8, 140
- Retinex, 137

Color constancy methods
- Bayesian methods, 162
- color-by-correlation, 161
- combination using output statistics, 162
- gamut mapping, 152
- gamut mapping combination, 157
- gamut mapping using N-jet, 157
- gamut-constrained illuminant estimation, 154
- gray-edge, 146, 148
- gray-world, 143
- max-RGB, 145
- physics-based methods, 150
- shades-of-gray, 146
- using high-level visual information, 169
- using natural image statistics, 163
- using scene categories, 167
- white-patch, 144

Color feature detection, 189

Color features, 82, 189
- boosting, 212
  - color distinctiveness, 214
  - function, 208
  - repeatability, 215
- Canny edge detection, 199
- circle patterns, 201
- color distinctiveness, 207
- color Gabor filtering, 245

Color in Computer Vision: Fundamentals and Applications, First Edition.
Theo Gevers, Arjan Gijsenij, Joost van de Weijer, and Jan-Mark Geusebroek.

Color features (*continued*)
  color tensor, 191–2
  corner detection, 199
  curvature detection, 201
  edge detection, 74, 108
  filterbanks, 253, 256
  Harris affine, 227
  Harris detector, 199, 204, 207, 216
  Harris-Laplace, 284
  illuminant invariant
    edge detection, 73
  optical flow, 202
  oriented patterns, 198
  star and spiral-like patterns, 200
  structure tensor, 191
    eigenvalue analysis, 198
    weighted, 192, 197
  symmetry detector, 200, 218
Color histograms, 61, 77
  density estimation, 61
Color image descriptors, 82, 120, 221, 271
  invariants, 229
  C-SIFT, 225, 230
  color ratios, 74
  discriminative power, 236
  H-SIFT, 230
  histogram
    O_1O_2, 273
    RGB, 273
    rg, 273
    hue, 273
    saturation, 273
  HSV-SIFT, 229
  information content, 236
  SIFT, 222, 273
    RGB, 273
    rg, 273
    C, 273
    gray, 273
    HSV, 273
    hue, 273
    opponent, 273
  W-SIFT, 230
Color image formation, 26
  diagonal, 272
  diagonal model, 139, 141, 143
  diagonal offset model, 142, 272
  full diagonal, 273
  highlights, 49, 52
  light color change, 273
  light color change and shift, 273
  light intensity change, 272
  light intensity change and shift, 273
  light intensity shift, 272
  linear model, 319
  local diagonal offset model, 142
  narrow-band filters, 71
  shadows, 49, 82
  specularities, 82
  von Kries model, 141
  white light assumption, 91
  white-balanced, 320
Color invariance, 49, 196
  blur, 74, 78
  body and surface, 55
  body reflectance, 53
  by learning, 117, 120, 123, 132
  coordinate transformations, 195
  derivative-based, 81
  derivatives, 193
  diversified ensembles, 113
  full-invariance, 82, 84
  H, 97
  histogram construction, 61
  histograms, 58
  illuminant, 71, 77
  illumination, 120
  irreducible color invariants, 54–5
  lighting geometry, 72
  multispectral images, 318
  noise
    estimation, 58
    propagation, 58
  noise propagation, 58
  optical flow, 203
  pixel transformation, 49
  pixel-based, 49
  quasi-invariance, 82, 101, 191, 193
  quasi-invariants
    shadow-shading, 103
  spectral derivatives, 84
Color names, 287
  basic color terms, 288
  Berlin and Kay, 287
  boundaries, 290
  calibrated data, 291
  evolutionary order, 289
  fuzzy logics, 291

parametric model, 291
PLSA, 307
properties, 288
psychophysical experiments, 292
relativist, 288
uncalibrated data, 304
universalist, 288
Color ratios
blur robustness, 74
color angles, 76
illuminant and object invariance, 73
illuminant invariance, 71
lighting geometry, 72
locally constant illumination, 72
Color saliency, 205
boosting, 207
saliency map, 207
Color spaces, 36
CIE standard, 37
HSI, 42, 210
HSL, 42
HSV, 42, 52
hue, 42
intensity, 51
Lab, 41
Luv, 41
normalized-rgb, 50
opponent color space, 40, 52, 88, 92, 106, 209
perceptually uniform, 41
RGB, 38
saturation, 43
spherical, 104, 208
XYZ, 36
Color vision, 13
assimilation, 21
center-surround, 17
chromatic aberration, 14
chromatic adaptation, 18
chromatic discrimination, 23
color deficiency, 23
color matching function, 36
confusion lines, 23
contrast, 21
contrast sensitivity, 22
dichromats, 23
L,M,S sensitivities, 15
LGN, 16
lightness, 20
luminous efficiency, 16
opponent cells, 17
receptive fields, 16
retina, 14
Retinex theory, 20
rods and cones, 14
spatial frequency, 22
spatial interactions, 20
stages, 14
trichromatic, 13
visual cortex, 16
von Kries adaptation, 20

Datasets
ALOI, 226
ALOT, 249
color constancy, 172
Barcelona set, 174
Bristol set, 173
Color-checker set, 174
Foster et al. set, 173
Grey-ball SFU set, 173
HDR Images, 174
SFU hyperspectral set, 173
SFU set, 173
Corel, 211
CURET, 258
Ebay, 314
Google data, 306
Mediamill, 279, 282
PANTONE, 109, 241
PASCAL VOC, 278–9
TRECVID, 279

Error propagation
polar angles, 326
Evaluation
color constancy
angular error, 176
color constancy index, 177
comparison, 182
perceptual significance, 180

Fisher criterion, 241
Focal colors, 291
Fuzzy sets, 300

Gabor filtering, 245
Gaussian color model , 84, 89–90
derivatives, 89

Gaussian color model (*continued*)
  differential invariants, 90
  Hering basis, 87
  RGB camera, 88
  spatial, 85

Hough transform, 202

Illumination, 65, 69
  black body radiator, 26
  candle light, 26
  correlated color temperature, 26
  diffuse light, 26, 140
  direction, 51
  fluorescent, 26
  halogen, 26
  intensity, 51
  white, 53
Image classification, 271
Image retrieval, 61, 77, 271, 273
  normalized average rank, 77
Image segmentation, 120
  region based, 264
  region detection, 120
    road detection, 129
    skin detection, 125, 132
  texture, 247
Image understanding, 120

LDA, 308

Machine learning
  classification scheme, 121
  classifiers, 114
  stages, 114
Material recognition, 249, 256
Michelson contrast, 72
Mondrians, 70
Monte Carlo simulation, 117
Multispectral images, 318
  error propagation, 325
    estimated uncertainty, 325
  hue distance, 324
  hue polar angles, 322
  K-means clustering, 328
  polar coordinates, 321
Munsell color array, 289

Noise estimation, 58, 61
Noise propagation, 58, 61

Object classification, 278
Object recognition, 61, 70, 77, 120, 132, 221, 271, 273
  bag-of-words model, 277
  color indexing, 70
  material classification, 249, 263
Optical flow, 203

Photometric invariance, 49, 196
Photometric Invariance from Color Ratios, 69
PLSA, 307

Reflection model
  ambient light, 31
  Beer-Lambert, 34, 95
  canonical illuminant, 35
  diagonal model, 34, 272
  diagonal-offset transform, 35
  dichromatic, 49, 140, 193
    differential structure, 193
  dichromatic reflection, 27, 29, 34, 101
  Fresnel reflectance, 27, 33
  interface reflectance, 27, 30, 33
  Kubelka-Munk reflection, 32
  Lambert's law, 29
  Lambertian, 28, 34, 36, 51, 53, 71, 90, 139
  neutral interface reflectance, 30, 33
  specularities, 31
  surface albedo, 28
  von Kries, 34
Region detection
  texture, 247
Retinex, 137
Retinex theory, 70

Scale space theory, 85
Sharpe Ratio, 117
Spectrograph
  Imspector V7, 320
Statistics of color images, 211

Taylor series, 59
Texture segmentation, 260
Trichromacy theory, 36

Video recognition, 279

Weibull, 164

Printed and bound by CPI Group (UK) Ltd, Croydon, CR0 4YY

16/04/2025

14658415-0004